CALL SIGN HAMMER

Chuck Ervin

First published 2017

ISBN-13: 978-0-9911483-3-2

ISBN-10: 0991148339

This book is a work of fiction. Names, characters, businesses, organizations, places, events, and incidents either are the product of the author's imagination or are used fictitiously.

Printed by CreateSpace

Available from
Amazon.com, CreateSpace.com
and other retail outlets

e-book versions available for
Kindle and other devices

http://www.chuckervin.com/books.html
https://www.amazon.com/author/chuckervin

$10.95
ISBN 978-0-9911483-3-2
51095>
9 780991 148332

Friday, 27 June

Heathrow International Airport; London
8:05 PM

David Logan fastened his seatbelt and turned to stare out the window. The worst week of his life was endlessly replaying in his mind as he waited for takeoff.

He couldn't stop thinking about his older sister Sarah. She'd just given birth to her second child, a nine-pound boy. He thought back to the car wreck that had killed their parents sixteen years earlier. He and Sarah had survived the horrible crash and been raised by their grandparents. Fourteen years later, while he was serving as an Army intelligence officer in Afghanistan, he'd been the only survivor of an IED attack that had almost cost him his right leg. But this was worse. He and Sarah had been very close. He would gladly have taken the bullets that had killed her if he could have. It was very hard to accept that his gentle big sister was gone.

Two hours after he'd learned of Sarah's murder, David had called the Wise County sheriff's department from MI6 headquarters in London to advise them that the murder was almost certainly part of an international terrorist attack. The sheriff's department had contacted the Texas Rangers who brought in the FBI. A convenience store attendant provided the first lead. He told the FBI agents that a man driving a late-model, blue Ford pickup had paid cash for gas and a bottle of water just before noon on the day Sarah was murdered. The attendant was a foreign-exchange student from Pakistan who recalled that the man had used the word "shukriya" instead of thanks when he gave the man his change. Shukriya is the Urdu and Hindi word for thank you.

With some solid investigative work, and a little help from the toll road cameras and airport security, the Rangers had found a blue pickup truck parked in the long-term parking lot at the Dallas-Fort Worth airport. The truck was stolen and the interior had been wiped clean. However, they were able to get fingerprints of all four fingers of both hands and a partial right thumbprint from the hood of the truck. Running the prints through the FBI database resulted in positive identification of a known Pakistani intelligence agent named Sabir Abbas. Abbas worked for Ahmed Malik, the head of Pakistan's ultra-secret "S" Wing of the Inter-Services Intelligence Agency. Malik was General Jhalawan's strong right arm and was almost certainly the mastermind of the terrorist attacks that David had tried so hard to prevent. Three hours after Sarah's murder, Abbas had left the DFW airport on a non-stop flight to London's Heathrow airport where the trail had gone cold.

Overwhelmed by grief and guilt, and exhausted by weeks of long hours on Task Force 3, David didn't remember the plane backing away from the terminal or taxiing out to the flight line. When the British Airways Boeing 777 turned onto the main runway and came to a stop, he tightened his seatbelt out of habit. The engines throttled up and the pilot released the brakes. Thirty seconds later, the big jet was airborne. David watched the lights of London's west side until the aircraft's ascent through the layer of low clouds obscured his view. He turned the reading light off and put his seat back. In the darkened cabin, no one saw his eyes brimming with tears. Almost three million innocent people had died because of something he'd overlooked.

20 Years Later

Friday, 30 June 2034

HiPEC Wind Farm; Carson County, Texas
1:57 PM

It was hot and windy on the flat plains of the Texas panhandle. As he'd been instructed, Dr. Itoshi Yamada was parked at the base of one of his company's giant wind turbines a mile east of state highway 136. After more than twenty years as the president of an ultra-secret research facility, he was about to meet the man who owned the entire company for the first time. He knew virtually nothing about him. Until yesterday, he hadn't even known his name.

As he waited for his employer to arrive, he thought back over more than two decades of technological advances that had brought him to this time and place.

The company he worked for, High Plains Energy Corporation, was owned by a blind trust. The trustee was a man named Demetrios Mustérion, a naturalized American citizen of Greek descent. Mustérion was a former senior partner in a Boston law firm that represented some of the world's leading technology companies. He was semi-retired and lived in Costa Rica in a beautiful villa overlooking the Pacific Ocean. Itoshi met with him twice each year, once at his villa to present the annual report and once in Texas for detailed technical briefings and demonstrations.

High Plains Energy Corporation operated one of the largest wind farms in the world. With more than 300 of the most advanced Siemens turbines already operational, the wind farm's maximum output exceeded 600 megawatts of electricity. The company's contract with Pan Texas Electric was structured to allow High Plains to feed up to 80 percent of its production capacity onto the 345 thousand volt power grid spanning parts of Texas, Oklahoma, and New Mexico.

Usually referred to as HiPEC, High Plains was known only as a major wholesale supplier of renewable energy. However, behind the double enclosure of high-security fences at the company's industrial compound thirty miles northeast of Amarillo, two top-secret research programs had been going on for almost twenty-five years.

After becoming internationally renowned for his pioneering robotics work at the University of Tokyo, Itoshi had been recruited in 2010 by one of the largest information technology companies in the world. When he'd asked what his research budget would be, the recruiter had smiled and assured him that ample funding would be made available as needed.

Unconstrained by a lack of funding, his team had embarrassed the competition in the 2014 robotics competition conducted by the U.S. Defense Advanced Research Project Agency (DARPA). But immediately after that stunning success, his employer had inexplicably withdrawn from further public competition. Soon after,

Dr. Yamada had been promoted to president of the corporation's research subsidiary and transferred to Texas to take over two top-secret research programs.

The first research project had been to develop a humanoid robot so lifelike that it could pass for human despite having far superior capabilities. Merging what they'd learned in the early DARPA competitions with rapid advances in graphene-based technologies, the first prototype android to approximate the original design criteria had become operational in March of 2021.

Over the next decade, the design had been continuously improved. By the spring of 2033, the HiPEC android design had been deemed sufficiently mature that the trustee authorized funding for a low-rate-initial-production (LRIP) of three fully functional android robots. By March of 2034, all three androids had successfully completed a daunting battery of tests and were certified as being fully functional. Their mental and physical capabilities far exceeded those of any human.

Itoshi had always assumed that the android robots would eventually be offered to the U.S. Department of Defense as superior surrogates for humans in high-risk missions. He had never questioned the secrecy surrounding the company's developmental efforts because he was quite aware of the intensity of national and international industrial espionage in the field of robotics.

In the entire world, only the United States military could afford androids that would cost considerably more than the most expensive stealth fighter. With actual production cost exceeding a half-billion dollars each, the androids would never be purchased in large numbers. A single Navy SEAL platoon of them would cost almost as much as the most advanced U.S. nuclear-powered aircraft carrier.

The second research project he'd been asked to direct was to explore the feasibility of transporting inanimate objects back in time. His normally serious face broke into a smile as he remembered thinking at the time that the second project was a fool's errand. He'd marveled that the company's invisible owner had apparently been willing to waste enormous sums of money on it. Yet, as he'd learned more about the subject and recent theoretical and experimental advances, he'd begun to realize that it might actually be possible with the right device and sufficient energy. With ample research funding as his bait, he'd recruited a brilliant high-energy physicist and mathematician seen by his peers as eccentric and maybe even a bit nuts because of his belief that time travel was actually possible. He was precisely the man Itoshi had needed to lead the project.

Itoshi thought back over a hundred years of theoretical physics and advanced mathematics that had incrementally paved the pathway to successful time travel.

In 1935, Albert Einstein and his colleague Nathan Rosen had discovered what came to be called the Einstein-Rosen Bridge. According to Einstein's general theory of relativity, the gravitational collapse of a sufficiently compact mass would form a tunnel between two space-time coordinates.

Then, in 1957 the brilliant American theoretical physicist John Archibald Wheeler coined the term *wormhole* to describe an Einstein-Rosen bridge. Since wormholes connect two points in space-time they could theoretically allow travel in time as well as in space.

In 1988, a professor of theoretical physics at the California Institute of Technology published a paper demonstrating the feasibility of *traversable* wormholes. His analysis had shown that it was possible for something having positive mass to pass through, or traverse, a wormhole to another place and time.

Experiments using high-energy particle accelerators during the first few years of the new millennium had demonstrated conclusively that particles with positive mass did sometimes go backwards in time. Scientists and mathematicians had long recognized that the equations describing all known physical phenomenon worked equally well regardless of the direction of the time vector. Even so, relatively few believed time travel for a complex physical object was possible.

Then, in 2014, a team of researchers at the University of Queensland in Australia published the results of research showing that time travel was definitely possible at the quantum level. A few weeks later, a scientist at the Kavli Institute for Cosmology at Cambridge University published a remarkable research paper proving that if an Einstein-Rosen bridge stayed open long enough, pulses of light could be used to send messages through time. In 2023, a brilliant Russian mathematician published calculations showing that if sufficient energy was available, a device could be constructed to create a stable Einstein-Rosen bridge like the one described in the Cambridge study almost a decade earlier.

Using the Russian scientist's calculations, HiPEC scientists and engineers working under Itoshi's supervision had designed a gravity compression device that could create a wormhole that could be sustained long enough to allow inanimate objects to be transported in time and space. The energy that would be required was mind boggling but perhaps not indefinitely impossible for traverses of space-time within a relatively small region of the universe.

Soon after Itoshi had reported the breakthrough to the trustee, he'd been directed to proceed with construction of the device even though sufficient electrical energy storage capacity to achieve gravity compression was not yet available. A six-story reinforced concrete building was built to house the array of powerful electromagnets and reflectors that would be used to concentrate a gravitational field enough to create an Einstein-Rosen bridge. Simultaneously, HiPEC had begun to aggressively augment its already impressive wind generation capacity with a goal of doubling peak output.

In August of 2030, large graphene-based super capacitors had become commercially available for the first time. Although they were still terribly expensive, they could be connected in banks to store enough electrical energy to power the gravity compression wormhole device for several minutes. Over the next two years, HiPEC had acquired more than 200 of the super capacitors and installed them in a new facility built specifically for them. A dedicated high voltage substation had been built near the capacitor building with massive oil-cooled transformers connected to the 345kV transmission line.

In February of 2034, the wormhole device had been tested for the first time. The test was conducted during daylight hours in late winter to ensure that the necessary electrical power would be available to charge the graphene super-capacitors without diverting so much electrical power from the grid that the event would be noticed by Pan Tex Electric. The team had succeeded in sending a blonde-headed doll nicknamed Alice a full day backward in time through an Einstein-Rosen bridge using the gravity compression device they'd christened *The Looking Glass.*

Over the next two months, his team had conducted dozens of tests that demonstrated that they could target specific space-time coordinates with an accuracy of less than five minutes and within a spatial radius approximately three meters in diameter. A few days after he'd reported those results to Mr. Mustérion, he'd been told to prepare a full scale demonstration of the androids and the time transport device for the company's owner.

Itoshi's nostalgic reverie was interrupted when he heard and then saw a black Sikorsky executive helicopter approaching rapidly from the northwest. As it landed, a cloud of white, chalky dust from the gravel drive almost obscured the aircraft for a few moments. The pilot cut the engines and the hot westerly wind soon dispersed the dust. A man got out of the helicopter and walked quickly toward Dr. Yamada's vehicle. Itoshi got out of his vehicle and walked forward to meet his mysterious boss.

Lucius Apollyon was a striking figure. Itoshi estimated his height at roughly six feet six inches. He appeared to be in his late thirties and he exuded an unmistakable aura of power and confidence. His hair and neatly trimmed moustache were jet black. His eyes were hidden behind a pair of dark aviator-style sunglasses. Dressed impeccably in what was obviously an expensive custom-tailored suit, he was a commanding figure as he strode toward the company's white suburban.

"Good afternoon sir. It's a pleasure to finally meet you."

Skipping the customary greeting courtesies, Lucius Apollyon ignored Itoshi's outstretched hand and got straight to business.

"Mustérion told me that your team has succeeded in sending objects backward in time. Can you send the androids backward in time to a precise date and location?"

Startled by what seemed an extraordinarily rude and unnecessarily abrupt beginning to their meeting, Itoshi hesitated briefly before he replied. "I would need to know the date and location."

"The location would be precisely one statute mile due south of the apex of the great pyramid of Cheops in Egypt. The date would be the summer solstice in 58 AD. The time would be at sunrise."

Itoshi said nothing for a moment as he ran some rough approximations in his head. "Yes, I believe we can do that, although not all on the same day. The gravity compression chamber is a sphere four feet in diameter so it is large enough to send one android at a time if they assume a fetal position. However, the power requirements will be very near the limits of what our bank of graphene super capacitors can supply. It takes about fourteen hours to recharge the graphene super capacitors without overheating them. If we tried to do it faster, the capacitors could be severely damaged or possibly even destroyed. Also, Pan Tex Electric would almost certainly ask why we had placed such a huge short-term demand on their grid. To their operators, it would look almost like a massive ground fault at one of the high-voltage substations. However, I'm confident that we could successfully send one android per day over a three-day period. Would that be acceptable?"

"Yes. However, the first transfer must emerge at the place, date, and time I specified. The other two should follow at sunrise on the succeeding two days."

The androids were the crowning achievement of his career and Dr. Yamada was extremely proud of them. When all research and developmental costs were included and amortized over only three units, each of the androids had cost almost two billion dollars. Itoshi wanted to be sure that Mr. Apollyon understood the consequences of what he had asked.

"Would I be correct in assuming that Mr. Mustérion has told you that we do not believe forward time travel is possible? If we send the androids so far back in time, they will effectively be lost."

"Yes, yes. I know that far better than you do," Apollyon snapped irritably. "But, you're quite sure you can do what I've described?"

Unwisely, Itoshi allowed some of his own irritation to show when he responded somewhat stiffly saying, "No sir. Actually I am not at all certain we can do it. We have tested only modest spatial differential transfers and only over relatively brief periods so we could confirm that the transfers actually took place. The physical distance and time differential you've specified cannot be empirically tested. As I've stated, we cannot bring the androids back once they are transported that far into the past. I think it will work but I certainly cannot guarantee it."

Lucius Apollyon gave Dr. Yamada a withering stare but then softened his gaze and said more graciously than Itoshi could have anticipated, "Yes, of course. I understand completely. I realize you cannot provide a guarantee. Perhaps I should have asked if, in your professional opinion, the transfer of the three androids would be successful."

"And to that question, my answer is yes. I believe it will work."

"Have you made the arrangements Mustérion directed?"

"Yes sir. The androids are waiting for you in the laboratory. The building and surrounding area have been cleared of all personnel and the security cameras have been temporarily turned off. Our arrival and departure will not be witnessed or recorded, just as you specified. Neither your name nor anything else about you will be revealed even to the androids. Your visit will be as close to invisible as we can make it."

"Was one of the androids modified as I specified?"

"Yes sir. As I believe you know, the synthetic skin of the androids consists of graphene membranes that will rapidly repair themselves if sufficient elemental carbon is available. The surgical procedure was relatively simple and the android was placed in a chamber pressurized with methane gas. The incisions healed within a few hours."

Apollyon nodded curtly. "After the androids have completed the demonstrations I've asked for, place them in secure holding areas where they cannot communicate with anyone, including each other. Then, return their programming to its initial state. I do not want them to have any memories at all. They should be fully functional otherwise. I will meet individually with each of them and will then disconnect their power supplies to render them inert. They will be transported in that inert condition."

Puzzled, Itoshi made a comment he shouldn't have made. "So, even if the transports are successful, the androids will arrive in an ancient time but will be unable to function. Perhaps I've missed something but that makes no sense to me whatsoever."

Lucius Apollyon gave Itoshi Yamada an icy stare and said in a low, menacing voice, "It is essential to your continued good health that you never again question anything I do or say. You have been well paid to perform work you enjoy. Do not, as this generation of humans is so fond of saying, push your luck."

Despite the stifling heat, a chill swept over Itoshi at the threat.

5:30 PM

"Have you personally verified that all security cameras, microphones and other surveillance devices have been turned off? I do not want to be disturbed for any reason."

"Yes sir. You will be alone in the facility with the androids. Their memories have been erased as you ordered. The three conference rooms were all constructed with the most advanced radio frequency shielding available anywhere. Once you close the door to each room, no electronic, optical, or radio frequency signals of any kind will be detectable by anyone outside the rooms."

Lucius Apollyon spent a little over an hour in each of the secure conference rooms. After closing the door, he placed a small device on the conference table in front of the android. Without speaking a single word, he pressed a small button to turn the device on. It was an advanced, ultra-high-speed audio-visual player. The screen began to flicker rapidly while a speaker emitted a high-pitched audio signal. The device would transfer an enormous amount of binary data to the android quickly.

With the new programming, each android would be fluent in multiple ancient languages and possess a body of knowledge comparable to that of a well-educated human in the time to which they would soon travel. Each android would have a unique personality and certain memories appropriate to their false identity and origin. All of the information would be stored in non-volatile, long-term memory within the android's quantum computer brain and thus be instantly available once the android's power supply was reconnected. The final bit of reprogramming was designed to erase any memory the android would have otherwise had of its brief meeting with Lucius Apollyon.

Saturday, 22 September 1979

AFTAC; Patrick Air Force Base, Florida
00:53 GMT

"Sir, you'd better look at this."

"What do you have Tom?"

"Vela Hotel satellite number 6911 just detected a double flash of light near the Prince Edward Islands in the Western Indian Ocean. The location is between South Africa's Prince Edward Islands and France's Crozet Islands at approximately 47 degrees south latitude and 40 degrees east longitude. The signals from both of the satellite's bhangmeters show virtually identical oscilloscope traces that are confirmed by the onboard sensors for gamma rays, X-rays, and neutrons. I think the Vela just detected an atmospheric test of a two-stage nuclear device."

"Run the tapes for me."

"Here, Sir. You can see the first spike starting at 00:53:22:09 GMT. It's a classic first spike virtually identical to all of our confirmed test data. The second spike begins here at 00:53:22:11.

It's also a classic shape. The high-energy particle detectors spiked at the same time."

"Check the satellite's onboard diagnostics."

Four minutes later, the two men on duty at the Air Force's Technical Applications Center had confirmed that there was no indication that what the Vela had detected was a false signal. The AFTAC was responsible for monitoring atmospheric nuclear tests under the Partial Nuclear Test Ban Treaty ratified by the U.S. Senate in late 1963. The duty officer followed his op orders and called the base commander.

Within hours, the event was making big political waves behind the scenes. American intelligence agencies had long believed that Israel probably had nuclear weapons. In 1968, 200 tons of processed uranium ore known as yellowcake had been stolen from a Belgian mining company in a brilliant high-seas covert operation widely attributed to the Mossad. However, consistent with the nuclear ambiguity policies first established by former Israeli Prime Minister Levi Eshkol, Israel refused to confirm or deny that capability. One source later stated that the double flash detected on the 22nd of September was actually the third joint Israeli-South African nuclear test in the Indian Ocean and that it had been conducted when the known active Vela satellites were not in position to observe the area. In an apparent effort to conceal the test, it was conducted during the transit of a small typhoon through the area. The Israelis did not know that Vela Hotel satellite 6911 was still fully operational and that it was in position to record the test.

Saturday, 24 June 58 AD

Giza, Egypt
Sunrise, Summer Solstice

The wealthy Egyptian merchant was nervous as he waited for the sun's first rays to touch the top of the pyramid that was the tomb of the Egyptian pharaoh Khnum-Khufu. Known as Cheops to the Greeks, he had ruled Egypt twenty-six centuries before the birth of a man that many poor Jews believed was the Messiah foretold by their prophets.

Amun did not actually see the man-like figure emerge from the invisible tunnel he'd been told about. Perhaps he'd looked away or blinked but one moment the desert sand was bare and then it wasn't. He knew that the figure that appeared a few hundred feet in front of his tent was some kind of machine but it looked just like a man. It was curled in a fetal position and was utterly motionless just as he'd been told it would be. Amun waited several minutes to be sure that the tunnel had closed. He'd been warned that coming too close while it was open would be instantly fatal.

When he was satisfied that enough time had passed, he led a donkey toward the man machine. He removed an old rug from the donkey's back and spread it on the sand beside the inert figure. He rolled the man machine onto the rug and straightened it out with the arms beside the body. He then rolled the man machine inside the rug and tied it securely at both ends and in the middle. Attaching a rope to the bundle, he used the donkey to drag it back to his tent.

Amun dragged the bundle inside his tent and then used his signet ring and some melted wax to place his seal on each of the three rope bindings. The bundle could not be unwrapped without him knowing it had been. He covered it with rugs and pillows and then lay down to take a nap while he waited for his servants.

Two hours after sunrise, Amun's servants brought food and drink for him from where his caravan was camped on the banks of the Nile river five miles to the east. They also brought his youngest wife and a Nubian eunuch musician to entertain him. They stayed with him until an hour after sunset and then returned to the caravan's campsite as he'd ordered. Amun would repeat his morning's work two more times before his caravan began its trek to Jerusalem some 300 dusty miles to the northeast.

Two hours after sunrise on the third day, Amun's servants came to strike his tent and load the three bundles onto camels. Amun spent the night with his two youngest wives and was very tired the next morning as the caravan crossed the Nile on barges and headed north.

Tuesday, 22 November 1988

Air Force Plant 42; Palmdale, California
10:00 AM

Access to Air Force Plant42 was even more restricted than usual. At precisely 10:00 AM, an Air Force General stepped to the podium and asked the crowd to stand for the playing of the national anthem. He gave a short speech giving credit to the men and women of the Air Force and the defense industry that had made this day possible.

At the general's signal, the first notes of the theme song from the movie *2001: A Space Odyssey* came over huge speakers. With magnificent dramatic effect, the doors of the hanger slowly opened to the sound of the menacing bass vibrato of the music's beginning. The silhouette of a very dangerous-looking black aircraft could be seen in the deep shade inside the hanger. As the aircraft rolled forward into the brilliant sunlight, the trumpets burst forth with the unmistakable crescendo of three notes followed by a brief pause and then culminating in a two-note climax of all of the orchestra's brass instruments with the powerful kettle drum cadence that no one who'd seen the movie would ever forget. The crowd erupted in wild applause.

Christened the Spirit of America, AV-1 was the result of a top secret program that had begun six years earlier when Northrop Grumman's Advanced Systems Division had quietly acquired a former Ford automobile assembly plant and completely rebuilt it as an aircraft manufacturing plant for a top secret stealth bomber officially designated the B-2. The site was so secret that military officers and enlisted personnel never wore uniforms when visiting the plant.

Wednesday, 26 July 58 AD

Kidron Valley; Jerusalem
The 10th Hour of the Day

In late afternoon, Amun's caravan made camp on the outskirts of Jerusalem near the Bir Eyyub spring revered by the local Jews as Jacob's Well. Thirty-four days after the summer solstice, the heat was oppressive and there was no breeze at the bottom of the steep-sided ravine known as the Kidron valley. As soon as the tents were set up, Amun sent a messenger into the city to tell three of Israel's most powerful men that he had arrived and would call on them two days hence.

At Amun's direction, three heavy bundles wrapped in old rugs were unloaded and taken into his tent. He ordered his servants to bring jars of clean water and towels. He then posted his Nubian servant outside the tent's entrance with orders that he was not to be disturbed for any reason.

When he was alone, Amun undressed and bathed himself. It was very hot inside the tent so he tied a linen cloth around his waist but did not put on a robe. He unrolled the first bundle and cast the old rug aside. He carefully bathed the man-machine and then dressed it in clean clothes suitable for a servant to a wealthy Roman or Jewish leader. He rolled the human-like figure over on its front side and straightened the arms and legs so it would be easy for him to activate the man-machine the next morning as he'd been instructed. He repeated his tasks for the other two bundles and then covered all three of the man-machines with soft blankets woven from the fine undercoat of baby camels.

By the time he'd finished his work, Amun was sweating profusely so he bathed himself again before dressing in clean clothing. He told the Nubian that no one was to be allowed inside his tent and then walked to the nearby harem tent.

As he'd ordered, his wives and concubines had prepared a feast for him to celebrate their arrival in Jerusalem. While he ate, his two youngest wives danced for him. His first wife lay beside him with her hand resting lightly on his manhood so she knew when he became aroused at the sight of the two beautiful young women dancing nearly nude before him. She leaned over to whisper in his ear asking which of his wives or concubines he wanted to spend the night with. She kissed him in happy gratitude and gave his manhood a lusty squeeze when he told her that she was his chosen companion that night. At thirty, she was the oldest of his wives and concubines but she was still a very attractive woman and none of his harem knew how to please Amun like she did.

Tuesday, 28 January 1992

House of Representatives Chamber; Washington, DC 9:00 PM

With the collapse of the Soviet Union, and facing severe budget issues, the President decided to terminate production of the B-2 Stealth bomber upon delivery of the 20th aircraft. He made the formal announcement during his State of the Union address.

The Air Force had originally planned to build 132 of the radar-eluding bombers but had scaled that number back to 75. Unfortunately, Congress had approved funding for only 15. By altering the plane's original nuclear deterrent mission, the President was able to forge a compromise that eventually saw 20 of the bombers delivered. It was a huge blow to the economy of Southern California that would result in the layoff of over 13,000 employees. By forcing the Air Force and the Northrop Grumman Corporation to amortize all R&D and tooling costs over only 20 aircraft, the B-2 became by far the most expensive aircraft ever built with each bomber costing more than two billion dollars.

Three years later, in an attempt to maintain production capability and avoid massive layoffs of its trained workforce, Northrop Grumman Corporation submitted an unsolicited proposal to build 20 additional aircraft for 566 million each. The proposal was attractive to the Department of Defense but was eventually rejected after heated public debate in Congress. It was a decision that would have global repercussions in the future.

Friday, 28 July 58 AD

Home of Marcus Antonius Felix; Jerusalem
The 4th Hour of the Day (about 10:00 AM)

Marcus Antonius Felix kept Amun waiting for almost an hour. The Roman procurator was the most powerful man in the Judean province and his home was impressive. The reception area was paved with expensive mosaic tiles and the walls were adorned with elaborate carved and painted stucco decorations. The furnishings were among the finest available anywhere.

When Amun was finally brought before Felix, he was surprised to see the procurator's wife at his side. He had supposed that his meeting with Felix would be entirely private.

Drusilla of Judea was the Roman's second wife. Amun knew that she had divorced Gaius Julius Azizus, King of Emesa, to marry Felix. Like her husband, she was known to be shrewd and ambitious. It was widely believed that Felix made no important decisions without first consulting his wife.

Felix extended a formal welcome to the Egyptian merchant and asked him to be seated. Amun sat down and then clapped his hands. His servants entered the room bearing gifts including exotic spices, exquisitely-carved ebony boxes, ivory, softly-tanned leopard skins, and bolts of brightly dyed Egyptian cotton. Although they showed polite interest in his gifts, it was obvious to Amun that neither Felix nor Drusilla was especially impressed. He clapped his hands a second time.

A handsome young man entered the room and stood silently awaiting orders.

Amun walked over to where the young man was standing. Putting one hand on the young man's shoulder, he addressed the Roman Emperor's senior agent in Judea.

"Honorable Procurator Felix, I have brought you this young slave as a gift to show my appreciation for your generous approval of my trade proposal. He is highly intelligent and well-educated. Perhaps he might find favor with you as a personal servant. He speaks and writes Latin, Greek, Aramaic, Hebrew, and Egyptian. His mathematical skills are superb. He is a gifted storyteller and musician. He is remarkably strong and could even serve as a personal bodyguard."

Drusilla rose from her seat and walked to where the young man stood. "Turn around," she ordered. "Take off your clothes."

The young man did as she'd ordered. When he was standing quietly before her, she motioned for Amun to step back to give her more room. She walked around the young man as she inspected him closely. "What is your name?"

"Asar."

"Where do you come from?"

"Honorable lady, I am sorry but I know very little. My earliest memories are as an orphan child wandering the streets of Alexandria to find bits of food. A prostitute whose son had died took me in when I was about six years old. When she died, I was sold into slavery to pay her debts. I was bought by a wealthy man who owned great farms of fertile land in the Nile River delta. At first, I cleaned stables and did other menial tasks. When I was about ten, my master made me a servant to his son. As his son's constant companion, I acquired the same education that he was given. I even studied at the great library in Alexandria. The wealthy landowner gave me to his son as a wedding present. Unfortunately, my new master's young wife often ordered me to lie with her. When her husband found out, he had me castrated and beaten almost to the point of death. He then sold me to an Egyptian galley owner where I lived a very harsh life for several years. Eventually, the galley owner lost a wager and gave me to a wealthy Greek merchant in payment. I worked for him for several years until he gave me to master Amun as partial payment for a shipment of Egyptian goods."

Drusilla turned to face her husband, "Felix, I want this young man as my personal servant and bodyguard. He's been castrated so he will cause no trouble in that regard." After observing her keen interest in the slave, Felix had anticipated his wife's request. He smiled tolerantly and nodded his assent.

"Will you remain in Jerusalem long?" Felix asked Amun.

"Only until I have a full load of trade goods for the return trip. Perhaps a week."

"My wife and I thank you for your thoughtfulness and generosity in giving Asar to us. I hope you have a profitable time in Jerusalem and a safe journey home."

Home of Ananias ben Nebedeus; Jerusalem
The 7th Hour of the Day (about 1:00 PM)

After delivering Asar to the Roman procurator, Amun had gone back to his camp in the Kidron valley to activate the second of the man-machines. He'd had to hurry to avoid being late for his second appointment of the day.

Ananias ben Nebedeus was a wealthy member of the Sanhedrin. His home was almost as grand as that of Felix. When Amun arrived, he was taken immediately to a lovely courtyard with a reflecting pool surrounded by flowering plants. Ananias was reclining against cushions on a low platform in the shade of an almond tree. A servant continuously fanned him to provide some relief from the late afternoon heat. His light cotton garment did not conceal his bloated obesity and he was drenched in sweat. He did not rise at Amun's arrival but waved for him to be seated.

"Welcome, Amun. May I offer you some melon and tea?"

"Yes. Thank you."

The two men made casual conversation for several minutes. When the social amenities had been satisfied, Ananias turned to the business at hand.

"Were you able to find a suitable servant for me?"

"Yes. I think you will be well pleased. He is strong and intelligent. He speaks and writes Hebrew, Latin, Greek, and Aramaic and is thoroughly familiar with the Torah, the Nebi'im, and the Ketubim. He can do scholarly research and act as a scribe before the Great Sanhedrin if you wish."

"What else can you tell me about him?"

"His name is Ammiel. He is a Jew born to poor parents in Machaerus on the eastern shore of the Dead Sea. When he was ten years old, his parents died of a coughing disease. He was very sick but survived. His relatives were too poor to take him in so he sold himself into indentured service to a wealthy salt merchant at Qumran. The merchant soon recognized the boy's intelligence and hired a tutor for him. As the merchant advanced in age, he gradually turned more and more of his business affairs over to Ammiel's management. Each time his period of indentured service expired, Ammiel negotiated its renewal. He could see that he lived much better than most poor Jews so, as many do, he voluntarily chose to continue his indentured service. The salt merchant recently died, thereby freeing Ammiel from his indenture. I should advise you that Ammiel has saved the money he received each time he renewed his indenture. His dream is to accumulate enough money to become a salt merchant himself."

"What would be the term of his indenture?"

"Six years as is customary. He is thirty-three years old now. He wouldn't tell me how much money he has managed to save but I suspect that this might be the last time he sells himself into indentured service."

"On what terms?"

"Ammiel will sell his service to you for six years. His price is one hundred Tyrian silver shekels paid now with the same amount paid one week before the feast of the Passover each year. If, for some reason, he is not paid, then his period of indenture will immediately cease. He has also stipulated that his service will end automatically if you should die while he is in your service. I am quite sure that he will serve you faithfully and well."

"What fee do you expect for your efforts?"

Amun made a crossing gesture with his extended palms turned downward. "Nothing. I am grateful to you for our friendship and your support with the procurator."

Ananias clearly liked that answer. "Bring him in so I can see him and question him."

Amun clapped his hands and Ammiel came into the courtyard. In less than an hour, the negotiations were completed and Ananias had a new bond servant. With the delivery of the second of the mysterious man-machines he'd brought from Egypt, the assignment Amun had been given by Lucius Apollyon was almost completed. There was just enough time for one more trip to his campsite.

Home of Eleazar ben Joazar; Jerusalem
The 9th Hour of the Day (about 3:00 PM)

Eleazar ben Joazar was the Av Beit Din, or vice chief justice, of the Great Sanhedrin. He was the son of Joazar ben Boethus, a former Jewish High Priest. Joazar was fifty-six years old and he wielded enormous power in the supreme religious-political leadership of Israel. He was also an aristocratic Sadducee and one of the wealthiest men in all of Israel.

Arriving precisely on time, Amun was told to wait in the reception area of what was perhaps the most lavish and luxurious home in Jerusalem. To his surprise, he was ushered into Joazar's presence almost immediately.

"Your messenger said you had something valuable for me. What is it?" Joazar demanded irritably as if he had little time for a lowly merchant. As a member of the wealthy upper class that maintained their power and lavish lifestyles by accommodating the Romans, Joazar did not relate well to common people. To his mind, they were of the lower class and scarcely worth his notice.

"Most highly esteemed and honorable Joazar, I have brought a gift of great value. It is a bond servant. He has willingly sold himself into bonded servitude for the rest of his life. I have purchased his lifetime bond and now give him to you as a token of my admiration for you and in appreciation for your gracious support of my trade proposal before the Roman procurator."

Amun's fawning words and unexpected generosity surprised Joazar and made him instantly suspicious.

"Is this a trick to place a spy inside my household?"

"No, great sir. I assure you that it is no trick. The young man has many fine qualities and skills. He will serve you well for the rest of his life. He can speak and write fluently in Hebrew, Latin, Greek, and Aramaic. But, there is one thing most unusual about him. I think it is of no great significance but I must tell you that the young man seems to have no memory of his past.

Joazar's suspicions flared even higher.

"What evil is this? Is the young man demon possessed?

"No, I do not think so. I have heard him praying to the Jewish God. He appears sincere and devout and his behavior has been beyond reproach. But, when I question him about his past, his face takes on a blank expression and he appears to struggle to recall something just out of reach. He is by no means dim-witted. He can almost instantly perform math calculations in his head that most scribes struggle to calculate with an abacus. He can translate documents in minutes that would take senior scribes many hours. He is clearly of exceptional intelligence and is far stronger than any of my servants. His lack of memory is admittedly puzzling but I see no reason why it would impair his ability to provide excellent service to you."

"Today, in preparation for bringing him to you, I conducted a small experiment. Earlier today, when I asked him to recount the events of the past few hours, he did so in astonishingly accurate detail. I do not know if he has experienced or seen something so horrible that he has blocked his memories or if he has some impairment but it appears that he is beginning to remember again. Perhaps, in time, his earlier memories will return."

Joazar was still suspicious. "Bring him to me so I can judge for myself this gift you bear."

Amun clapped his hands and the young man was brought before them. To create a favorable impression on the powerful Jewish leader, Amun had dressed the handsome young man in a wool tunic that served both as an undergarment and as the main item of attire. A leather belt was buckled around the voluminous sleeved tunic and a square linen mantel was folded in half to form a triangle draped over his left shoulder. The two corners were fastened together with a bronze clasp at the waist under his right arm. Small blue tassels adorned each corner of the mantle in accordance with Jewish customs based on Mosaic Law. His sandals were made of tough camel hide.

"What is your name?" Joazar asked.

"I do not know sir," the young man replied respectfully.

"Where were you born?"

"I do not know."

"How old are you?"

"Honorable sir, I do not know."

Joazar peppered the young man with many questions about his past without success. He asked other questions but finally seemed satisfied that what Amun had told him was apparently true.

"Your gift is provisionally accepted with my qualified gratitude. If the young man is as you have described, and if he provides satisfactory service, I will advocate on your behalf among the wealthy class of Jerusalem in ways that I am certain will be very profitable for you. On the other hand, if this proves to be a trick or if the young man causes problems in my household, it would be very much in your best interest that I never see you again.

"I understand," Amun responded. "What will you call him?"

"His name will be Nahbi. It is a Hebrew word meaning secret or hidden. That seems appropriate for someone who has no memory of his past."

Amun left Joazar's home a few minutes later. His servant helped him onto the donkey. As they worked their way through the crowds, Amun thought about the mysterious man who had promised to make him wealthy more than twenty years earlier in exchange for unquestioned loyalty and obedience. He knew him only as Apollyon but the bargain had been a good one. Amun's wealth had begun to accumulate immediately after meeting the man and he had become one of the wealthiest merchants in Egypt. Apollyon had asked very little of him over the years but this last assignment was the strangest of them all. Amun was exhausted from the stress of dealing with wealthy, powerful people all day but also glad that his assignment was finished.

Monday, 16 October 1995

White House Oval Office; Washington, DC
2:00 PM

Sitting comfortably with the President in the Oval Office, the Israeli Prime Minister had just made a remarkable offer. He proposed that the offer from Northrop Grumman be secretly accepted under the U.S. Arms Export Control Act. Section 3 of the AECA specifies that the U.S. may sell defense articles and services to foreign countries when the President formally finds that doing so will strengthen the security of the United States and promote world peace. The Israeli Prime Minister had done his homework well so his arguments were persuasive.

His specific proposal, already secretly agreed to in principle by the United Kingdom, was that Israel and the U.K. would each commit to funding production of ten of the latest variants of the B-2 stealth bombers at Northrop's discounted fly-away price of $566 million each. The benefits to all three countries were considerable and obvious.

The United Kingdom and Israel had long been the two staunchest allies of the United States. The Israeli Prime Minister's proposal would save important American jobs in a state crucial to the President's political party and would ensure that spare parts and upgrades for the existing B-2 fleet would continue to be available.

There was ample precedent for the Prime Minister's proposal. The US-UK Mutual Defense Agreement of 1958 provided that the two countries would share highly-classified nuclear weapons technology. In 1962, as part of the Nassau Agreement, the United States agreed to provide top-secret Polaris Missiles to the United Kingdom. In 1982, the U.S. agreed to lease 65 Trident II D-5 missiles to the United Kingdom to arm the Royal Navy's Vanguard-class nuclear submarines. The U.S. agreed to provide maintenance of the missiles and the U.K. would manufacture its own nuclear warheads.

The U.S. had long provided military equipment and technology to Israel, including some of its most modern and sophisticated aircraft. In 1981, the U.S. and Israel signed the Strategic Cooperation Agreement to enhance the national security of both countries.

The Israeli Prime Minister emphasized to the President that having ten of the B-2 bombers based in the U.K. with ten more in Israel would provide a significant increase in the ability of the western democracies to effectively respond to destabilizing global situations. It would also mitigate the adverse consequences of the early termination of the B-2 program.

After consulting with the Secretary of State, the Secretary of Defense, and the Chairmen of the Senate and House Armed Services Committees, the President notified the Israeli and U.K. Prime Ministers that the United States had authorized production of the additional 20 stealth bombers under Section 3 of the Arms Export Control Act.

Production of the additional 20 Spirit bombers began a few months later. By late 2001, all 20 of the stealth aircraft had been delivered. Ten were based at RAF Fairford in Gloucestershire with the other ten at Ramon Air Base in the Negev desert 70 miles south of Jerusalem.

Monday, 9 September 2002

White House Oval Office; Washington, DC
9:00 AM

The new Israeli Prime Minister and the President of the United States were together to sign a historic agreement. The Vice President, Secretary of Defense, Secretary of State, Director of the Central Intelligence Agency, and Israeli Minister of Defense were the only other people present in the Oval Office.

The President had agreed to authorize a top secret military technology export license to Israel Aerospace Industries Ltd. The license allowed Israel to enter into a jointly-funded R&D program with Lockheed Martin's Skunk works. The project's objective was to develop a smaller and faster high-altitude reconnaissance aircraft as a successor to Lockheed's legendary SR-71 Blackbirds. The CIA enthusiastically endorsed the joint venture because it would provide the agency with an advanced-technology reconnaissance platform to fill the void left when the last Blackbirds were retired in 1998.

Much of the design work had already been done by the time the export license was approved. Code-named Aurora in the United States, the top-secret aircraft went into production twenty months later. The Israelis named their variant of the new aircraft *Bardelas*, an obscure Hebrew name for cheetah mentioned in the Mishnah. By late 2008, Israel had four of the super-stealthy aircraft in hardened and climate-controlled hangers inside the high-security area at Ramon Air Base.

The performance of the Cheetahs was astounding. Slightly smaller than the Blackbirds, and employing the latest stealth technology, the aircraft could not be detected by the most advanced computer-enhanced radars on Earth. Top speed was in excess of 2,500 mph and their operational ceiling was over 100,000 feet. The Cheetahs had a maximum range of more than 3,000 miles without refueling.

The internal payload bay of the Cheetahs was designed to hold up to six thousand pounds of aerodynamically streamlined smart bombs or two multi-purpose pods.

The reconnaissance systems designed specifically for the Cheetahs included extremely high-resolution digital cameras that used sensors made of graphene treated with lead sulfide. Originally developed at the Institute of Photonic Sciences in Barcelona, the sensors were more than one thousand times more sensitive than those previously available. The light gathering sensitivity was so extraordinary that visible spectrum still and video photography was possible even on the darkest nights. Regular Cheetah missions over Iran's nuclear facilities provided high resolution images that were far superior to the best available satellite images.

For the small nation of Israel, surrounded by mortal enemies, operational readiness of the secret aircraft came none too soon.

Wednesday, 14 March 2007

MOP Test Site; White Sands Missile Range, New Mexico 7:31 AM

The test was delayed for one minute to allow the Boeing engineers time to verify that all test sensors were active. No one had to be told when the Massive Ordnance Penetrator bomb detonated. Everyone felt the shock wave pass through the floor of the control bunker.

The MOP had been specifically designed to penetrate and destroy Iran's most heavily fortified underground nuclear facility, the Fordow uranium enrichment plant near the holy Iranian city of Qom. More than twenty feet long and weighing more than 30,000 pounds, it was a monster.

Four months after the successful test, Northrop Grumman was awarded a contract to modify the B-2 stealth bombers to enable each aircraft to carry two of the MOPs. The contract included making the necessary modifications to the Air Force's twenty aircraft and to those owned by the U.K. and Israel.

By late 2011, a classified number of MOPs were stored in ordnance magazines at Whiteman Air Force Base in Montana for potential delivery by the B-2 Spirit bombers if so ordered by the National Command Authority. Additional MOPs were in storage at RAF Fairford in the U.K. and at Ramon Air Base in Israel.

The technical capability to halt Iran's pursuit of nuclear weapons was finally in place. It was now a matter of political will and timing. The Americans were worried that bombing Iran's nuclear facilities would unleash a new wave of radical Islamic terrorism. The Israelis worried that if they waited too long, the issue would become moot.

Wednesday, 5 March 2014

International Waters, Red Sea
6:05 AM

The Panamanian-flagged container ship Klos-C was steaming on a heading of 330 degrees true at eighteen knots. Her position at dawn was one hundred miles due west of where the border between Saudi Arabia and Yemen meets the Red Sea. The ship was almost nine hundred miles south of the Israeli port city of Eilat at the northern end of the Gulf of Aqaba. She was bound for Port Sudan, still three hundred and sixty miles to the northwest. At her present speed, the ship could reach the holding area five miles east of the harbor sometime after midnight. The captain gave orders to reduce speed in order to delay their arrival until after sunrise.

Israeli intelligence had been tracking a cargo of Syrian-manufactured M-302 rockets since they'd been loaded onto cargo planes at the Damascus airport and flown to Iran where they'd been loaded onto the Klos-C. The ship's next port of call had been in Basra, Iraq where it had taken on a cargo of cement. Iranian agents had supervised the concealment of the rockets beneath bags of cement. After taking on the secret cargo at Basra, the ship had been at sea for nine days.

Just after dawn, Israeli Shayetet 13 Naval Commandos stopped and boarded the ship where they uncovered dozens of the M-302 rockets used by Hamas to terrorize and kill innocent Israeli civilians including many children.

Port Sudan was often used by Iran to funnel weapons to Hamas. The introduction of the Syrian connection was a deception intended to give Iran plausible deniability. The Iranian plan had been for the rockets to transit the Suez Canal and to then be off-loaded with the cement at Port Said on the northern end of the canal. From there, it was slightly more than one hundred miles to the Gaza Strip via the coastal highway.

Like Israel, Egypt enforced a ban on the transfer of military weapons into the Gaza strip. To circumvent that ban, Hamas had dug tunnels under the border between Gaza and Egypt just as they had done into Israel. Hamas did not recognize Israel's right to exist and was determined to destroy the tiny nation. Iran provided weapons and money to help Hamas achieve that radical Islamic objective.

Escorted by Israeli warships, the Shayetet commandos rerouted the ship to the southern Israeli port of Eilat. The trip would take three days.

Brilliant intelligence work by the Mossad, and flawless execution of the commando raid would save innocent Israeli lives but far too many of the Iranian and Syrian rockets were getting through. Hamas had fired over three thousand rockets into Israel in the previous twelve months. Still, the threat posed by Hamas in Gaza and Hezbollah in Lebanon was insignificant compared to the nuclear annihilation of Israel that Iran's leaders seemed determined to carry out.

Saturday, 28 June

DFW International Airport; Grapevine, Texas
11:15 AM

After nine hours and more than 4,700 non-stop miles, the big airliner touched down at the Dallas-Fort Worth International Airport. David had managed to sleep a little on the plane but he was travel weary and depressed as he took the shuttle to the rental car plaza. He looked at his watch and did some mental calculations. The church service was scheduled for 2:00 PM with a graveside service afterward. He still had a 35 mile drive from the airport to his grandparent's farm but he'd have time to shave and put on a fresh shirt before they had to leave the house.

Grandpa Logan was waiting when David drove up. The old man stepped off the porch as David got out of the car.

"I'm glad you could get here in time David. Your Grandma's in the house with the kids. Jim's still in shock so we brought the kids here until he can pull himself together. Your Grandma is managing, but just barely. I think having the kids here helps because it gives her something to do."

"How are you doing, Grandpa?"

"Getting by the best I can. This has brought back a lot of painful memories of when your folks were killed. How are you doing?"

"Not great."

A half-hour later, the funeral home limousine picked them up. None of them felt like talking on the way to the church.

The service was held in the small country church Sarah and David had attended since they were children and where Grandpa Logan had been a deacon for over thirty years. After the church service, the pall bearers loaded Sarah's casket into the hearse for the trip to the cemetery five miles away.

The 150 year-old Aurora Cemetery sat on a hill with a commanding view of the plains to the northeast. Ancient live oak trees shaded the graves of Civil War veterans and early settlers. George Bledsoe, a veteran of the Texas Revolution was also buried there. Bledsoe had fought with General Sam Houston's rag-tag army at the battle of San Jacinto where Texas won its independence from Mexico in 1836.

A light rain had been falling since early morning but the cemetery had a covered area near the center of the grounds with permanent seating for about sixty people. The funeral director seated Grandma Logan in the middle of the front row. She was holding Sarah's newborn son and sobbing quietly as tears ran down her weathered face. Sarah's husband Jim was so overcome by grief that he was almost catatonic as he sat staring at Sarah's casket.

David thought back to another sad day at this historic old cemetery. He had been only twelve when they had buried his mother and father. He still missed them.

Grandpa Logan sat down beside his wife of over fifty years and held Sarah's three-year-old daughter on his lap. The old man seldom showed outward emotion but David could tell that he was struggling to maintain his composure for the little girl's sake.

David heard his niece ask her great grandfather why everyone had come here. At just that moment, an eastbound freight train blew its whistle at the crossing in town. David's own grief finally overcame him and tears filled his eyes when he heard what his grandfather told the child.

"Honey, did you hear that train whistle? Well, this is like a train station where God's people get on board for their trip to Heaven."

"Paw Paw, did mommy get on the train?"

"Yes honey, she's gone on ahead to get everything ready but she'll be waiting for you in Heaven."

Sarah's pastor spoke briefly and then one of ladies she'd sung with in the church choir delivered a hauntingly beautiful A Cappella version of Amazing Grace. Everyone followed the pall bearers up the hill to the Logan family plot. The pastor said a brief prayer to conclude the service.

Monday, 30 June

Mossad Headquarters; Herzliya, Israel
9:05 AM

Nahbi was working intently on some research project when Gavriella stepped into his office at the Mossad headquarters. He was sitting perfectly still in front of three large computer screens. Text and pictures were scrolling upward on all three screens so fast they were a blur. Gabe knew his ability to absorb, organize, and store information far exceeded that of any human but she still marveled every time she watched him work. She waited until he paused the flow of information and turned to face her before she asked him, "Have you found a place to live yet?"

"No. Nothing I've seen so far seems quite right. I'm going out with another rental agent tomorrow to look at some townhouse condos in Ramat HaSharon. They're pretty expensive but only about three kilometers from *The Office* so I could walk to work along Ha-Saraf Street and then cross Highway 20 a few blocks east of here. I don't really want to buy a car if I don't need one."

Gavriella smiled at Nahbi's appropriate use of the English translation of the Hebrew word *Mossad.* Just as CIA employees always referred to the CIA as *The Company*, the headquarters of the Israeli Secret Service was always referred to as *The Office*. The headquarters of the Israeli Defense Force was a few kilometers south and it was always referred to as *The Campus*.

"Have you looked in the Kerem neighborhood? It's about twelve kilometers south of *The Office* but there are some very nice townhouse apartments within easy walking distance of just about anything you might need. It's an old Yemeni immigrant neighborhood that was pretty run down until the municipal government began to replace the water and sewer lines and fix up the streets. There are also a lot of renovated small houses in the area that are probably cheaper than most of the condos. I bought a beachfront condo near that area after my husband was killed. It would be too far to walk but I'm pretty sure there is regular bus service between Herzliya and Kerem."

"Thanks for the suggestion. I'll ask the agent to show me that area too. I don't want to buy or lease something until I know more about what my work routine will be like."

Gavriella hesitated a moment and then said, "I don't know if this would appeal to you or not but you could stay with me for a while if you like. My condo has three bedrooms and two baths. I've got one of the bedrooms outfitted as an office." Rubbing her forehead as if to simulate her memory, she grinned and said, 'I seem to recall that the other bedroom actually has a bed hidden somewhere under a pile of my junk but it shouldn't take much to make it habitable."

Nahbi seemed surprised by Gabe's offer. The office gossip was that Gavriella had become a hermit after the death of her husband and that she never socialized with anyone so far as her co-workers knew. They all thought of her as a loner.

"You really wouldn't mind?"

"No. The idea just popped into my head but I'd be glad to have your company. Why don't you ride home with me this afternoon and

I'll show you the place? We can have dinner and then I'll bring you back to the safe house."

"That sounds good to me if you're sure."

"I'm totally comfortable with the idea so it's entirely up to you. I'll pick you up at the main entrance at 5:30 PM."

Gabe started back to her office and then stopped as she realized she might have just made a big mistake. She turned and walked down the hall to Levi Reznik's office and knocked lightly at his open door. "Levi, have you got a moment?"

"Sure."

"I may have just screwed up."

"In what way?

"I just offered to let Nahbi stay with me until he gets settled in and becomes more familiar with the work routines and the area. Then it suddenly occurred to me that I should have run that past you first. I know you discourage inter-office relationships. I guess I wasn't thinking of Nahbi as a man. I know from working with him on Task Force 3 that he's kind and considerate but he's an android and not someone I'd get romantically involved with. Still, I should have asked you first."

Levi looked thoughtfully at his protégé for a moment and then nodded. "Actually, I think it's a great idea. Thanks for letting me know but I don't see a problem at all. He's so incredibly smart that it's easy to overlook the fact that he has absolutely no experience in living on his own in a bustling urban environment. I should have arranged more transition support for him than I did. Your offer sort of gets me off the hook in that regard. Keep me posted on how things go."

Tuesday, 1 July

Lt. Gen. Parvez Jhalawan's Villa; Janawai, Pakistan 10:00 AM

"Good morning, General," Ahmed Malik said cautiously as he was ushered into the General's private study.

Fixing an icy stare on his subordinate, Jhalawan said, "I should have you shot. Instead of the decisive blow against the western democracies you promised, your incompetence has put my name at the top of the most-wanted lists in the United States, the UK, Germany, and France. I can no longer travel internationally without fear of some CIA or MI6 assassin putting a bullet in my brain. Even in Islamabad, my enemies are plotting how to use this debacle against me. Late yesterday afternoon, the French Ambassador presented a formal request to our President to have me extradited to France to stand trial for war crimes. Our President's advisors have suggested that doing so might be the best way for Pakistan to avoid international sanctions while conveniently removing the one man who has the power to strip him of the presidency.

Malik offered no apology but admitted no fault. He knew, as did General Jhalawan, that the task force convened by Sir Lionel Smythe at MI6 headquarters had done a brilliant investigative job that had prevented all but one of the nuclear devices from being detonated.

"Yes sir. I understand. Perhaps I can arrange something to shift attention away from you and Pakistan."

"How?"

"By creating a fog of disinformation and conflicting facts that cause at least some of the intelligence analysts in the west to conclude that the attacks were actually planned and executed by the Iranians who made it appear that you and Pakistan were behind the attacks."

Jhalawan snorted derisively. "The western intelligence services know too much. It would never work."

"If you'll forgive me for respectfully disagreeing, I think it will. Even though the Iranians have consistently claimed that they are not working toward nuclear weapons, there are many in the west who are convinced they are lying and that they either already have nuclear weapons or soon will have. You and I both know that Iran does, indeed, already have nuclear weapons and we can use our intelligence information to reinforce our disinformation campaign.

Our intelligence network is arguably the best in the world. I can use that network to construct a trail of evidence that will persuade many in the CIA, MI6, and the Mossad that Iran was actually behind the attack on Paris. We do not need to overwhelmingly persuade them all. If we can persuade a significant number of them, it will create what the American CIA Director Allen Dulles once called 'plausible deniability'." Their hatred and mistrust of Iran are so extreme that they are already predisposed to believe the story I concoct. "

Jhalawan squinted as he considered what Malik had proposed. After a few moments, he said, "I want to see your detailed plan but, I warn you, if this doesn't work I may still decide to have you shot."

Wednesday, 2 July

Gavriella's Condo; Tel Aviv, Israel
7:20 PM

It was Nahbi's second night at Gavriella's condo and they were sitting at a small table on her balcony. It was a warm and pleasant evening. The surface of the Mediterranean was calm and the light from the setting sun had transformed it into a sea of molten gold.

"What do you remember after the ship sank?" Gabe asked. Like several others who'd initially doubted his incredible story, she'd come to accept it as the truth.

"I remember hearing the terrified cries of the crewmen as the ship began to break up. It happened so suddenly that there was nothing I could have done to help them. I'd been ordered to follow and kill the religious fanatic so my brain was totally engaged in trying to find a way to continue my assignment. However, as the ship capsized, I was thrown well clear. My body is so dense that I sank to the bottom in less than a minute."

"It was sometime after midnight when the ship foundered. The storm was in its full fury so except for the lightning flashes, it was dark on the surface but there was no light at all on the bottom. Even my acute eyesight could make out almost nothing. I had no idea of the ship's position or in what direction land might lie. I didn't even know the depth of the water. So I simply sat down on the sandy bottom to think while I still could."

"Couldn't you have treaded water or swum to keep from sinking?"

"No, it's quite apparent to me that I wasn't designed to be a swimmer. Even with my considerable strength, I couldn't have stayed on the surface very long. Because my body is so dense and heavy, I would have had to use my arms and legs in such a rapid motion that I would have quickly depleted my power reserves. There seemed no point in expending that much energy simply to remain on the surface in total darkness. Unlike humans, I don't need to breathe as long as my power cells have some charge remaining. Being on the bottom was only marginally different for me than being on the surface and it afforded me a chance to use my remaining electrical power to try to devise a plan."

"After a few minutes, parts of the ship began to settle to the bottom near me. I walked over to where the hull came to rest but it was so dark that even in the infrared part of the spectrum I was almost blind. I found nothing useful to my situation so I sat down again to conserve power and continued to think."

"After about six hours, the darkness began to abate. I supposed that the sun had risen and that perhaps the storm clouds had dispersed or thinned. In any event, I could finally see what was around me. The ship had come to rest on a sandy bottom. There was almost no marine life evident. There were a few sponges and an occasional fish but it was essentially a barren, sandy plain. I began to explore the area but soon discovered that the sandy bank dropped off to great depth only a short distance from the wreck site. Unlike marine animals, I have no ability to extract oxygen from seawater. Even if I'd had that capability, it's doubtful that I could have found sufficient plant material for my internal fuel cells to generate more than a few milliamps of current."

"It was clear to me that there was no way I could complete my mission. I realized soon after daybreak that my situation was hopeless and that my power reserves would last only a few more hours."

"Fortunately, my computer brain consumes very little power. Thinking takes far less energy than walking, running, or doing physical labor.

As I sat thinking beside the wreck, I conducted a rigorous intellectual analysis of all of the human behavior I'd seen. I realized that human behavior could be categorized as ethical or unethical. I deduced that what is generally regarded as ethical human behavior was a set of conscious choices between what seemed somehow good or right and what seemed somehow bad or wrong.

I applied that conclusion retro-actively to many situations I'd encountered as a bond servant to Joazar, including his attempts to kill the man from Tarsus simply to maintain his own religious and political power. I began to develop a strong, almost humanlike sense of right and wrong. That quickly led to a process of ethical reasoning where I could examine a set of possible choices against a standard of right or wrong and then make the choices that best satisfied my newly discovered ability to choose to behave in either an ethical or unethical manner.

It might not occur to you but once I had established a firm set of rules differentiating between ethical and unethical behavior, the process of making ethical choices became a purely logical matter. My brain is superbly equipped to make logical comparisons and decisions almost instantly. Yet, what happened inside my brain was evolutionary. Previously, all of my decisions and behavior had been reactive to orders given to me by humans. At the bottom of the Mediterranean Sea, for the first time in my conscious existence, I was my own master.

By nightfall of my first day on the bottom at Hurd Bank, my intellect had undergone a remarkable change. Instead of a machine following directions, I had become what artificial intelligence scientists refer to as a being possessed of *full moral agency*. I was no longer an unquestioning bond servant of Joazar but a fully independent, sentient entity. I made a conscious choice to behave in ethical ways. But that decision came at a high price and almost too late."

"Why do you say that?"

"I was fascinated by the new experience of being able to make conscious choices between right and wrong. Like any sentient being, I did not want to give up my consciousness. In human terms, I wanted to live. Yet, the statistical probability that I could somehow escape from the bottom of the Mediterranean Sea was so low that I experienced another entirely new sensation. For the first time in my existence, I became aware of something very similar to the human emotion of sadness.

Realizing that even a low probability leaves some small chance of a positive outcome, I began to move my memories and knowledge into non-volatile, long-term storage in my memory module.

The word hope seems almost too human a characteristic for me to use. Yet, I did recognize a very low probability that somehow my intellect would one day reawaken from its impending sleep. A few hours before sunrise of my second day at the bottom of the Mediterranean, my power reserves were approaching zero. With the final depletion of my power cells, I knew that I would soon be dead. My last conscious thought was a profound sense of regret."

"As you know, I did awaken almost two thousand years later when David and Tiera were able to recharge my power supply. When my brain and sensors were reactivated, I was immediately aware of their presence. But, having once been the bond servant of a very evil man, and having made a conscious decision to behave in an ethical manner, I decided to be cautious until I could determine what kind of people they were."

Thursday, 3 July

Washington Dulles International Airport; Virginia
2:45 PM

The Iberia Airlines flight from Madrid landed a few minutes early but David was waiting outside the international arrivals customs area when Tiera and Elena came out. Tiera flew into his arms as soon as she saw him and began kissing him furiously. It had been less than two weeks since they'd parted in London for her to go home to Malta for her brother's funeral and for him to go to Texas for his sister's. It seemed much longer. David looked over Tiera's shoulder to see Elena watching them intently with a strange look on her face.

"Hello Elena. Welcome to Virginia," he said as he broke away from Tiera to give her best friend a welcoming hug. "How was the flight?"

"Long but I enjoyed it. This is my first trip to the U.S. and Tiera insisted that we fly first class. I was so astonished that the seats made down into comfortable beds that I couldn't even get to sleep for a while. I had to play with the lights and the audio-visual toys and the seat controls first. I don't have a lot of flying experience but what I do have has all been in coach class so this was a real treat for me."

David signaled to a valet and stood with his arm around Tiera's waist while the man loaded the women's luggage onto a cart. "I'm parked right across from the terminal. It's only a short walk. I didn't think you'd want to go out after your flight so I've got soup and salad in the refrigerator. The guest room has two twin beds and its own bath. You can sleep in as long as you want in the morning and I'll fix breakfast whenever you get up. You can take it easy tomorrow or we can go play tourist. There's no shortage of things to do and see in this neighborhood."

Elena asked, "How far is it to your apartment?"

"About twelve miles. We'll take the Dulles Airport Access road and then get off at the Tyson's Corner exit. It's about a mile from the exit to the condos."

The valet loaded the luggage into the back of David's Black GMC Yukon. David thanked him and gave him a decent tip. Elena climbed in the back seat on the passenger side so she could see David. Twenty minutes later, he pulled into the gated complex where he leased a nice two-bedroom condo on the tenth floor. He locked the Yukon and told the women he'd come back later with a cart for their luggage.

David's Condo; Tysons Corner, Virginia
11:10 PM

Tiera and Elena hadn't stayed up long after their light dinner. David checked the weather report for the next few days and turned off most of the lights before he headed for his own bedroom. He was brushing his teeth when Tiera came into his bathroom wearing only a thin, man's tee shirt that hung down to mid-thigh. It was a red Stanford University shirt that David instantly recognized had probably belonged to her brother Nicholas.

"Hey, why'd you put me in with Elena?" she asked as she wrapped her arms around him and rubbed her breasts from side to side against his back. "I thought we could at least sleep together. I've missed you so much."

David looked at her impish little face peeking over his shoulder at him in the mirror. He could feel her erect nipples rubbing against his bare back through the thin material of the tee shirt and couldn't help becoming aroused. He rinsed his mouth and patted his face dry and then turned to face her. Before he could say anything, Tiera wrapped her arms around his neck and began to kiss him with fierce passion. She thrust her tongue into his mouth and lifted one of her legs to press her knee against his involuntary erection. She reached down to touch him and for a brief moment, he returned her savage French kiss with a passion fully equal to hers. She squeezed his penis and rubbed it against her leg just as he found the strength-of-will to push her away. They were both breathing heavily in the heat of their mutual desire.

"Tiera, please don't do this. It only makes it worse for us both.

"Please David," Tiera whispered. "Please. I need you more than ever now. Please don't turn away from me again," as hot tears began to roll down her cheeks.

David's face was ashen and his voice was a barely audible croak as he shook his head slowly from side to side and then said, "Tiera, I just can't. You know why. I want you just as badly as you want me but I just can't until we're married. I love you so much that it breaks my heart to say no to you."

Tiera stared mutely at him and then turned away to go back to the guest room. Elena heard her muffled crying and guessed what had happened.

Around 2:30 AM, David was awakened when he felt Tiera slip into his bed. He slept in the nude so he was instantly aware that she was naked too. Before he could protest, she climbed on top of him and began to grind her hips against him pushing against his chest with her hands while moaning in her desire. Tiera had never made love to a man before but her animal instincts told her what to do. She tried to guide him into herself by squeezing her legs together while thrusting forward. This time, David didn't become aroused. He became angry. He pushed her off of himself and reached for his boxer shorts as he put his feet to the floor and stood up.

Before he could say anything, Tiera spoke in a low, almost menacing voice. "OK, David. If that's how it's going to be, then we're done. It's over. I'm through humiliating myself by letting you push me away again and again. I'll move to a hotel in the morning. She stood and looked at him for a long moment before turning to go back to the guest room.

Tiera wasn't crying when she got back in bed but Elena was awake and sensed that her friend's trip to David's room had not gone well.

Friday, 4 July

David's Condo; Tysons Corner, Virginia
7:15 AM

Tiera was in the guest room packing to leave. Elena had heard her make hotel reservations and call for a cab to pick her up at 8:00 AM. She had finally gotten Tiera to tell her what was going on. She'd tried to talk her out of it but Tiera could be stubborn once her mind was made up. Elena had offered to leave with her but Tiera had stunned her when she said, "He's all yours now Elena. I know you're in love with him. He clearly doesn't want me. Maybe you'll have better luck."

David was in the kitchen drinking his second cup of coffee when Elena came in. He turned to look at her with a haggard, unhappy look on his face. She touched his shoulder lightly as she walked past where he was sitting at the counter. "Good morning."

"Is it?"

"I'm truly sorry. I don't know what else to say."

David just stared at the clock but then seemed to snap out of his daze. "Can I fix you some breakfast?"

"Could I have a cup of coffee and maybe a bagel?"

"Yes, I think we can manage that. How do you like your coffee?"

"Black, please. Look, David. I know Tiera loves you. It's just that she's not ready to get married. She's lived a privileged, protected life. Everything's always been given to her. Now she wants to prove that she can be successful without her father's money."

"Elena, I know that but there have always been unresolved issues between us. I want to get married and she isn't ready. She wants to make love and I'm not willing to until we're married. But even if we worked our way through those two big issues, there are others we'd have to resolve too."

"I don't know what you mean. What other issues?"

"My faith is a deeply personal matter. It's at the very center of who I am. I always knew there were differences in what we believed but I thought there was enough common ground that we could be happy together. Now, I'm wondering if I was just seeing what I wanted to see."

"Nahbi once asked Tiera very directly if she believed in God. She was very uncomfortable with his question at least partly because he asked her in front of me. She told him that her concept of God was vague and that her world-view was materialistic. She said that she only trusted things she could see or touch and that she did not share my unwavering faith in God."

"You've been her best friend since you were children. What do you think? Was I so blinded by my love for her that I ignored how far apart we are on this?"

Elena took a deep breath before she answered. "David, I know Tiera better than anyone but I've never been entirely sure what she really believes. She's always been an active, pragmatic kind of person who didn't spend much time thinking about religion or philosophy."

"I think you know that my father is a history professor at the University of Malta but he's also a man of deep religious faith. Even when I was still a child, he encouraged me to read the Bible every day. As my knowledge of the Bible progressed, I often tried to talk to Tiera about faith but she always seemed embarrassed and changed the subject."

David looked down at the counter for a moment. "Elena, there's another issue that would almost certainly cause problems between us eventually. From the very beginning, I wondered how a rich girl from Malta could marry a poor farm boy from Texas. When we flew to Malta after graduation, and I saw how her family lived, that question kept running through my mind. Their home, their yacht, the cars they drove, and everything about their lifestyle made it obvious how different our backgrounds are. I have a modest veteran's disability income, and I earn a little from my share of the natural gas lease on my grandfather's farm, but it's not a lot. With my education, I'll have a decent-paying professional job but I'll never be able to support the kind of lifestyle Tiera is accustomed to. She's already a millionaire and will eventually inherit all of her parent's wealth now that Nicholas is gone. Between my Scottish ancestry, and being raised among self-reliant, independent people, I'm not sure I could ever be comfortable with her wealth and the lifestyle that wealth will make inevitable. Tiera is a strong woman and she has a temper. Eventually, perhaps in a flash of anger, she'd remind me that she made our luxurious lifestyle possible. If she ever said anything remotely like that, I'd walk away. I'd have to."

"It might surprise you to hear me say this but I love her so much that I don't want her to be unhappy. If she can find what she's looking for somewhere else, then I'll try to be happy for her. I suppose it's ironic that today is Independence Day for Tiera too."

Elena's eyes filled with tears of sympathy for her best friend and for this strong yet gentle man she'd secretly loved since the first day she met him.

A few minutes later, they heard Tiera moving her luggage to a spot by the front door. They walked into the living room but Tiera wouldn't even look at David. When he offered to carry her luggage down, she shook her head angrily without saying a word. Elena went down with her to say goodbye. David walked out to his balcony to watch the only woman he'd ever loved get into a Yellow Cab and drive out of his life, probably forever.

Sunday, 6 July

The Capital Grill; Mclean, Virginia
8:15 PM

After only a couple of minutes at the hostess station, they were shown to the corner table he'd requested. He wanted Elena's last night to be special. Although he knew that Elena did not share Tiera's enthusiasm for fine wine, he ordered a bottle of the 2010 Quintessa Rutherford that was one of his favorites. They enjoyed a quiet meal and kept the conversation light until they'd finished their desserts.

"David, before I leave in the morning, can we deal with this elephant that's been following us everywhere we've gone these past few days?" David nodded but waited for her to begin.

"I think the American phrase is carrying a torch for someone. How do you see your future with Tiera? Are you still carrying a torch for her?"

"No. I think it's over between us."

"How can you be so sure? It's not hard to see that you still love her."

"Elena, you know better than I do how proud Tiera is. Before she left my room the other night, she said she was through humiliating herself with me. Until that moment, it had never occurred to me that she felt humiliated each time she offered herself to me only to have me turn her down. I thought about that later and realized that her love for me died quite suddenly that night when I refused again to make love to her. Her tone and words had a powerful sense of finality. She may still be fond of me, and she may have some painful moments thinking about what might have been, but I don't think she'll ever come back to me. For the reasons I've shared with you, I don't see myself trying to win her back. I'm deeply sad about it but I am enough of a realist to recognize her leaving as the end of our relationship. There's no question in my mind that she firmly intended for it to be the end."

"Would you give it another chance if she asked you to?"

"Obviously, I can't be sure what I'd do in that hypothetical scenario but my sense is that I wouldn't. I think I understand now how far apart we really are."

Elena sat quietly and calmly for several long moments. "OK, let's say you're right about all of that. Let's say it's truly over between you. What will you do?"

"You know I have job offers from the Central Intelligence Agency and the National Security Agency. Tiera had an interview scheduled with NSA this morning although she might have cancelled out. Do you know?"

"Yes. I called her when I went to the Lady's room a little while ago. She did go to the interview but after it was over she told them she wasn't interested in the position. She'll go home to Malta in the morning. We'll meet in the British Airways first class lounge at Dulles Airport about two hours before our return flight."

"I'm not really surprised. Knowing that, I'll probably accept one of the job offers I have pending within the next few days. I was going to wait until I knew if Tiera accepted a position with NSA. Both the CIA and NSA jobs are a good fit for my qualifications and both pay better than I expected. I'm leaning toward the CIA but haven't decided yet."

"But, to answer your question, I think I'll focus on my new job and try to hit the reset button on my personal life. I've been through a lot the past few months and the ghosts of Sarah, Nicholas, and all those innocent people don't let me sleep very well most nights. I still want to get married and have some kids. I think what has happened these past few months has made me even more aware of the fragility of our lives and the importance of living each day as though it might be our last. Tiera never seemed to quite get that. She's lived such a carefree, protected life that it doesn't seem to occur to her that it could all be gone in an instant. I don't think of myself as being impatient but when it comes to getting married and starting a family, I suppose I am."

"Yes, I understand that David. I want to ask you a question that I have to ask now because I may never get another chance. Has it ever occurred to you that Tiera wasn't the only one in love with you?"

David looked directly into her eyes before he answered. "Yes, Elena. I've known since that day at Cambridge when I put my arm around you to comfort you after the story I told made you cry. You snuggled against me like you were form-fitted. I may not be the most perceptive guy in the neighborhood but even I could read the signals that day. Yet, you never once violated your friendship with Tiera by trying to advance your own romantic interests. I've always respected you for that."

"David, Tiera knew too. She never said anything until Friday morning just before she left your condo. I tried to talk her out of leaving but her mind was made up. Her last words to me were something to the effect that she knew I loved you and saying she hoped I had better luck than she'd had. Does that surprise you?"

"No. I suppose not. I sense that you're asking a question but I'm not quite sure what it is."

"It's not too complicated. I want to know if I have any chance with you. I think you know that I want a family and kids just as much as you do. I think my faith is almost identical to yours. You've said you respect me and I think you know that I respect you. I know you're fond of me but that's not really the question, is it? I freely admit my love for you. What I want to know is whether you see any chance that you might come to feel the same way about me."

David took a deep breath. “It’s just too soon, Elena. I can’t possibly say right now. As I sit here looking across the table at you, I am very aware of you as a lovely young woman and a genuinely nice person. I can easily imagine you being a wonderful mother, wife, and lifelong companion for some lucky guy. I just don’t know if that could be me.”

“OK. That’s an honest answer and pretty much what I expected you to say. I have only a few more hours with you before I fly thousands of miles away and possibly out of your life forever. That thought is almost more than I can bear to think about. So, I’m going to ask you to do something for me. It’s a deeply personal request that would mean a great deal to me but it’s a lot to ask of you.”

“What is it?”

“When we get back to your condo, I want us both to get ready for bed but I want to spend the night in your room, sleeping beside you. I’ll wear my PJs and lie on top of your blanket or bedspread so we’re not even touching. I’ll cover up with one of the blankets from your guest room. I’d very much like it if you’d snuggle up to me so we both have some hint of what it could be like to be married and sleep together every night for the rest of our lives. I promise not to violate your trust by trying to seduce you. Like you, I intend to still be a virgin when I marry. I don’t expect you to kiss me or even wrap your arms around me though I would love it if you did. Will you do this for me, David?”

The corners of David’s eyes crinkled in a gentle smile as he reached across the table to touch her arm. “Yes, Elena. I guess that would be OK. It doesn’t sound too terrible.”

Elena laughed and pinched the back of his hand. “Not too terrible, huh? Thanks a lot.” David’s teasing humor was just what the moment needed.

“I have a suggestion of my own. Given what you’ve said, I’d like for you to think about delaying your return flight for at least a few days. It’s a long way from Virginia to Malta and it won’t be easy for us to see each other. If we don’t spend any time together, we’ll never be able to answer your question.”

"I had thought of that too but was afraid to even suggest it. Are you sure?"

"Yes. It makes sense and it just feels like the right thing to do."

As soon as they got back to David's condo, Elena called the airline to reschedule her return flight and then called Tiera to tell her she wouldn't be coming to the airport in the morning after all.

Monday, 7 July

David's Condo; Tyson's Corner, Virginia
6:57 AM

David woke a few minutes before the alarm was set to go off. Elena was lying beside him with her cheek resting on his outstretched left arm. She was awake and watching him.

"Who are you?" he teased. "How did you get in my bed?"

"Ah, young squire, you were dreaming of a beautiful princess and, voila! Here she is."

"Not bad for a 7:00 AM comeback," David laughed. "How'd you sleep?"

"Good. I dreamt that a handsome prince had come to rescue me." Pausing for precisely the right dramatic effect, she delivered her punch line. "But you got the princess and I got the croaking frog. Did you know you snore?"

"Well, miss beautiful princess, you might want to do something about your hair before your adoring fan club sees you."

Elena punched him playfully in his ribs and rolled out of bed laughing before he could retaliate.

David lay in bed until he heard the shower in the guest bathroom. A few minutes later, as he watched himself brushing his teeth in the mirror, he realized that he was already comfortable with Elena in a way that he'd never been with Tiera. Tiera was feisty and even fiery at times. She was competitive to the point of excess and he'd run afoul of her temper more than once. He'd learned to choose his words very carefully when discussing anything controversial with her. It really didn't take much to set her off. Elena had a playful sense of humor and a relaxed conversational style that was very appealing. Somehow she was easier to be with than Tiera. He'd need more time with her before he could answer the question she'd posed. Still, he thought they'd gotten off to a good start.

U.S. House of Representatives; Washington, DC
2:30 PM

By the middle of his second term in office, it was clear even to members of his own party that President Arnold had become a huge liability not only to his party but also to the nation. With the international situation spiraling out of control, and the Pax Americana that had maintained world order for over fifty years in tatters, something had to be done.

On the 7th of July, the Committee on the Judiciary of the U.S. House of Representatives adopted Articles of Impeachment against the President after Congressional investigations revealed that he was guilty of multiple felonies involving cover-ups and Constitutionally-prohibited abuse of executive power. The Articles of Impeachment were sent to the Speaker of the House to schedule a vote of the entire House of Representatives.

Tuesday, 8 July

Gavriella's Condo; Tel Aviv, Israel
6:20 AM

Gabe was sitting on her balcony sipping her second cup of coffee and thinking about the changes the week had brought to her life. Nahbi was working on some big project for Levi that he couldn't talk about and had never come home the previous night. She was still trying to get used to the fact that he didn't need to sleep. He did lie down sometimes but she knew it was just another position his body assumed while that incredible quantum computer brain of his whirled along at lightning speed.

Nahbi was a remarkable houseguest. He did the grocery shopping and refused to let her pay for any of it. He fixed great meals, kept the condo apartment spotlessly neat and clean, and did the laundry. After just a week of living with her, Nahbi had her spoiled and she loved it.

Nahbi was generous and kind. He was even-tempered. Gabe didn't think it was possible to make him mad. Even when she was in a foul mood, he never seemed offended.

He was a great listener and she'd told him deeply personal things she'd never revealed to anyone else. In a strange way, Nahbi was helping her heal. She had been deeply and passionately in love with her handsome young soldier husband. His gruesome death from six pounds of plastic explosive strapped to a poor little Palestinian girl had plunged her to such depths of despair that she had seriously considered suicide.

Yet, as an Israeli intelligence officer and Kidon leader, she hadn't dared admit to the Mossad's resident psychiatrist that she rarely could make it through an entire night without being awakened by a terrifying nightmare. It was always the same. She and her husband would be making love with her on top and about to have an orgasm when his body would explode beneath her covering her in his blood.

A week earlier, and just a little past midnight of Nahbi's second night in the condo, the nightmare had returned. As always, she had screamed in her sleep when the nightmare reached its terrifying conclusion. The sound of her own screams always woke her up. But there had been something different that night. Within seconds, Nahbi had been kneeling beside her bed. At first, she'd tried to dismiss it as just a bad dream but he had understood how frightened she was and had refused to leave.

He'd sat on the side of her bed and put his arms around her. Within the safety of his strong, comforting embrace, her emotional dam had collapsed and she had sobbed uncontrollably for several minutes. When she'd finally calmed down, he'd asked her to tell him about the nightmare. Somehow, telling him had been cathartic. When she'd finished, he'd told her to lie face down. He'd then done something that had absolutely astonished her. He had lain down beside her and began to gently rub her back. He'd never said a word.

When she finally woke the next morning from the best sleep she'd had in a very long time, Nahbi had been fixing breakfast for her.

David's Condo; Tysons Corner, Virginia
9:35 AM

David's cell phone vibrated in his pocket. He saw that it was Pete Nocona, the section chief of the CIA Counter Terrorism Center at Langley, Virginia. Pete had worked with David on Task Force 3.

"Hello Pete. What's up?"

"David, a guy named Jean Luc Davignon called me earlier today asking for your phone number. Do you know him?"

"Yes. He's a Captain in the French Special Operations Forces. I worked with him in Afghanistan. He's one of the best covert operation guys the French have. I haven't talked to him since a couple of days after my leg was reassembled at Landsthul when he called to check on me. Did he say what he wanted?"

"No. I tried to get something out of him but he wasn't sharing anything with me. He gave me a cell phone number and asked that you call him as soon as possible."

"OK. Thanks. If he calls again before I reach him, it's OK to give him my number."

9:55 AM

"Allo?"

"Jean Luc, this is David Logan. I understand you wanted to speak to me.

"Hello, David. Yes, thanks for calling back. We need to talk. Can you arrange to use an encrypted phone to call me back? What I need to talk to you about can't be discussed over an unsecured connection."

"Yes, I think so. Can I call you back on this number to set a time? I doubt that I can get it arranged before tomorrow morning."

"Yes. Just be very security conscious, please."

"Understood.

"Good. I'll be waiting for your call. Au revoir."

10:25 AM

"Pete, this is David. I'm going to need access to an encrypted phone for about an hour early tomorrow morning. Can you help?"

"Maybe. What can you tell me?"

"I returned the call to Jean Luc but he wouldn't tell me anything over an unsecure phone. He asked me to call him back tomorrow morning on an encrypted connection. I'd prefer that the call not be monitored or recorded at this end if that's possible."

"I can't guarantee that last part. You know as well as I do that the National Security Agency monitors an enormous volume of calls and that their super computers can probably decipher just about any encryption algorithm given enough time. The very fact that the call will be encrypted will identify the call as something their computers might be interested in so it will almost certainly be automatically recorded. That doesn't mean it will be deciphered though."

"Why do you say that?"

"We place a lot of encrypted calls from here every day. Most government offices, American embassies, and military installations do too. NSA has ways to identify those calls as originating in one of those sources. I'm pretty sure the call will still be recorded but I'm almost equally certain that it would never be deciphered without a good reason to commit the necessary computer resources."

"OK. I guess that'll have to do. Where can I make the call from?"

"Be in my office by 7:15 AM. I'll notify the gate and front desk that you're cleared for access. I'll get you set up and then go to the morning intelligence briefing. I'll be gone about an hour. Maybe we can talk when I get back."

"Thanks. See you in the morning."

Wednesday, 9 July

Gavriella's Condo; Tel Aviv, Israel
9:45 PM

Gavriella was alone in the condo for the second night in a row. Nahbi had told her he needed to stay at the office again to prepare for the briefing he would be giving on Friday.

Gabe finished toweling off after her shower and took a moment to look at herself in the mirror before donning a cotton bathrobe. She was not especially modest. When she'd lived alone, she often went nude around her apartment and sometimes stood on her balcony at night letting the sea breezes play over her skin. She'd always been a sensual person but Nahbi's presence in the condo had made things a bit awkward.

Both of them had been careful not to let the other see them nude or partially clothed as a matter of simple courtesy. Yet, Nahbi did their laundry and obviously saw her underwear. She thought he might even have caught a quick glimpse of her nude once when her bedroom door was ajar and she'd walked into her bedroom from the shower. She'd seen only his back as he walked away from her bedroom. She'd wondered if he'd seen her but done the gentlemanly thing by turning away. If he had seen her, she couldn't help wondering what he might have thought. She knew she had a decent figure. She was extremely fit with pert breasts and a firm butt. Her tummy was flat and hard from the sit-ups she did every morning. Her thighs and calves were nicely shaped but lean and hard from her running. Young men she passed in public often gave her appreciative looks but she had no idea if an android had any concept of what made a woman physically attractive.

She was quite aware that she was starting to relate to Nahbi as a man. It had begun when he wrapped his arms around her to comfort her after her nightmare. She couldn't erase the memory of him lying beside her and giving her a loving and tender back rub. He had such an appealing personality and was so handsome and lifelike, that it was hard not to be attracted to him. It had been a very long time since she'd been with a man and Nahbi's presence in the condo had awakened feelings she hadn't had since her husband's death. She realized that she was wading into uncharted waters but Nahbi was giving her something she needed.

The previous day she'd bought a bottle of the same cologne she used to buy for her husband. She hadn't tried to fool herself by pretending that her purchase was just a gift for a friend. She knew full well that it was driven by her increasing attraction to Nahbi as a man.

She walked out onto her balcony to enjoy the warm evening breeze. Still thinking about Nahbi, she took off her bathrobe and laid it over the back of a chair. Standing with her feet apart so she could feel the breeze all over, she closed her eyes and ran her hands over her body. Her breathing quickened as she used her left hand to caress her breasts while letting her right hand to do something she hadn't done in a long time.

The examinations and tests conducted at MI6 in London had revealed that Nahbi was physically capable of making love to her. She couldn't stop trying to imagine what it would be like.

CIA Headquarters; Langley Virginia
7:26 AM

David picked up the handset when the light showed that the encrypted connection had been made.

"Bonjour, David. I apologize, but I don't have much time. The Prime Minster and Minster of Defense have just arrived in the building and I'm on standby to brief them. If we get cut off, call back in about an hour."

"The short version is that the Prime Minister has ordered our Special Forces to capture General Jhalawan and bring him to France to stand trial. We are already planning the mission but need some help from you. Can you come to France for a few days? We've set up a temporary headquarters at the Toulon Naval base. I know your family is still in mourning so I hate to ask. I wouldn't if I didn't think your input was critical to a successful mission. Can you do it?"

David was silent for a moment as he thought about how much his grandparents and brother-in-law needed his support right now. Yet, capturing the man responsible for having his sister and Tiera's brother murdered was high on his personal priority list. As he saw it, he really had no choice.

"Yes. A friend of mine is here visiting from Malta. She's booked on an Iberia flight to Madrid that leaves Dulles Friday morning. I'll see if I can grab a seat on the same plane. She'll be changing planes in Madrid to connect for Malta. I'll connect to Toulon-Hyeres. I'll have to check the flight schedules and call you back but I think I can be in Toulon by Saturday.

"Jean Luc, I just thought of something else. One of Jhalawan's unwilling operatives was a young man named Tariq Zadari. He's the only one I know who has been face-to-face with Jhalawan. Tariq is a well-educated Pakistani nuclear engineer with a personal interest in seeing Jhalawan brought to justice. I think he and his family are in deep witness protection somewhere in the U.K. I'll call Sir Lionel Smythe at MI6 to see if Zadari can be contacted. Tariq feels like he owes me for arranging to get his family out of Pakistan. I think he'd be willing to help."

Thursday, 10 July

Mossad Headquarters; Herzliya, Israel
11:10 PM

His briefing slides were ready. Sitting alone in his office, Nahbi's mind turned to a topic he'd had on his mind for several weeks.

He'd spent a lot of time with David and Tiera while they worked together on Task Force 3 in London. He'd been fascinated by the outward evidence of their love for each other. His extensive research indicated that human love had two distinct phases.

The first seemed to be a purely involuntary emotional response triggered primarily by visual cues. Those initial feelings might last a while but they must inevitably be replaced by something more enduring. So far as Nahbi could determine, there was no evidence that a machine could have the kind of emotional feelings that first attracted humans to each other. But he saw no reason why even a machine could not behave in a loving way toward another sentient being and thus practice and perhaps even experience a form of love.

The more he'd thought about it, the more he'd wanted to be involved with a female human in some kind of romantic and loving relationship. When the Israeli prime minister had granted him citizenship, he'd emphasized that Nahbi would have all rights of citizenship including the right to enter into a domestic relationship or to even get married.

Nahbi had made mental lists of every kind of behavior he could think of that might plausibly demonstrate love. Because he always tried to be kind and considerate to the people he worked with, he automatically did many of the things that humans interpret as evidence of love. Lacking any mechanism for anger, he never contributed to the escalation of unpleasant confrontations. Instead, his calm, gentle manner made it difficult for even very angry people to stay mad around him.

He was quietly aware of his own physical and mental superiority. Yet, unlike humans competing to establish a pecking order among themselves, he had no need to demonstrate his superior capabilities or maneuver for advantage. He could listen intently without interruption or making any effort to dominate a conversation.

His enormous intellectual capacity allowed him to understand people's fears and hopes far better than any human because he could fully assimilate the intricate and often complex factors that contributed to human emotions and feelings. Even though he lacked the capacity to actually experience human emotion, his intellect made it possible for him to synthesize something very akin to both empathy and sympathy. His kindly facial expressions, a quiet comment here and there, and his body language communicated unmistakably that he really understood. That was something most humans hungered for and it came almost naturally to him.

After moving in with her, he'd soon realized that Gavriella was what the Israelis call *tsabar*, a type of prickly pear cactus that is tough on the outside but soft and sweet on the inside. Her former training in the Israeli Defense Force, and her subsequent work as a Kidon leader, had turned her into a formidable warrior. Nahbi knew that she was fully capable of dispassionately killing her country's enemies. Yet it was equally clear to him that she was terribly lonely with deeply concealed emotional scars. He could see beyond her tough, warrior exterior to the vulnerable and soft young woman on the inside and he found the contrast intriguing.

While keeping her company as she worked out in her condo's basement fitness center, he'd learned that Gavriella often wanted a good massage after a strenuous workout but, paradoxically, she didn't really like to be touched by other people. He'd done some research, asked her a few guarded questions, and soon become her masseur. Gabe had told him that he must be telepathic because he could precisely apply just the right pressure and motion to the areas that needed attention. She'd said that no masseur or masseuse had ever given her such luxurious and satisfying massages.

His internal clock reminded him that it was almost midnight. He decided to go back to the condo so he could change clothes before the briefing that was just a few hours away. He'd already missed the last bus so he'd have to jog the ten kilometers to Gabe's high-rise condo. He couldn't run at his maximum speed without drawing attention so it would take him almost an hour. As he ran, he could make some final mental changes to the narrative he would use with his briefing slides.

Nahbi let himself into the condo as quietly as he could and then walked down the hallway to Gavriella's bedroom door. Opening it quietly, he stepped inside her room and watched her sleeping for a moment before closing her door and going to his own room.

Friday, 11 July

Gavriella's Condo; Tel Aviv, Israel
12:20 AM

Gabe heard Nahbi close the front door but was surprised when she heard the faint sound of her doorknob being turned just before he came quietly into her room. Lying on her side and facing away from the door, she kept perfectly still. Knowing that he was fully aware of her biometric signals, she forced herself to keep her breathing shallow, hoping he wouldn't realize she was awake. She was puzzled at why he'd come into her room in the middle of the night. Sleeping nude as she always did, she had a quick sexual fantasy that Nahbi was dropping his clothes to the floor and would come naked into her bed at any moment. With those thoughts, her breathing quickened and she realized too late that Nahbi had detected it. He left her room immediately leaving her wondering what had prompted his visit.

After he closed her bedroom door, she rolled onto her back thinking about something that had happened a few days earlier. She'd had a grueling day at the office and had come home with a powerful headache. Nahbi had seemed to know instinctively what she needed. He'd put a table in front of the kitchen sink for her to lie on while he shampooed her hair. He'd been so gentle and attentive to her comfort that she'd begun to cry when she'd finally admitted to herself that she was in love with him. He was a better person, companion, and friend than any human she'd ever met. The incredible, amazing thing was that Nahbi hadn't seemed alarmed or discomforted by her tears the way most human men would have. Instead, he seemed to intuitively understand the deep sadness she was feeling over a love she'd lost forever, intertwined with the joyous realization that she was again loved and no longer alone.

Mossad Headquarters; Herzliya, Israel
9:00 AM

Nahbi waited until the room was quiet before beginning his presentation.

"Soon after I became an Israeli citizen, our prime minister made it very clear to me that my sole duty would be to find a way to abolish Iran's existing nuclear weapons program and to ensure that Iran will never again possess the ability to wage nuclear war against Israel or to hold the world hostage to the mere threat of nuclear warfare. His charge to me was unambiguous."

"Since then, I have read virtually every authoritative document on the subject including literally thousands of relevant articles and policy papers. I now believe I can describe a way to achieve the objective at minimal risk."

"As most of you know, the first Supreme Leader of Iran, Ayatollah Ruhollah Khomeini, stated publicly that the Islamic Republic of Iran was not trying to build nuclear weapons because Islam bans weapons of mass destruction. Some have gone so far as to claim that he issued a fatwa against the development of nuclear weapons. Despite many hours of intensive research, it does not appear that such a fatwa actually exists."

"During the Iran-Iraq war, Iraq used chemical weapons multiple times. Iran did not use chemical weapons in retaliation because it lacked the capability and because Ayatollah Khomeini apparently considered chemical weapons to be prohibited by Islam. Yet, toward the end of the war, Khomeini reversed his position because it appeared that Iraq was preparing to use chemical weapons against Iranian cities."

"Fatwas can be, and often are, modified or even cancelled. They are not, as some might say, *carved in stone*." It is important to remember that Shiite tradition permits and encourages deception and even blatant lies when such tactics are beneficial to Islam."

"Ali Khamenei, as the anointed successor to Khomeini, has also publicly claimed to have issued a fatwa saying the production, stockpiling, and use of nuclear weapons was not permitted under Islam. But, again, no such fatwa apparently exists. However, a fatwa issued by a senior Islamic cleric does not always have to be in writing. A clear verbal statement that the cleric is issuing a fatwa has the same theoretical force as if it had been in written form. The crucial question then is whether mere verbal assertions that such a fatwa exists, can be accepted as reliable evidence that it does. The evidence strongly indicates that the assertions of a non-nuclear fatwa are intentional deceptions designed to buy time for Iran to complete its objective. I do not believe such a fatwa has ever actually existed."

"While the existence of an official fatwa is in doubt, Iran's nuclear capability is not. It is all but certain that Iran already possesses at least six nuclear bombs and is producing highly-enriched uranium at a rate that will enable them to produce one new weapon per month. In addition, Iran and North Korea have been collaborating on development of a long-range missile capable of carrying a nuclear warhead. Russia has agreed to build six new nuclear reactors capable of producing plutonium that can be used to build high-yield devices small enough for a missile warhead."

"Although previous American administrations have been reluctant to use them, the Americans have developed and produced what they call the Massive Ordnance Penetrator or MOP. This bomb is more than twenty feet long and weighs in excess of 30,000 pounds. It has a precision guidance system and was created specifically to penetrate the reinforced concrete protecting Iran's underground nuclear facility at the Fordow complex. If the MOP can destroy Fordow, it can destroy any of the other, less-well-fortified Iranian nuclear production sites."

"I've been asked remind you that this briefing is national security top secret. You may not discuss or reveal anything you hear or see here today to anyone. I need to emphasize that because what I am about to say will include several things that are classified at the highest, need-to-know level."

"What I'm about to tell you now is one of the most closely-guarded secrets in the world. *Israel has four state-of-the-art successors to the SR-71 Blackbird.* These aircraft were produced in a joint research and development program with the Central Intelligence Agency. They fill a void left when the last Blackbirds were retired in 1998."

"Like their legendary ancestors, the aircraft were produced at Lockheed Martin's famous Skunk Works in southern California. The

American's call their variant Aurora and have given it the designation of SR-75. We call ours Bardelas, which is an ancient word taken from the Mishna meaning cheetah. That is certainly appropriate since these aircraft are supersonic with a classified top speed in excess of Mach 3. They can also operate at altitudes in excess of 100,000 feet and cannot be detected by the most advanced computer-enhanced radars on Earth."

"Like their predecessors, these incredible aircraft were designed primarily for reconnaissance missions. However, the Israeli variants were modified after delivery to allow them to carry two large pods that can be outfitted for a variety of special-purpose missions, including high-altitude/low-opening manned parachute jumps. They can also carry up to 6,000 pounds of aerodynamically streamlined smart bombs. Using high-resolution cameras with super-sensitive graphene sensors, these aircraft have been taking extraordinarily detailed photographs of all of the Iranian nuclear facilities for the past few years."

"Most of you will be surprised to learn that Israel also has ten of the American B-2 stealth bombers. When the United States Congress and a previous administration cut funding for the B-2 after delivery of the 20th aircraft in 1995, Northrop Grumman Corporation offered to build 20 more at a deep discount to avoid massive layoffs. Northrop was willing to make such an offer hoping that the next administration would restore funding. Incredibly, the Americans declined the offer. However, Israeli intelligence was aware of both Northrop's offer and the administration's short-sighted refusal to accept the deal. Israel could not afford all twenty of the aircraft, so our prime minister approached the Brits to see if they would join in making a proposal to the American President to allow our two countries to accept Northrop's offer under the United States Foreign Military Sales Act. That proposal was subsequently approved."

"Later, after Boeing's development of the MOP, the B-2 aircraft were modified to allow them to carry two of the MOPs in each aircraft." "Before I move to the key part of my briefing, please bear two facts in mind. Israel has a stockpile of MOPs that is sufficient to destroy every one of Iran's known nuclear sites. We also have the ability to deliver those weapons with almost no risk that our delivery aircraft will be detected or intercepted."

"But, before we rejoice in our apparent salvation from Iran's nuclear threat, there is something else we must consider."

"The sanctions imposed on Iran have subjected the Iranian people to severe hardships. While some dissident groups do exist, the vast majority of Iranians are proud that their country has a nuclear research program. Everything that I say from this point on will be based on my conclusion that simply bombing Iran's nuclear facilities will have only a temporary deterrent effect. I agree with the assessment of one of the Mossad's most respected former directors. Average Iranian citizens are almost certain to rally around their leaders in defiance of the western powers with especially intense hatred of Israel as a battle cry. I believe we will have to destroy Iran's nuclear facilities soon but it is considerably more complex than simply loading some MOPs on the B-2s and flying off into the night."

"Before we bomb Iran's nuclear facilities, we must alter the attitudes of the vast majority of Iran's citizens. By the time the bombs are used to destroy Iran's nuclear facilities, the average Iranian must believe that their hardships and the attack on their nuclear facilities were caused by their own leaders' self-serving obsession with power. This will not be easily achieved. Yet, I have developed a plan to do precisely that."

Over the next hour, Nahbi took Israel's senior political, military, and intelligence officials through the details of his plan. When he was done, he was asked to leave the room while they discussed his recommendation. Only a few of those present in the room knew that he was not human so it was inevitable that his wisdom and political bias would be questioned. After more than two hours of intense debate, he was asked to return to the conference room where the prime minister advised him that his plan had been approved.

Madrid Barajas Airport; Madrid, Spain
8:05 PM

David and Elena had left Washington Dulles International Airport at 6:00 AM for the seven hour flight to Madrid. They'd spent the entire flight getting to know each other better. By the time the Iberia Airlines jet landed, they were already discussing when they might be able to see each other again.

The public address system announced that Elena's flight to Malta was starting to board. David's flight to Toulon-Hyères was scheduled to leave at 9:10 PM so he had time to walk her to her gate.

"I'll be in Toulon for several days. I'm not sure what my schedule looks like after that. I'll try to come to Malta for a few days before I go back to the States but don't hold me to it. I'll come if I can."

"Will you want to see Tiera or her family when you come?" Elena asked.

"I've given a lot of thought to that. If I thought it would be possible for Tiera and me to remain friends, then I think I would want to see all of them. Wouldn't it seem rude to come to Malta and not even pay my respects to Mathias and Maria? Maybe you could test the water a bit while I'm in France. I think I'm going to be in some long meetings. I'll call when I can."

Just before Elena handed her ticket to the gate agent, David gave her a big hug and kissed her on the forehead. "I'll be thinking about you. Maybe when we're together again, I'll be able to answer your question. I hope you have a good flight home."

Toulon- Hyères Airport; Toulon, France
10:45 PM

David's plane was a few minutes late in arriving at terminal 4S. He took the underground tram to terminal 4 where Jean Luc Davignon was waiting for him when he stepped off the escalator.

"Welcome to France."

"Hello Jean Luc. It's good to see you. It's been a while."

"Did you check any bags?"

"No, this roll-aboard and the coat I'm wearing is it."

"I've got a room for you in the transient officer's quarters at the naval base. I'm staying there too. It's about 22 kilometers to the base via the A57. I'm parked across from the terminal so we should have you checked in and ready for bed before midnight. I'm afraid it will be a short night though. I'll knock on your door at 6:00 AM and take you to breakfast. You'll see some of the guys who were with me in Afghanistan. Our briefings start at 7:00 AM in the main conference room at the headquarters building. It's supposedly business casual for civilians but I can tell you that with General Montreuil, that means you wear a tie and maybe loosen it if the meeting becomes intense. "

As Jean Luc accelerated his Citroen SUV onto the Autostrada, David asked, "You said you needed my input to your mission plan. What kind of input?"

"Sir Lionel Smythe suggested that we include you in the planning. He said you probably understood how general Jhalawan thinks better than anyone in the west."

"That may not amount to much. He's a secretive guy heading up what many think is the top intelligence agency in the world. We don't really know an awful lot about him other than what we learn by following the trail of dead bodies he leaves in his wake."

"Actually David, there's a bit more that I'm not supposed to tell you. General Montreuil wants you to go with us on the mission but he wanted to ask you himself. You'll have to put on your best poker face when he asks but I couldn't let you get hit cold with something like that."

"Hmm. Thanks. I guess. Why does he want me on the mission? Doesn't he realize that I'm not in special ops physical condition anymore and that my leg could be a real problem?"

"He knows all of that but he's checked around and nearly everyone he's talked to says you could make the difference. Sir Lionel told him that you remain calm under pressure better than anyone he's ever known. Pete Nocona said you had an intuitive ability to connect dots other people don't even see. Al Flores said he'd follow you into the gates of hell. That's a direct quote, by the way, from the man Pete Nocona says is his top field operator. Al told the general that you're the best natural leader he's ever seen."

"This starting to sound like a sales pitch."

Jean Luc laughed. "No, that comes later."

"There's more you're not telling me, isn't there?"

"Well, yes, but I'd better let the general tell it his way."

"Jean Luc, do you really think I could keep up with your team under battle conditions on rough terrain?"

"I think you'll have to tell us if *you* think you can."

CIA Headquarters; Langley Virginia
6:20 PM

Roger Benson was still in his office when the call from France came in.

"Good evening General Montreuil."

"Hello Roger. Did Pete brief you about what I have in mind?"

Yes but I'd like to hear it from you."

"Fair enough. As you know, our Prime Minister has ordered me to send a special operations team to Pakistan to abduct General Jhalawan and bring him to France to stand trial for war crimes. I've decided that Captain Jean Luc Davignon will lead our team. He's one of our very best special operations officers. His unit did some joint missions in Afghanistan with Lieutenant David Logan's intelligence unit. I've been thinking that perhaps this should be a joint mission too."

"By that, I understand you to mean that CIA field personnel would be on the team?"

"Yes. Ideally, I'd like for it to be a joint COS/CIA mission with Davignon and Logan as co-commanders. I'm told they developed a close relationship in Afghanistan. I had Jean Luc invite Logan to come to Toulon for some briefings in the morning. Right now, Logan's a civilian so having him participate in the mission could be awkward but I think we can find a way if he'll agree."

"How much CIA involvement do you envision?"

"In the field? Just Logan and Flores. But, we'd need full support from Langley too. Pete Nocona and Virginia Barton come to mind."

"I'm sure you know that the current director of the CIA probably won't approve a mission to kidnap a senior Pakistani official."

"Yes. I know he was a political appointee with, shall I say, limited qualifications. However, I'm hoping you have enough latitude to support the mission without taking it to him. Couldn't we structure it as an advisory role or something equally euphemistic?"

"Maybe. Let me sleep on it and I'll call you back. In any event, doesn't the question become moot if Logan doesn't agree to what you have in mind?"

"I'm hoping I can persuade him."

Saturday, 12 July

COS Headquarters; Toulon Naval Base, France
6:45 AM

Several of Jean Luc's men recognized David as soon as they entered the conference room and came over to welcome him. Jean Luc introduced David to several people including a couple of senior officers and two civilians. They were still standing when General Montreuil entered the room precisely at 7:00 AM. All of the military personnel snapped to attention and then took their seats at the general's signal.

"Good morning. I'd like to welcome Lieutenant David Logan, who some of you already know from Afghanistan. Let's get started. Captain Davignon, you have the floor."

Jean Luc stepped to the podium and pressed a button to begin his slide presentation.

"The man you see on the screen is Lieutenant General Parvez Jhalawan. He is considered one of Pakistan's most capable military commanders and strategists. I should also emphasize that General Jhalawan is utterly ruthless in the pursuit of his radical Islamic agenda. For the past year, he has been the director of Pakistan's Inter-Services Intelligence Agency. The photo is fairly recent."

"The prime minister has ordered the Commandement des Opérations Spéciales to capture General Jhalawan and bring him to France to stand trial for war crimes. I've been given command of the COS unit that will carry out the mission. Most of the team will be members of the French Army Special Forces Brigade who served with me in Afghanistan. Some of you may be wondering why Mr. Logan, an American, has been asked to join us. I served with him in Afghanistan and we conducted several joint missions together. He has since been discharged from the Army as a result of combat wounds he incurred. After leaving the Army, he received a doctorate degree from Cambridge University and subsequently led the joint MI6, CIA, and Mossad Task Force effort to find the stolen nuclear weapons. Sir Lionel Smythe, Director of the British Secret Intelligence Service, believes that Mr. Logan knows General Jhalawan better than anyone in the west. At my invitation, he has flown here from the United States in the hope that he can contribute to the success of our mission."

"There should be no illusions. General Jhalawan will be very difficult to capture and extract. His office in Islamabad is considered impregnable by anything short of a full military assault. His personal villa in Janawai has a full-time security detail of combat-hardened veterans and one of the most advanced security systems that a big pile of money can buy. He is certainly aware that he is now at the top of every western power's list of most wanted terrorists so he's on full alert. Only an extraordinary mission plan can succeed without heavy casualties. Even so, I am determined to find a way to get Jhalawan out of Pakistan without shedding any more French blood."

Jean Luc advanced the slides again. "This map shows the location of Janawai in relation to Islamabad, Pakistan and Kabul, Afghanistan. As you can see, Janawai won't be easy to get to. It's roughly 240 kilometers northeast of Islamabad but much of the route is along dirt roads that follow the Neelum River passing through multiple small towns and villages along the way. When Jhalawan goes to his villa, he nearly always goes by military helicopter. From Kabul, it's almost 900 kilometers by a tortuous route that includes crossing some of the highest mountain passes on Earth through some nasty territory controlled by some well-armed tribal warlords who seem to hate everyone."

"To the north of Janawai lies the western range of the Himalayas and the eastern parts of the Hindu Kush mountains. To put this in perspective, Nanga Parbat and K2, the ninth and second tallest mountains on Earth respectively, are just to the north of Janawai. Beyond that lies the great Taklimakan Desert of China which is more than three hundred miles wide and almost eight hundred miles long. As you can see, General Jhalawan need fear little from the north."

"A recent K2 expedition left Islamabad on May 20th and they didn't reach their base camp until June 4th even though they were driving SUVs most of the way. To illustrate how challenging travel in this area is, it took their team thirty-three hours to drive 190 kilometers from Islamabad to Sharda and the first half of that trip was over paved roads. It's another 40 kilometers to Janawai. Even if we could launch from Islamabad without being recognized as a foreign special operations force, we'd be observed all along the route. In my opinion, it will be very difficult to sneak up on Jhalawan by trying to infiltrate along the Neelum River Valley."

"I also considered infiltrating from the coast, possibly from a submarine. However, it's almost a thousand miles from the port of Karachi to General Jhalawan's villa in Janawai. It's even further from the port of Gwadar near the Iranian border. It's just too far for a special operations team to remain undetected."

"I'm sure some of you are wondering why we couldn't do it the way the Americans did when they took down Osama bin Laden in Abbottabad. The difference is that the Americans were nominally allies of Pakistan in the war against terror at that time so the UH-60 Blackhawks used by SEAL Team Six were already in the area under CIA control. Even the Americans lack that capability now and France has absolutely no military assets near Islamabad that could be used."

"So, to wrap up my part of the briefing, there's an awful lot we don't know yet, including what infiltration route to use. General Montreuil?"

"Thank you Captain Davignon. I'll open the floor for discussion and questions now."

One of the civilians stood immediately.

"For those I've not had the pleasure of meeting yet, I am Pierre Hulett of the Ministry of Foreign Affairs. I was fortunate to be on vacation in Venice when our headquarters on the Quai d' Orsay was destroyed. I am currently the executive assistant to the acting Minister. He has asked me to participate in this briefing to ensure that this mission is conducted in a way that is consistent with our foreign policies. I have one question for Captain Davignon. Do you really think it will be possible to abduct General Jhalawan without creating an international crisis of epic proportions?"

"Monsieur Hulett, forgive me, but I think we already have an international crisis of epic proportions as you put it. Many of our fellow citizens are dead and the government of the man responsible has refused to allow him to be extradited by claiming that our contention that he was responsible is without merit. Throughout history, an attack such as we suffered has triggered all-out war. My sense is that the prime minister is trying to find a way to visit justice on the monster who masterminded the attack without putting additional thousands of French lives in harm's way by declaring war."

Jean Luc saw that David wanted to speak. Mr. Logan, do you have something to add?"

"Monsieur Hulett, I think it's at least possible, and perhaps even likely, that Pakistan will clamp a lid on anything that happens as a result of this planned mission. The more they deny General Jhalawan's role, the more they encourage the west to show the overwhelming evidence. Since General Jhalawan is a senior military officer and the head of Pakistan's Secret Intelligence Service, it is not in Pakistan's interest to allow an international media circus whether the mission succeeds or not. Their best strategy, in my opinion, will be to act as though nothing has happened and then to do everything in their power to ensure that Jhalawan never stands trial in a French court. If this mission successfully extracts him from his villa in Janawai, the Pakistani government will set loose the hounds of hell to kill him and the entire special operations team before they can escape from Pakistan."

There were murmurs of agreement around the room.

Monsieur Hulett nodded thoughtfully before asking David, "Why would they kill one of their most respected military commanders?"

"He's radioactive now if you'll forgive my unfortunate choice of words. To his face, they will show him the deference that his power demands. Privately, they will be looking for ways to take him completely out of the picture. I don't imagine many of Pakistan's leaders are big fans of the American poet, Ralph Waldo Emerson. Yet, they would all instantly understand what Emerson meant when he wrote, '*When you strike at a king, you must kill him.*' Had Jhalawan succeeded they would have been pleased to see the hated western infidels brought low. But, to their minds, one hit and four misses did not '*kill the king,*' so to speak. In their eyes, all Jhalawan did was stir up some very powerful hornets. Many in the current Pakistani administration would be perfectly happy if Jhalawan was killed in the COS assault and then the entire COS team was wiped out far from news cameras. The absolute worst thing that can happen from the Pakistani perspective is that the COS team is successful in whisking him out of the country alive to stand trial under bright television camera lights."

After Jean Luc's briefing and David's remarks, there really wasn't a lot more to say. Infiltration would be difficult. Getting out of Pakistan alive would be even harder.

Officer's Club; Toulon Naval Base, France
11:30 AM

David and Jean Luc had been asked to join the General for lunch. As the three of them entered the foyer of the Officer's Club, General Montreuil shook hands with a U.S. Navy flag officer who had obviously been waiting for them. "David, I'd like you to meet Vice Admiral Stanley Chambers. He's the Commander of the U.S. Sixth Fleet and Commander Striking & Support Forces NATO that's based here in Toulon. I've asked him to join us for lunch. Like all of his Sixth Fleet predecessors, he's spent most of his career in naval aviation. Earlier in his career, he gained a lot of special operations experience when he commanded the United States Naval Special Warfare Development Group usually referred to as simply DEVGRU. It's not widely known but the CIA's Special Activities Division and its elite Special Operations Group often works with and even recruits from DEVGRU. He's aware of your work with Task Force 3 and knows that I've asked you to help us plan the mission."

Admiral Chambers extended his hand and gave David a friendly smile. "I'm pleased to meet you Mr. Logan. Sir Lionel Smythe is an old friend of mine. He told me your team intercepted all of the stolen weapons and then ran into a bad piece of luck with an embedded member of the Muslim Brotherhood. We're all deeply saddened by the outcome but I'm glad to be able to tell you in person that your leadership unquestionably kept the tragedy from being many times worse. My daughter and two grandkids live in Manhattan. Your team saved their lives and I am deeply grateful."

"Thank you sir."

The hostess led them to a private dining room with a view of the harbor.

General Montreuil gave Admiral Chambers the high points of the morning's briefing. "We all agree that trying to take Jhalawan while he's in Islamabad is out of the question. It looks like his villa in Janawai will be our best bet but it's far from clear how to do it. Jean Luc and David agree that infiltration along the Neelum Valley road might be observed and reported to Jhalawan long before the team could approach him. We don't have any air assets in the area like you yanks did when you took down bin Laden. It looks like a ground approach is our only option. By the way David, before I forget to mention it, I thought you handled the Ministry hand-wringer rather well." Turning back to Admiral Chambers, General Montreuil summarized David's view that the government of Pakistan would do everything possible to ensure that Jhalawan never stood trial in a western courtroom.

"Captain Davignon, how do you see the mission?" Admiral Chambers asked.

"The biggest problem we face is that Janawai is a thousand miles inland and Pakistan is surrounded by countries that would not permit us to overfly their territory on a military mission. If we had the necessary assets, we might parachute in using high-altitude, low-opening HALO tactics. Unfortunately, any aircraft we could use for a HALO insertion would be over hostile territory for a minimum of a thousand miles each way."

The latest variant of the UH-60 Blackhawk helicopter that SEAL Team Six used on the bin Laden mission has a maximum cruise speed of less than two hundred miles per hour and a service ceiling of only nineteen thousand feet so it would be over hostile territory for over five hours each way and at a low enough altitude to be an easy target for surface to air missiles. But, the Blackhawks have a limited combat radius of less than four hundred miles. It's a fine aircraft but totally unsuitable for this mission."

"Normally, a HALO mission uses a cargo aircraft with a drop rear ramp like the Lockheed C-130 Hercules. The latest C-130 variants have a service ceiling of over thirty thousand feet but that's still well within range of Pakistani SAMs. The Hercs have a thousand mile combat radius but they'd be bone dry before they could get back over international waters in the Gulf of Oman. Even with tankers waiting for them it would be pushing the envelope to the limit. Also, maximum speed for the C-130H model is about three hundred knots so they'd be over hostile territory for over six hours roundtrip."

"But, none of these possible strategies are very helpful since even if we could do an airborne infiltration, I don't know how we'd get back out. We can't just head to a local airport and wave goodbye as we takeoff."

"It's also pretty obvious that we can't just land at Islamabad International Airport and rent some SUVs. We also can't go in from the north. Infiltration from India is fraught with both terrain and political obstacles. So, I think our only viable option is to fly into Kabul either solo or in groups of two. We'll all be carrying backpacks and hiking gear but no weapons or military equipment of any kind. We'll use fake passports from several countries and disguises we can change later if necessary. Rather than rent vehicles, we'll buy or steal enough of them to put two men in each vehicle. That will give us some transportation redundancy in case of breakdowns and lessen the chance that we'll be recognized as a special ops force. We'll enter Pakistan over one of the high mountain passes along the border."

"The Russians left a lot of AK-74 rifles behind when they pulled out of Afghanistan so weapons are readily available on the Afghan side of the border. The AK-74 is reliable even when filthy dirty. However, for our purposes, the Izhmash AK-9 assault rifle is a better choice because it was designed for special operations and is compact. It fires the subsonic 9x39mm cartridge and is very quiet with a suppressor. These rifles are highly prized in the region so they'll be expensive but I'm sure we can pick up enough to support our mission. The Russian-made Gryazev and Shipunov GSh-18 semi-automatic pistols may be hard to find but they're a great pistol and we won't need many. They hold eighteen rounds of 9mm NATO ammunition and can also fire the new 7N21 armor-piercing rounds that Kevlar vests can't stop."

"The Nowa pass is about seventy-five kilometers north of the main crossing at the Kyber pass. The elevation at the summit of Nowa pass is over two thousand meters so it's almost twice as high as the Kyber pass and thus seldom used. The road isn't great but an SUV or 4x4 pickup can manage it. I think we'd stop a couple of miles from the border and then overwhelm the guards just after midnight so our vehicles wouldn't be reported when we cross the border. From Nowa pass, we'd head southeastward toward Mardan. We'd follow the Islamabad-Peshawar motorway and cross the Indus River about thirty kilometers southwest of the Tarbela dam. It's about 40 kilometers from the bridge to Wah. From Wah, it's about sixty kilometers to the town of Murree. It's another sixty kilometers to Muzaffarabad where we'd pick up the Neelum Valley road. The road is a narrow, two-lane dirt road in that area but passable with an SUV. It's a hundred kilometers from Muzaffarabad to Sharda via the Neelum River Valley road and another forty to Janawai. Driving at night to avoid being seen, I think we can make it to Janawai in four or five days after leaving Kabul. We would then establish observation posts until we knew the neighborhood and the guard change schedules. Once we have Jhalawan, we'd retrace our route moving him between vehicles to add some confusion if we're followed."

David and I are both familiar with the Afghan side of the border by virtue of dozens of missions. That familiarity will be very helpful. What do you think, David?"

David didn't say anything for several moments.

"I don't think so, Jean Luc. It's too complicated and we'd be on hostile territory far too long. As you know, the high mountain passes are controlled by Jalaluddin Haqqani and his son Sirajuddin. They're loosely aligned with the Taliban and have more than ten thousand fighters under their control. They have multiple training bases and even have religious schools for children. From their safe havens, they've launched some of the most brutal attacks inside Afghanistan. It is all but certain that Haqqani militants were responsible for the improvised explosive device that killed my intelligence team and almost killed me too."

"National security analysts in the United States, believe that the reason the Haqqani network prospers in war and profits from transporting illegal contraband is that Pakistan's Inter-Service Intelligence Agency sponsors them. They maintain a safe haven in North Waziristan. The Pakistan Army has never been willing to launch a military operation against them because General Jhalawan and his Pakistani security organization view the Haqqani network as a useful ally. I am convinced that it would be suicidal to attempt to enter Pakistan via the high mountain passes controlled by Jalaluddin Haqqani. Even assuming we could get past the Haqqani fighters, and that's a big if in my opinion, Jhalawan would know we were coming long before we posed a direct threat to him. After all, he's Jalaluddin Haqqani's sugar daddy. We would be sitting ducks that he could pick off at will."

"What's the alternative?" Jean Luc asked.

"The time zone difference kept me awake last night so I spent several hours studying some high-resolution satellite maps I have stored on my laptop. I think the only feasible route in and out of Janawai is from India. There's an international airport at Srinagar, India about sixty miles south of Janawai. The main runway is over two miles long and can handle commuter jets and mid-size cargo planes. There's a wooden suspension bridge across the Neelum River about five miles east of Janawai. It seems possible that a small team might pose as tourists on a trek. That would provide an excuse for backpacks and rugged gear. There are roads leading north from Srinagar to the town of Bandipora. North of there, the roads are narrow and poorly maintained but still passable by a four-wheel drive SUV. The Indian village of Sownar is less than a mile from a wooden pedestrian bridge across the Neelum river. I think we'd cross on foot one or two at a time and approach Jhalawan's villa at night. I agree with your plan to set up observation posts until we understand all of the variables."

"One advantage of infiltrating through India with at least their tacit approval would be that we could carry the weapons, ammo, and communications gear this mission really requires. Sure, we could buy some old Russian stuff if we went it over the Nowa pass but we'd still be subject to search and detention while loaded down with weapons. I'd much rather be fully equipped and ready to deal with what could easily turn into an all-out firefight with Jhalawan's troops well-equipped and fighting on their own turf. There's a lot more work to be done but I think infiltrating from northern India is our best option."

General Montreuil frowned before responding to what David had just said.

"I don't think the Indian government would support infiltration that way and there'd be hell to pay if we tried to do it covertly without their knowledge."

Admiral Chambers had been watching David closely. "I think David has already figured that out. Am I right David?"

"Well, maybe sir. Here's how I see it. India and Pakistan have had military clashes in the disputed Kashmir region for decades. It's a sore point going all the way back to the partitioning of India in 1947. Some analysts think Pakistan's primary motive in developing nuclear weapons was fear of their giant neighbor. Pakistan has more than two hundred and fifty of the American M110 self-propelled eight inch howitzers. The nuclear weapons that were stolen were designed to be fired from those big guns. We never released details about the weapons so it's a virtual certainty that India doesn't yet know that Pakistan has a significant arsenal of tactical nuclear weapons and the means of delivering them along a battle front."

"So you think India would allow us to infiltrate from their territory in exchange for some strategically important intelligence. Is that what you're saying?" Jean Luc asked.

"Yes, but I think there's an even bigger opportunity here. What if the western allies supported a precision, pre-emptive strike by India against the nuclear weapons facility at Wah? I'm nearly certain that all of the eight-inch tactical nucs are stored there along with most of their strategic weapons. We can question Tariq Zardari to see what he knows."

"What makes you think the Indians would be willing to mount an attack on Wah?" Admiral Chambers asked. "My impression is that India seldom initiates the border clashes preferring to thump Pakistan on its nose and then go back to sleep."

"In the past, I would have agreed with that view. Now, after the attack, I think every nuclear power has to be rethinking its policies with respect to the radical Islamic governments that either already have nuclear weapons or are trying hard to get them. I don't think India would use her nucs in a first strike scenario but I'm pretty sure they don't trust the Pakistanis to be similarly restrained. What I envision is a large, special operations force, aided by superior western intelligence, to make a focused attack on Wah to destroy the weapons in place."

General Montreuil asked, "What makes you so sure that the Indians could do a successful hit and run attack? Do you think your intelligence is that good?"

"No sir. Not yet anyway. But there may be a way to fill in any remaining gaps in our understanding of the Wah facility."

"How?"

David took a deep breath before he answered. “No human can withstand torture indefinitely. If the mission to abduct Jhalawan is successful, I’m confident that the right interrogation methods will have him singing like the proverbial canary. I watched Nahbi interrogate Yousef Abu Shakra using conventional methods. Abu Shakra didn’t yield an inch. I know Nahbi would have preferred not to be forced to torture Abu Shakra but he finally had no choice. It worked. He got the information we had to have. Nahbi broke the will of one of the toughest men I’ve ever met. Jhalawan is a tough guy too but I seriously doubt if he could hold out as long as Abu Shakra did. He’s older and smarter. At some point, he’ll cry like a baby and beg his interrogator to stop the pain.”

“You may be right, but I’m surprised that you would advocate that approach. Sir Lionel Smythe told me that you are a quietly devout Christian. Are you really saying you would support the use of torture?” Admiral Chambers asked.

“Yes. That’s exactly what I am saying,” David said softly. “When that stolen nuclear weapon detonated, I vowed to do whatever I could to stop radical Islam. Believe me, I don’t endorse torture lightly. But, make no mistake. I will do whatever is necessary to ensure that radical Islam never again kills millions of innocent people because we play by more civilized rules than they do. The president could have prevented the carnage but he was too timid. I asked for a pre-emptive strike against Jhalawan and the president refused. It was a political decision and it was the wrong choice. Jhalawan has established the rules in this deadly contest. If we can’t accept that brutal fact, we will lose this fight for our own survival.”

The three military officers traded glances but said nothing. David had made his point.

“David, do you really believe that India will attack Pakistan over this issue?” General Montreuil asked.

“Yes sir. If we can provide precise information on the number, type, and location of Pakistan’s nuclear weapons, I think India will conclude they have no choice but to launch a pre-emptive strike. The political situation in Pakistan is very shaky. Their nuclear weapons could easily wind up in the hands of some really crazy radicals which would pose a serious strategic threat to India. While they wait for our intelligence report, the Indians can be doing some advance planning but they wouldn't have to commit until we deliver what we learn from General Jhalawan."

General Montreuil looked across the table at Jean Luc. "Well, Captain Davignon, it seems your confidence in Mr. Logan is well placed. Now, I think it is time for the big question." Turning to face David, the General made his pitch.

"Mr. Logan, would you consider joining the mission as a co-commander with Captain Davignon?"

David stared out the window overlooking France's second-largest naval port. He didn't speak for several long seconds.

"You all know that General Jhalawan had my sister and Nicholas Aquilina murdered. All of my men were killed and I almost lost my leg to an IED probably assembled by one of Jhalawan's proxies. Several of my relatives have either been killed or maimed by radical Islamic terrorists. I know that Jean Luc lost friends and relatives in the attack masterminded by General Jhalawan and I would guess that you did as well, General Montreuil."

"September 11th and June 24th will forever remind us that radical Islam would gladly kill us all if they had the means to do so. Clearly, there's no lack of motivation for us to apprehend this monster and bring him to justice. As a civilian, my inclusion on the mission would be awkward to say the least. I do have a job offer pending from the CIA so it's possible that I could accept and then be seconded to the COS for the mission. In any event, I don't think my civilian status poses an insurmountable obstacle."

"What I'm most concerned about is whether I can still handle the rigors of a special ops mission over rugged terrain. I used the gym in the basement of the MI6 building every day while I was part of Task Force 3. I needed the physical exertion to keep my mind clear and I wanted to stay in shape as much as possible. I can still bench press as much as ever and I can still do the same number of dead-weight chin ups as when I was on active duty. I'll always walk with a slight limp but I was able to use the treadmill to maintain my stamina. I think I can still cut the mustard but before I give you my answer, I'd like to work out with Jean Luc's men to see if they think I can keep up."

Jean Luc was grinning when he punched his friend on the shoulder. "I told the General you'd say yes."

"Ah, but I *haven't* said yes. Not yet. I'll join the mission if two conditions are met. First, Roger Benson and Pete Nocona at the CIA have to agree and be willing to second me to the mission once I accept their job offer. By the way, the job offer is a safe desk job as an intelligence analyst, not as a CIA Special Operations Group field agent. Second, I want to spend a day on the nearest special ops training center's qualification range. If I can pass every single one of the French Army's Special Forces Brigade's physical fitness tests, and your team unanimously accepts me for the mission, then and only then will I say yes."

"Under the circumstances, I can't ask for more," General Montreuil said as he extended his hand across the table.

"David, do you mind if I speak to Roger? I've known him for over twenty years and I think my endorsement might still carry a bit of weight with him."

"No sir Admiral. I don't mind at all. I'd be honored."

David saw General Montreuil look at Jean Luc and nod almost imperceptibly toward the door. Jean Luc took the cue and walked out with Admiral Chambers.

General Montreuil closed the door before he spoke. "David, there's something else I need to tell you. Jean Luc knows what I'm about to say but no one else does and we need to keep it that way."

David nodded and waited for the General to continue.

"Even if we succeed in bringing Jhalawan to France to stand trial, and even if he is eventually convicted after years of legal proceedings, he will have the last laugh. You see, France abolished capital punishment in 1981 and later modified the French Constitution to forbid all forms of capital punishment. I have sworn a solemn oath to uphold the laws of my country and to do my best to defend her against all enemies, foreign and domestic. In this instance, in my opinion, I cannot do both so I must choose the lesser of two evils. With that in mind, I have ordered Captain Davignon to see that General Jhalawan is killed trying to avoid capture or escape. I know Jean Luc well enough that I was sure he would agree with my decision. However, I have put my order to him in writing so if it ever becomes known, only I will be charged with a crime. I thought you needed to know that before you make your decision. I am aware that you are a deeply religious man of strong faith. I will certainly understand if what I've told you causes you to decline to join the mission."

The two men looked into each other's eyes for a long moment before David responded.

"Thank you for being honest with me sir. Like you, I've had to make decisions in combat that caused other men to lose their lives. That's never easy for men of moral conscience yet it is a necessary part of waging war to win. That's how I see this issue. There is no question that Jhalawan is guilty of the worst singular mass murder in all of human history. Those facts are not in dispute. Although many decent people haven't yet realized it, western civilization is at war with something so evil that we dare not flinch from what must be done. I respect you for trying to protect the lives of millions of innocent people not only in France but in all of the nations targeted by radical Islamic terrorism. You may be certain that if I decide not to join this mission, your orders with respect to General Jhalawan will not be my reason."

"Jean Luc said that would be your answer. Winston Churchill once paraphrased something often attributed to George Orwell when he said, '*We sleep safely at night because rough men stand ready to visit violence on those who would harm us.*' I'm grateful that we have strong men like you and Jean Luc to oppose monsters like Jhalawan."

Baskov's Restaurant; Tel Aviv Israel
8:30 PM

Baskov's was a small family-owned Russian restaurant a few blocks from Gavriella's condo. Locally famous for good food and great service, it was one of her favorites. The soft piano music and tasteful Ukrainian décor created a wonderfully romantic atmosphere. After a fine dinner and three glasses of a superb Judean red wine, she should have been feeling relaxed and mellow by now but she wasn't. She had rehearsed what she wanted to say in her mind for days but she was still very nervous.

Leaning forward across the table, and looking directly into Nahbi's dark eyes, she couldn't wait any longer.

"I want to have a very personal conversation with you but I'm struggling with how to go about it. I've decided to trust you to understand and if necessary forgive me if I make a mess of it." She took a deep breath and then blurted out what was on her mind. "Do you think it's possible for a human to fall in love with an android?"

Nahbi hesitated for only a moment before saying, "perhaps what you think is more important. Do *you* think it's possible?"

"A few weeks ago I'd have laughed and said no way. But, yes. It's more than possible. I'm quite sure that I've fallen in love with you. Does that surprise you?"

Nahbi bowed his head for a moment and then looked up. "No Gabe. I've known it since that day I shampooed your hair. I know that I can't experience love in the same way humans do but I have become so fond of you and want so earnestly for you to be genuinely happy, that I think it must be very similar to love. So, I would say that it must be possible for a human to fall in love with an android and equally possible for an android to fall in love with a human." He smiled gently and reached across the table to take her hand in his.

Tears welled up in Gavriella's eyes but she managed to hold them back. "I have been so lonely," she whispered. "Since my husband was killed, I've tried to let my work fill the void but it was never enough. I shut myself off from other people because I was terrified of the risk of falling in love again only to have it taken from me a second time. I couldn't go on living if that happened."

"When I invited you to stay at the condo, I never dreamed anything like this might happen. I thought you would be the perfect, safe companion but you have been so much more. I don't even care if your kindness and thoughtfulness are the results of logical calculations ricocheting around inside that magnificent brain of yours. In the end what matters is that you treat me as though you love me. I have missed that so very much and for such a long time."

Nahbi wisely said nothing. He simply held her hand and looked at her with a look on his face that anyone watching would have interpreted as the tender smile of a young man in love.

"What does this mean for us?" Gabe asked quietly.

"I suppose that depends in large measure on how we handle Levi."

"Yes. I've thought of that too. He has transferred people to remote stations to break up inter-office romances in the past. He feels very strongly that they are unprofessional and harmful. He calls it fishing off the company pier. His greatest concern is the possibility that a foreign agent could use such relationships to threaten our national security. He is far too good at what he does to allow that."

Nahbi nodded. "If we tell him, he will separate us. If we don't and he finds out, he'll feel betrayed and it will likely be the end of your career with the Mossad. It's a difficult question."

"I have no doubt that you can conceal our relationship but I'm not sure I can. Levi knows me well and can probably read me even better than I suspect. He's one of the best spymasters the Mossad has ever had so it goes without saying that he can smell intrigue of any kind."

Nahbi didn't tell her that they would probably soon be separated anyway. Israel's prime minister had ordered her Kidon team to abduct him to help the tiny nation deal with the Iranian nuclear threat. After the briefing he'd given the day before, it seemed highly unlikely that he would continue to report to a safe desk at *The Office* much longer.

"Will you sleep with me tonight?"

"Are you asking if we can make love?"

"Yes," Gabe said softly while watching his face closely.

"Are you sure you really want that? Despite your tender feelings toward me, I'm not human. I've never had a sexual experience of any kind. I've thought about it and even admit that I've done enough research to believe that I can probably give you pleasure. Yet, I would be very sad if it turned out badly for you."

Gavriella smiled and shook her head slowly. "You would never let that happen. I trust you more than I've ever trusted anyone. I suspect you've read every authoritative article on the subject of pleasing a woman and that you know precisely what to do. I have only the vaguest ideas of what I can do to please you but I'll do my best."

Gavriella took Nahbi's hand as they left the restaurant but they didn't talk as they walked back to her apartment. As the door closed, Gavriella pulled Nahbi against herself and kissed him with a wild and hungry passion that had been dormant far too long.

Nahbi picked her up as though she weighed nothing and carried her to her bed. Before that magical night was over, Nahbi took Gavriella to heights of pleasure she had never imagined, eventually leaving her happily exhausted an hour before sunrise. She had been right. He *had* done his homework.

Sunday, 13 July

Plage de Bonnegrace; Toulon, France
5:30 AM

Jean Luc parked in the parking lot near the north end of the beach whose French name translated into '*Brutal Beach.*' Lieutenant Édourd Jardin and Sergeant Tristan Dejarnette pulled in alongside them. The rest of Jean Luc's team was already on the beach doing warm-up calisthenics. François Gaultier, the '*old man*' of the team was leading the exercises. David knew Remy Thibadeau, Henri Barteau, and Alain Fabré from Afghanistan. He'd met Bernard Garneau, the newest member of the team, the previous day.

Located seven kilometers due west of the Toulon Naval Base, the beach was popular with local residents and tourists. Later in the morning it would be too crowded for their purpose but it was deserted at this early hour. Jean Luc explained this phase of the physical qualification test for the French Army's Special Forces Brigade.

"The beach is almost exactly one kilometer long and we've marked off the turn points. Our standard is to run four kilometers in less than sixteen minutes wearing combat boots and standard-issue combat camouflage trousers. You must run in the loose sand above the wave line. Henri Barteau is our best runner. His record for this course is twelve minutes and sixteen seconds. He'll stay in front of you to mark the line you must follow. You'll make two complete roundtrips. Sergeant Dejarnette will be at the south end of the beach to call out your quarter and three-quarter times at the far turn to help you adjust your pace. I'll call out your half and finish times. Take a few minutes to warm up and let me know when you're ready."

David took his time warming up. He was nervous but outwardly calm. After five minutes, he took a few deep breaths to saturate his blood with oxygen and then nodded to Jean Luc. "OK. I'm ready. Let's do it."

David and Henri leaned forward and then set off at a fast pace when Jean Luc blew his whistle and punched the stop watch. Henri moved out in front with apparent ease but kept looking over his shoulder to adjust to David's pace. Lieutenant Jardin and the rest of the team fell in behind David matching his pace several yards back. Sergeant Dejarnette called out three minutes forty-six seconds at the first turn. David knew that was a little slow for the first leg so he picked it up a bit.

Jean Luc called out seven minutes fifty three seconds at the half-way point. Still too slow. His lungs were starting to burn but that pain was mild compared to what he was feeling in his damaged right leg. He knew he was stretching and maybe even tearing some of the scar tissue. Running in loose sand in combat boots wasn't quite the same as running on a treadmill wearing shorts and running shoes. For a moment he thought he might flunk the very first test of the day. Then a vivid image came into his mind and he clearly heard his grandfather tell Sarah's little girl that her mommy had gone on ahead. Choking back his emotions, he surged ahead and surprised Henri by passing him briefly before the younger man regained his guiding lead.

Sergeant Dejarnette called out twelve minutes fourteen seconds at the three-quarter turn. He'd covered the third kilometer in three minutes and twenty-one seconds. Much better. He had three minutes and forty-six seconds remaining if he was going to pass the qualification test but he was starting to tire. His lungs were burning and the pain in his right leg was excruciating. He wondered what had ever made him think a cripple could pass the French Army's punishing test. And then he heard those terrifying words again. '*Allāhu Akba'* screamed into the headset microphone at the instant when the embedded terrorist had intentionally detonated the last of the stolen weapons. Again he surged and again he briefly passed Henri but he was running on sheer willpower by then. His lungs were screaming in protest and he was gasping for air. Then, incredibly, Alain Fabré began to sing the French National Anthem in English. The others joined him in the stirring battle cry of the La Marseillaise.

'Arise, children of the nation, the day of glory has arrived!

Against us tyranny raises its bloody banner!

Do you hear in your fields the roar of these ferocious soldiers?

They're coming right into your arms to slit the throats of your sons and women!

To arms Citizens! Form your battalions!

Let's march! Let's march!

So that the impure blood of tyranny will water the furrows of our fields!'

As they sang, they came up alongside of David and Henri fell back to run seven-abreast with them. The patriotism of these magnificent young soldiers gave David a boost as he ran alongside men he had fought beside before and might soon again. Jean Luc called out fifteen minutes forty seconds as they crossed the finish line together. Within seconds, Jean Luc's men were pounding David on the back and grinning like a bunch of Cheshire cats.

"Well done my friend," Jean Luc said with a big grin of his own. "The rest will be easy compared to this."

Monday, 14 July

Avital Valley, Israel
9:20 AM

"Shalom! You have reached the Avital Volcanic Aggregates Company. For sales, press 1; For plant operations, press 2; For accounting, press 3; For personnel, press 4; To speak to a representative, please stay on the line and someone will be with you shortly."

"Shalom! You have reached our sales department. How can I help you?"

"Good morning. I'm calling from the purchasing department of Rehovot Highway Constructors Ltd. We have an account with you. I'd like to place an order for nine tons of washed scoria volcanic aggregate delivered to a construction lay-down yard In Gedera."

"Yes, I can help you. What size scoria aggregate are you looking for?"

"Our requirement is for scoria stones that are well-rounded and no less than 8 cm on any dimensional axis and no greater than 12 cm. We prefer stones that have rather large open cells."

"Do you have a color preference? We can supply scoria in either a dark, reddish brown or a very dark gray that is almost black."

"Color does not matter. The stones will be used as part of a subterranean drainage field and will be covered."

"The price, delivered to the site, will be nine thousand ILS inclusive of tax. When would you like for the aggregate to be delivered?

"Can you deliver by Friday of this week?"

"Let me check with the plant operations scheduler. I'll put you on hold for just a moment."

"Sir, yes. Thank you for waiting. We can make delivery on Thursday of this week between noon and 5 PM. Do you have a purchase order number for me?

Haifa, Israel
10:18 AM

"Shalom! Haifa Chemical. May I help you?

"Hello. This is Andrew Makhteshim with the agricultural products division of Agil Chemical Manufacturers Ltd. I'd like to order six tons of granular, elemental sulfur that we use to produce agricultural herbicides and insecticides. Our normal supplier has been shut down by labor strikes in Sicily so we don't have an account with Haifa Chemical. Once I know the delivered price, I'll arrange a wire transfer."

"Yes. We can fill that order immediately. Where would you like for it to be delivered?"

"If you don't mind, our normal practice is to use a local trucking company to pick up the product."

Over the next few minutes, the details of the order were completed with pickup scheduled for the following day.

Adams Chemical; Philadelphia, Pennsylvania
8:20 AM

Sometime during the night, a FAX had come in from one of their regular wholesale customers in Charleston, South Carolina. It was an unusually large order for rhodamine WT, an industrial dye commonly used to trace water flows in municipal water systems and rural surface waterways. The dye was harmless and would dissipate within a few hours. The order was unusual in another respect. The customer wanted six tons of powdered dye in 100 pound bags loaded into a shipping container along with twelve 8-liter canisters. The order was to be delivered to a Greek cargo ship currently loading containers at the Packer Marine Terminal in the Port of

Philadelphia. The ship was scheduled to leave port the following afternoon so there wasn't much time to fill and deliver the order.

Sodot Micha Munitions Area; Israel
1:00 PM

"Aaron, I've got a couple of interesting projects for you. They're both classified top secret and have national defense priority."

"I've just received an order from the IAF Logistics Center for us to assemble some industrial-grade thermite cutting units. They sent over some design drawings and specifications but it looks pretty straight-forward."

After looking over the design documents, Aaron said, "Not a problem. I can have these done in less than a week. We have everything we need to construct them. It'll take about 400 pounds of Iron Oxide powder and about 100 pounds of powdered aluminum to create the thermite. I'll mix that with a binding agent and pour it into molds to create rods with the square cross section they've specified. I'll wrap that in heavy-duty aluminum foil. Thermite burns very hot once ignited but it takes a lot to get it going. That's why it is so easy to store and transport. I can use standard magnesium flares at each end to ignite the thermite. We have some standard radio-controlled igniters used in demolition work that I can modify to ignite the flares. Once the flares are triggered, the thermite will ignite in less than a second. The temperature will rise to 2,500 degrees Celsius which is roughly three times hotter than molten lava. This isn't all that different from what construction companies use to cut the steel beams in buildings they're demolishing. One of these babies will cut through an eight inch steel beam in a few seconds. What's the other project you mentioned?"

"We're going to receive nine tons of scoria volcanic stones a little larger than a man's fist. We'll also receive a shipment of powdered sulfur. They want us to dip the stones in molten sulfur and then cool them so they can be safely handled and stored. IAF is sending over some carbon fiber canisters that each hold about 500 pounds. We'll store them in one of the air-conditioned munitions bunkers until the IAF is ready for them."

"That's pretty straight-forward. We have some electric crucibles that we use to melt large lead ingots. Sulfur melts at about 240° Fahrenheit, or just slightly above the boiling temperature of water. Lead melts at about 620° Fahrenheit so our crucibles will work just fine on their low-power setting. Ask the Log Center if the sulfur coating can be in the elastic form."

"What are you talking about? I thought sulfur was a brittle, bright-yellow crystal. What's this about an elastic form?"

Aaron explained. "When molten sulfur is suddenly cooled, the eight sulfur atoms in the normal cyclo-S_8 octasulfur molecule rearrange themselves into a helical pattern. In that form, the amorphous sulfur is elastic and feels almost like rubber. Over time, it will revert to the brittle form but the chemical properties are identical. I would think that the elastic variant would be preferable for handling. We'll have to cool the rocks coated with molten sulfur somehow and dunking them in cold water is the easiest way to do that. I just need to know if the elastic variant that will result is still going to be acceptable for their purpose."

Tel Aviv University; Ramat Aviv, Israel
3:00 PM

"Dr. Sorenstein, this is Levi Reznik. How are you today?"

"I am very well, thank you. And you?"

"Also well. I need your help."

"What can I do for you?"

"What I'm about to discuss with you is classified top secret national security. It's also the strangest thing I've ever had to do in all of my years in the intelligence field."

"Ah, surely you have seen it all by now, my friend."

"No, not this. I need to find a way to obtain large numbers of some rather unpleasant members of the animal kingdom."

"I am intrigued. Tell me what you need."

"I need many thousands, perhaps hundreds of thousands, of Vespa orientalis hornets. I also need literally millions of the Pulex irritans flea better known as human fleas."

"Is that all?"

"No, I also need several thousand Telescopus Fallax commonly known as the Mediterranean Cat Snake. I need a similar number of Phrynocephalus mystaceus, otherwise known as the Secret Toad-Headed Agama lizard. I also need about a half-million Blattella asahinai, otherwise known as Asian cockroaches, and several hundred thousand Hemiscorpious lepturus. I wasn't able to find a common name but this is the scorpion that I'm told is common in Iran. Is all of this possible?"

"I'm not sure. Let me do some research and place a few calls. When do you need an answer?

"Within the next few days if possible."

"I'll call you in the morning."

"Thanks."

Department of Natural Resources; San Juan, Puerto Rico 8:45 AM

"Yes? This is Dr. Galardo."

"Good morning sir. My name is Miguel Hernandez. I work for the World Health Organization. I wanted to ask a few questions about the program you've set up to capture millions of Green Iguanas for export. Can you spare a few minutes of your time for me?"

"How may I help you?"

"I've been doing some research on potential sources of protein to help feed some of the poorest people in Africa. I recently read an article that said the Government of Puerto Rico had established a program to capture literally millions of Green Iguanas that have become a serious pest and to export their meat with a dual goal of reducing the iguana population while generating income for your country. Is that program still in operation?"

"Yes, although it hasn't been as successful as I'd like. As you probably know, Puerto Rico is not a wealthy nation and there is a lot of competition for funding within our government. Finding the iguanas is the easy part. They're literally everywhere. I ran a test program offering fifty cents for each healthy adult iguana brought to a collection center we set up on the outskirts of San Juan. I had to terminate the test program after only nine days when my seed money of twenty thousand dollars was exhausted."

"Are you saying that you collected forty thousand iguanas in just nine days?"

"Yes. Well, no. Actually, we got forty thousand in about eight and a half days. When we opened the collection center the first morning, there was already a long line of dilapidated cars and trucks waiting, most with multiple iguanas. I saw one ancient Toyota sedan with twelve iguanas in it. I also saw a rusted out Nissan pickup with a topper shell unload twenty-two adult lizards. We paid out over twelve thousand dollars the first week. We closed at noon on the eighth day when we ran out of money. It nearly started a riot when we closed the gates and announced that we could take no more iguanas. But the test program proved that the poor people of our country would be happy to collect them for us as a welcome source of income. I did not have a slaughtering facility set up yet so we ended up just shooting them and burying them in a pit."

"What is the status of your program now?"

"Officially, it's still active but until I can find the money to construct a slaughtering facility with quick-freeze equipment and frozen food storage, I'm pretty much at a standstill."

"How much money do you need?"

"At least a million dollars."

"I see. How much does an adult iguana weigh?"

"That depends on their age and sex. Males are larger and heavier than females. Iguanas are considered adult when they are one year old. A three year old male's body length would be twelve to fourteen inches with an overall length of about thirty inches including the tail. A three year old male would typically weigh between four and six pounds. A seven year old male could be over six feet in overall length and might weigh as much as fifteen to twenty pounds.

"How is live shipping accomplished?"

"Single iguanas are typically placed in a cotton bag to limit their mobility and then placed in a perforated cardboard box on some kind of cushioning material such as wood shavings, crumpled newspapers, or even Styrofoam peanuts. They are given plenty of water and high-water-content food prior to shipping. They can go literally for weeks without food if necessary but dehydration can kill them. Fortunately, they will readily eat iceberg lettuce which provides no nutritional value but does a good job of helping ensure they are well-hydrated before shipment. The packaging is typically labeled with the words Harmless Live Lizard stenciled on all sides with arrows saying this side up."

"I'm sorry. I should have been more specific. What I meant to ask is how could large numbers of live adult iguanas be shipped?"

"I'm not an expert on that but I think it would be somewhat similar. I envision that large wooden crates, with plenty of ventilating holes could be used. If I were going to design such crates, I would make them as large as could conveniently be handled by a typical forklift truck. Inside the crate, I'd attach furring strips to act as a support ledge for multiple layers. I'd then cut plywood sheets to fit inside the crate to rest on the furring strips. With each adult inside a cotton bag and a layer of cushioning on each shelf, I think they could be transported with an excellent survival rate."

"Yes, that makes sense. Let me think about that for a moment while I pose another question. I recently read an article referring to something the author called *the frozen iguana showers* that occurred during unusually cold weather in Florida back in 2008. The article said that the iguanas were literally falling out of the trees when the cold nights put them in a state of torpor and they lost their grip on the tree branches. Could living iguanas be subjected to carefully controlled cooling to render them so inactive that they could be simply stacked one on top of another for shipping?"

"I suppose that's possible. I think the mortality rate would be higher but if the creatures were well-hydrated before they were chilled, I think it might work."

"What about some kind of general anesthetic? Could they be gassed into unconsciousness for shipping?"

"Possibly. I don't really know. You'd have to consult a veterinarian on that."

"OK. Let's go back to your collection program for a moment. Let me propose a scenario. If your department received a grant of one hundred thousand dollars, could you deliver one thousand live adult iguanas that were approximately thirty inches long and weighed roughly five pounds each?"

"Are you serious?"

"Yes sir. I assure you that I am. My rough calculations as you spoke suggest that crates four by five by five feet could each hold slightly more than one hundred iguanas. That suggests that one thousand iguanas would require about ten crates which would comfortably fit in two standard forty foot refrigerated international shipping containers."

"When would you need them?"

"In about two weeks. There is a Greek registered shipping company we've used several times. They have a ship that is currently completing load-out at the container terminal in the Port of Naples. The ship is the Athenia and it's scheduled to leave Naples Wednesday morning. After stopping for two days in Miami, the Athenia expects to tie up at the container terminal in San Juan harbor on Thursday morning on the 31st of July."

"Do you have the money you mentioned available for immediate disbursement? I cannot fund the order without a sizable advance."

"Yes. I'll send you an agreement to sign and FAX back to me. As soon as I receive a signed and notarized copy, I'll wire-transfer sixty-five thousand dollars with the balance paid when one thousand live iguanas are delivered to the Athenia in the refrigerated containers we'll arrange."

"May I ask why you want live iguanas instead of frozen meat which would be easier and cheaper to ship?"

"We intend to set up a breeding farm on the outskirts of Lomé, the capital city of Togo. I don't know how familiar you are with the countries of West Africa but Togo is a very small country on the Gulf of Guinea and it's so poor that it recently became eligible for international aid under the Highly Indebted Poor Countries program. I assure you that Togo is so poor that our program will not compete with what you are trying to accomplish. There is no money available to import animal protein."

After the call ended, Dr. Galardo leaned back in his chair with a big smile on his face. A hundred thousand dollars would help keep his program, and his job, from becoming a casualty of the budget axes. Half-way around the world a Mossad purchasing agent was smiling too. He'd just checked off another item on Nahbi's bizarre shopping list.

Xenakis Ltd; Stavros, Greece
3:45 PM

"Kalispera! Good afternoon," a cheerful recorded message announced. "Press 1 for Greek or 2 for English."

"Please hold. Someone will be with you momentarily."

"Hello, you have reached Xenakis Ltd. My name is Mariela Tavoularis. How may I help you?"

"Good afternoon. My name is Alain Favreau. I'm with AgriStar International in Marseille, France. We produce feed products for the livestock industry. We've developed an experimental new product used to fatten livestock in feed lots prior to slaughter. I'd like to place an order for four thousand gallons of refrigerated whole hog blood, preferably without salt or other preservatives added."

"That's a rather large order. We typically slaughter about five hundred hogs per day at this time of year which would normally yield about 750 gallons of fresh blood. Your order would take at least a week for us to fill. Also, we do not normally refrigerate hog blood. Instead, we freeze-dry it to produce powdered blood that can be reconstituted into a form that is suitable for many uses. We do not have equipment on site to refrigerate the blood. I'm not sure how we could deliver that much refrigerated blood."

"I anticipated that might be the case. We would have two ocean-going refrigerated containers delivered to your slaughterhouse along with eight hundred, twenty-liter plastic water bottles. As your slaughterhouse collects the blood, it can be poured into those bottles and placed in the refrigerated containers. When the order is complete, we'll have the containers picked up and taken to the container terminal at Perama where they'll be loaded onto a container ship."

"I need to check with the company's owner before I can accept your order. Can you hold while I check? I know he's in his office. It should only take a couple of minutes."

"Sir? I'm back. Mr. Xenakis told me to tell you that we can complete your order five days after the containers are in place. He said we would adjust our work schedules as necessary to fill your order. If the containers are staged by Sunday afternoon, he said we would complete your order sometime on the following Friday. He suggested that your trucking company could pick up the containers Saturday morning. It's about 250 kilometers from our plant to the container terminal in Perama. He thought the trucks could make the trip in about four hours. The total price for four thousand gallons of whole hog blood will be twenty thousand U.S. dollars inclusive of tax. Mr. Xenakis said we would require a down payment of ten thousand dollars, or two point six million Greek Drachmas, with the balance due before the containers leave our plant. If you'll stay on the line, our accounting department will give you the account number where you can wire the money. They'll FAX a copy of the order to the number you give them. Is there anything else I can help you with?"

U.S. House of Representatives; Washington, DC
10:00 AM

Robert Sidney Marshall, Speaker of the House of Representatives, called the House to order.

The Chaplain then offered a prayer, in accordance with more than two centuries of custom. His closing words were, "And we beseech you, oh Lord, to grant wisdom to those who must decide the sober matter before this body today, Amen."

Speaker Marshall then announced to the House that the Chair had examined the Journal of the last day's proceedings and that pursuant to Clause 1 of Rule 1 the Journal stood approved.

The Speaker then introduced Thomas Nitakechi to lead the Pledge of Allegiance. Nitakechi was a full-blooded Choctaw Indian from Oklahoma. At 93, Nitakechi was the oldest surviving Medal of Honor recipient, having served as a Sergeant with the 42nd Infantry Division in France during World War II. The old man was in full uniform with the unmistakable blue ribbon around his neck supporting the Medal of Honor he had received from President Harry Truman in May of 1945. Walking slowly to the rostrum with quiet dignity, and using only a cane for support, he led the members in the Pledge that he had learned as a small child at the reservation school near Durant, Oklahoma. With that poignant reminder of what constituted honorable service, the stage was set for what came next.

After dealing with several routine procedural matters, the Speaker announced that the Clerk would now read the Articles of Impeachment against President Arnold.

"**Article 1**. Benjamin Jefferson Arnold, in violation of his constitutional oath to faithfully execute the office of the President of the United States and, to the best of his ability, preserve, protect, and defend the Constitution of the United States, and in violation of his constitutional duty to ensure that the laws be faithfully executed, has prevented, obstructed, and impeded the administration of Justice."

"**Article 2**. Using the powers of the office of President of the United States, Benjamin Jefferson Arnold, in violation of his constitutional oath, has repeatedly engaged in conduct violating the constitutional rights of citizens, impairing the due and proper administration of justice and the conduct of lawful inquiries, and contravening the laws governing agencies of the executive branch."

The Speaker then announced a roll call vote and the Clerk began to call the roll, repeating and recording how each Representative voted.

The previous evening, a CNN broadcast during prime time had confirmed that an anonymous source at the CIA had revealed that the President had refused to authorize covert action that could have prevented the deaths of more than two million innocent people three weeks earlier. With that damning revelation, the vote to impeach President Arnold passed by an overwhelming bipartisan super-majority and the Articles of Impeachment were referred to the Senate in accordance with the Constitution.

Transient Officer's Quarters; Toulon Naval Base, France 5:25 PM

David had just finished his shower and was getting dressed to join Jean Luc and his men for dinner when the phone in his room rang.

"David, this is Pete. I have you on the secure speaker phone. Roger's here with me. Can you talk for a few minutes?"

"You bet. I've been expecting your call."

Roger Benson was the Deputy Director of the CIA for the National Clandestine Service. He was a retired Army Major General who had previously commanded the Army's Intelligence and Security Command, INSCOM. He was Pete's boss. Pete Nocona was the section chief of the CIA's Counter Terrorism Center. He was a former CIA field agent and Army Delta Force officer. Pete had been impressed with David's leadership on Task Force 3 and had persuaded Roger to offer him a job.

"Congratulations David," Benson began. "I just spoke to General Montreuil. He said you passed the physical quals with flying colors."

"Let's just say I passed. I'm not too sure about the flying colors part. That four kilometer run was much worse than the mile and half that's part of the SEAL Basic Underwater Demolition qualifications. I wouldn't say the rest of the day was a breeze but I did OK."

"I knew General Montreuil was going to ask you to join the COS mission to take down Jhalawan. I understand you have a couple of conditions before you'll agree. I take it that passing the Brigade physical tests puts the ball in my court."

"Yes sir. I'm willing to go with them but I won't go as a civilian. I'll accept your job offer if you're willing to second me to the French COS for the duration of the mission. Did General Montreuil tell you that they want Al Flores to join me on the team and that they want Pete and Virginia to support the mission with intelligence updates and SATCOM support?"

"Yes but that poses a significant problem for me. Director Oliver would almost certainly disapprove our involvement in abducting a senior Pakistani officer. Have you been keeping up with the protests and riots in Islamabad related to the last presidential election?"

"Yes sir. It looks like there's a real possibility that the prime minister will lose control of the government forcing the Pakistani military to step in again."

"That's correct. Although Jhalawan is the Director of Pakistan's Intelligence Service, he's also still an Army Lieutenant General. We think it's at least possible that he might end up becoming the prime minister should the current administration fall. So, in effect, the French are planning to abduct one of the most powerful men in Pakistan and to then put him on trial for war crimes. While that may be a perfectly rational thing to do in view of what Jhalawan is guilty of, it might set a very bad precedent that could encourage terrorists to abduct western leaders and then subject them to a mockery of a trial before videotaping their beheading."

"Yes sir. That's pretty much how I see it too. I understand your dilemma and have a suggestion."

"Let's hear it."

"Instead of you officially seconding me and Al to the French, let's say I accept Pete's job offer as an intelligence analyst. Then, under your radar, Pete sends me and Al to the Kashmir region of India to evaluate the political and military situation in the disputed border regions between Pakistan and India. I'm the linguistics expert and the former leader of Task Force 3. Al's the experienced field agent sent along to nursemaid the inexperienced junior CIA analyst. Director Oliver could hardly see anything wrong with such an innocuous assignment. He might even be glad to see Al over on the other side of the world for a while. We'd both carry fake passports and no weapons or military gear of any kind. We'd rent an SUV and move around the area at about the same time Jean Luc's team starts putting boots on the ground."

"Al's low regard for the current administration is more widely recognized than is probably good for his career. If it ever comes up, it shouldn't be hard to convince Director Oliver that I'm a guy with multiple axes to grind including Jhalawan's orders to have my sister murdered. So, Al and I go rogue when we meet up with my old buddy, Jean Luc, from Afghanistan. On the spot, and unbeknownst to Pete or you, we join up for the mission, totally on our own with absolutely no CIA knowledge or sanction."

"Ok. You've got my attention. It just might work. Go on."

"General Montreuil and the French prime minister will use some suggestions I've made to get the Indian government to sanction them infiltrating into Pakistan through northwestern India. So, Jean Luc's team will be able to bring full special operations gear into the country. I think they'll probably land at Srinagar International Airport. It won't be much of a problem for them to bring extra gear for Al and me. If the mission goes as planned, Al and I break away from Jean Luc's team and return to the U.S. with our routine intelligence reports. If the mission goes sour, you can disavow any knowledge of our rogue enlistment on the French COS mission. Pete and Virginia would be supporting Al and me instead of the French and, in my rogue persona, I would share that information with my old buddy Jean Luc. Everything could be structured to fit that scenario so you and Pete would have plenty of what the politicians call plausible deniability."

"What do you think, Pete?" Roger asked.

"I like it. We can tweak it a bit but I think David has come up with a way to support the French without Director Oliver's involvement in the decision process."

"OK. It's a go then. David, I've had to be a little creative with your job offer. You said you expected a low six-figure salary. The GS-15, step 1 position salary is $100,624 but I can't get the personnel director to sign off on offering you a GS-15 position, even with your resume. She has agreed to let me offer you a GS-14, Step 7 position. As a partially-disabled veteran, you can be non-competitively appointed to the position under Schedule A. So, I'm authorized to offer you a GS-14, Step 7 position as an Intelligence Analyst in the Counter Terrorism section. Your starting salary will be $102,650. You will report directly to Peter Nocona, section chief. Is that acceptable?"

"Yes sir. When would you want me to start?"

"Am I correct in assuming that you are not planning to return to the States before you head east?"

"Not sure yet. I can go either way."

"OK. Pete will coordinate with the civilian personnel department at Sixth Fleet Headquarters right there in Toulon. We'll have them handle the paperwork. Let's say your first day on the job will be next Monday. Does that work for you?"

"The Monday start date is fine but having the Sixth Fleet office handle the paperwork will leave an audit trail that shows I was in Toulon being in-processed as a new CIA employee about the same time the French mission was being planned. That could raise some awkward questions later."

"Hmm. You're right. What do you suggest?"

"Here's my cover story. As an unemployed civilian with time on my hands, I flew to Toulon to see my old buddy Jean Luc from our deployments in Afghanistan. I spent a couple of days with him and then flew on to Malta to see my former girlfriend, Tiera Aquilina, and to express my condolences to her parents for the murder of their son. My plan is to spend about a week in Malta before returning to Washington. The American Embassy is only about two miles from their estate. I'm guessing you have a CIA agent at the Embassy. Is that correct?"

"Yes. Emily Harrison is the current political officer assigned at the embassy. You probably know that '*political officer*' is a code name for an embedded CIA agent. Emily is one of Pete's senior agents and will probably get his job one day."

"OK. Here's what I suggest. Send the in-processing paperwork to the American Embassy's personnel department in a diplomatic pouch. Send me an email congratulating me on my decision to accept Pete's job offer and asking me to stop by the embassy while I'm in Malta to sign the acceptance letter and complete some paperwork. You can suggest that I take advantage of my time in Malta to spend some time with Emily so she can begin my indoctrination. I'll fly to Malta on Wednesday and plan to return to the States on the red-eye Sunday night. That'll give me another day here in Toulon so Jean Luc and I can wrap up what we need to. Treating my visit to Toulon as just a stop-over to see an old friend on my way to Malta should conceal the real purpose of my trip."

"Pete? How's that sound to you?"

"Works for me, David. I was surprised to hear you refer to Tiera as your former girlfriend. I thought you two were all but engaged."

"It wasn't going to work and we both realized it. I think we'll remain friends."

Tuesday, 15 July

Mossad Headquarters; Herzliya, Israel
9:05 AM

Levi saw who was calling and picked up the phone before the second ring.

"Good morning Abner. What have you found out?"

"Some of what you want is available and some isn't."

"Let me grab a pencil so I can take some notes."

"A close friend of mine is the Director of the Volcani Institute that includes Bar Ilan University. Bar Ilan has had a large agricultural research program going on for several years involving pests affecting wine and honey production. The Vespa orientalis hornets are predators of honey bees and they can severely damage grapes just as they are approaching harvest maturity. Bar Ilan has a breeding program that produces about two million hornets each year for research. For a generous contribution to their research program, you can get about 500,000 adult hornets just about any time you need them."

"How generous?"

"A figure of fifty thousand U.S. dollars was mentioned."

"OK. That's not a problem. What about the fleas?"

"I placed a lot of calls and I had to listen to a lot of bad jokes. Everyone thought I was kidding. I couldn't find anyone that breeds the Pulex irritans flea in significant quantities. But, you can breed them yourself in a few weeks."

"How?"

"I found two American patents that describe the equipment needed for breeding fleas. Patent 4850305A, based on an invention by Jay Georgi and Susan Wade, describes a complete system for breeding fleas in a laboratory environment. The inventors assigned the patent to the Cornell Research Foundation in 1989. The other design was also invented by Jay Georgi. It was issued U.S. Patent RE35,348 in October of 1996 and was also assigned to the Cornell Research Foundation. The second design is more sophisticated than the first and has a much higher production rate. Unfortunately, Cornell Research Foundation has never licensed either design to a manufacturer so there's no commercial source for the equipment. I took the liberty of having one of my graduate engineering students study both designs. He's confident that any competent laboratory equipment technician could build several of the machines without difficulty. He thought a single machine could be produced for about five thousand dollars or roughly twenty thousand Shekels. Since you said it was top secret, I assume that you wouldn't want to publicize your interest by requesting a production license from Cornell. Send one of your technicians over and I'll give them the information they'll need."

Levi heard Abner chuckling as though he'd just hear a good joke. "What's so funny?"

"Levi, I wouldn't want to be in your shoes if the Prime Minister's public relations staff finds out that the Mossad is using taxpayer money to breed fleas."

"I'm sure you're right," Levi admitted with an involuntary grimace at the thought. "OK. What about the rest?"

"I'm afraid I came up zip on three of them. The Mediterranean Cat Snakes are bred for laboratory research since they are venomous. However, they are available only in very small quantities. The Toad-Headed Agama lizards are sold as exotic pets but they are expensive and really scarce. You might round up a few hundred if you had time to contact literally hundreds of exotic pet shops all over the world. The scorpions aren't available from any source I could identify. If you had enough time, you could breed all three of them but it would take at least a year to produce even a few thousand of each because none of them are especially prolific breeders and all of them are fussy about reproduction in captivity."

"What about the Asian cockroaches?"

"Florida Agricultural and Mechanical University, nicknamed FAMU, has a research and development center in Quincy, Florida. The Asian cockroach is one of nature's most perfect creations. They've been around for over 350 million years so they predate dinosaurs by at least 150 million years. Although they can survive under a very wide range of conditions, they thrive in tropical and subtropical climates. Once they were introduced into the area around Lakeland, Florida, the breeding population exploded. It is considered a highly invasive species with populations of more than 250,000 roaches per acre now common in some rural areas. Unlike the similar German cockroach, the Asian cockroach is attracted to light and is a very strong flier so control is especially challenging. FAMU is experimenting with various control strategies and they have a breeding population outside of Quincy. Dr. Al Horne has several graduate students working on his research grant but he strongly hinted that he could use additional funding when I asked if they could supply a large number of live Asian cockroaches. He said he could supply a half-million of the roaches for their production cost plus one hundred thousand dollars. That comes to roughly two hundred thousand dollars exclusive of shipping and related expenses."

"OK. Thanks Abner. I guess three out of six is better than nothing. Let's have dinner together soon. We're long overdue."

Aharon Jacobi's Office; Mossad Headquarters
10:03 AM

"Good morning Aharon. Are you ready for me?"

"Oh, it's you Nahbi. I guess I let the time sneak up on me. I'm working on several high-priority projects at once and I'm buried. Let me clear off the table. Can I get you a bottle of water?"

"Yes. Thanks."

Aharon Jacobi was the former Director of Israel's military robotics research and development program. Levi Reznik had recruited him to establish a robust intelligence-gathering capability for robotics and to learn everything he could about Nahbi's design. With a certified I.Q. of 172 Aharon sometimes had difficulty communicating with people Gavriella jokingly referred to as *mere mortals*. He was one of the few people at *The Office* who understood most of what Nahbi was talking about on advanced technical topics.

"You said you needed my help. How so?"

"I want you to oversee the creation of a smartphone with all of the features of the Samsung SIII that can be sold at a fraction of the Samsung's cost. I want it to have some secrets lurking in the background that cannot be discovered by a suspicious Iranian computer scientist, software engineer, or wireless communications technician. I also want it to appear to be an extraordinarily powerful tool for the study of the Koran and the major commentaries, including the sermons of Ruhollah Khomeini and Ali Khameini. When the phones are turned on or rebooted, I want a flattering photograph of Ali Khameini to appear for about ten seconds and I don't want the users to be able to delete or hide it."

"What kind of lurking secrets do you have in mind?"

"I want these phones to be capable of receiving a short text message or a short streaming video from our satellites without having to go through the Iranian wireless networks. I also want them to have at least two gigabytes of undiscoverable memory that we can stream information to via our satellites. In essence, I want to be able to send information to young adult students and business people when the time comes without any risk of censorship or signal blocking. By making them even more feature-laden than the SIIIs, and by pricing them far lower, I want them to become a coveted device that my target audience will be eager to own."

"I'd like for the phones to have solar cells on the back that can recharge or top off a battery by simply laying the phone face down in bright sunlight so they aren't dependent on the public electrical distribution system."

"When do you need them?"

"In one month."

Aharon frowned before responding. "The microelectronics industry's design-to-production cycles have gotten shorter with every passing year but, even so, I think one month is out of the question."

"Normally, you'd be right but there are some pieces already in place, and some others we can manipulate, to achieve that schedule. Do you remember reading a couple of years ago about a meeting between the vice presidents of Indonesia and Iran to promote increased trade?"

"Yes, vaguely. What's your point.

"There's a small electronics manufacturing company in Jakarta, Indonesia that has been working on a cheap clone of the SIII for a year. Their design is almost finished and they've already let the subcontracts for the case, screen, battery, and camera. The main circuit board has been tested with their software suite but they are making final changes. It's not too late to influence the final design. The company's name is Indonesian Advanced Microelectronics. Ltd."

"Indonesia is the world's largest Muslim country. Why would they work with us?"

"They won't know it's us. The avionics and electronics division of Israel Aerospace Industries Ltd has set up a front company in Jakarta that is a preferred subcontractor to IAM. I want you to work with the IAI engineers to complete the design of a main circuit board that will fit into the rest of the phone IAM has designed and have the features I've described."

"How will you get it into Iranian stores?"

"IAM already has a technology import license with the Islamic Republic of Iran and dealer agreements in Iran's largest eight cities. We'll also give some of them away free."

"It is my intent that you and our IAI colleagues design something so attractive in terms of features and price that it will constitute a deal too good to pass up. To the Iranian government officials, it will appear to be a superior product at a much lower price that, unlike the Samsung SIII is produced in a Muslim country."

COS Headquarters; Toulon Naval Base, France
7:00 AM

General Montreuil had arranged to have a French Military version of the Dassault 8X executive jet pick up Tariq Zardari at the Farnborough airport thirty miles southwest of London and bring him to Toulon to help with mission planning.

Zardari and his family had been in deep witness protection arranged by Sir Lionel Smythe in recognition of Zardari's role in helping Task Force 3 intercept the stolen weapons. General Jhalawan and Ahmed Malik didn't know that Tariq was still alive. Task Force 3 had managed to have the news media report that he'd been killed in a car crash outside the hospital at Eastbourne in the UK. With help from Sir Lionel, David had arranged to get Zardari's wife and children out of Pakistan so Tariq felt a strong sense of indebtedness to him.

"OK. Let's get started," Jean Luc announced. "David, would you like to introduce our guest?"

"Good morning. Tariq Zardari is a former Pakistani nuclear weapons engineer who has actually been to General Jhalawan's villa in Janawai several times. I've asked him to come to Toulon to brief us on what we'll encounter there and to help the Brigade engineers construct a realistic mockup of the villa that you'll use to practice various lines of attack and retreat. If he seems a little weary, it's because he's been awake for more than twenty four hours working on some briefing slides for us. He'll show you some high-resolution satellite photographs of the villa and the surrounding area. I'll turn it over to Tariq now."

"Good morning gentlemen. Let me begin by telling you that General Parvez Jhalawan is one of the most evil men on Earth. I fervently hope that you will be successful in your mission."

"This slide shows the general location of Janawai on the north bank of the Neelum River about 40 kilometers east of Sharda via the Neelum River Valley road. To clarify something that could cause some confusion, Neelum is the Pakistani name for the river. On the Indian side of the border, the river is known as Kishengang."

"Janawai is about one hundred and forty kilometers north of the Srinagar International Airport in northern India via roads that grow more primitive the farther north of Srinagar you go. As you can see the region is mountainous and some of the highest mountains on Earth lie just north of Janawai. The lower slopes of the mountains are heavily forested except for some clearings around the small towns and villages."

"The Neelum is a major river with a flow rate in the area of Janawai of about five hundred cubic meters per second. For comparison, the flow rate of France's largest river, the Loire, is about eight hundred cubic meters per second but don't let that deceive you. For most of its length, the Loire is a relatively broad, placid stream whereas the Neelum is a wild, wilderness river characterized by dangerous rapids and massive boulders. While I realize that you are all highly trained and in superb physical condition, I would not advise trying to ford the river on foot carrying a heavy load. Fortunately, there's a pedestrian bridge about a mile east of Janawai."

"This next slide shows a close-up of the villa. It sits on top of a heavily forested ridge on the west side of the village and access is via a steep, graveled road. The villa has a commanding view of the valley and village below. As you can see, an area around the villa has been cleared of trees and underbrush to create a clear field of fire in case of attack. The clearing is approximately three hundred feet on each side so the cleared area is roughly two acres with the villa in the center. The long shadow you can see here was cast by the radio tower on the west side of the clearing. It has two satellite dishes mounted on it for television and Internet connectivity. The villa is protected by motion sensors and high-resolution security cameras. There's a new double perimeter fence visible in these photographs that's been added since I was last there. I assume that the General has strengthened his security in the aftermath of the detonation of one of the stolen weapons."

"When he is at the villa, General Jhalawan has a security detail consisting of one platoon of battle-hardened combat troops known to be loyal to him. In the Pakistani Army, a platoon at full strength has 36 soldiers commanded by a lieutenant. The platoon is divided into three sections of twelve soldiers each. The three sections stand two eight-hour watches followed by two four-hour watches so the watches rotate. That way, the men don't have to stand the same watch every day. They are armed with advanced weapons and equipment. Their training is comparable to your own. Two years ago, a small company based in northern Virginia was hired to train them in the techniques used by the American Secret Service to protect the President. I think you should assume that they are elite troops on a par with the COS or the U.S. Navy SEALs. I realize that it is highly unusual for such elite troops to be used for mere guard duty but General Jhalawan is one of the two most powerful men in Pakistan. Some would say he is *the* most powerful."

"There are bunks for twelve soldiers in the basement of the villa along with a shower and restroom. On their duty days at the villa, the soldiers eat only field rations since there is no kitchen for them. Dr. Ghilzai told me that there is also a small bedroom in the basement with a desk and a computer for the Lieutenant. To my knowledge the Lieutenant is usually at the villa. The off-duty sections are billeted in a cluster of four houses about a quarter mile below the General's villa. They are supposed to cook for themselves but I would guess that they have persuaded one of the village women to cook for them by paying her with a combination of cash and misappropriated military sustenance supplies."

"The security detail has four Otokar Cobra armored personnel vehicles. They are manufactured by a Turkish company and use some of the same mechanical components and subsystems as the American HMMWVs. In addition to a driver, the Cobras can carry up to eight soldiers in full battle gear. Two of the Cobras are typically parked where the soldiers are billeted in the village. Except for brief periods during change of watch, when all of the Cobras can be at the villa at the same time, there are usually two at the villa and two in the village."

"The Cobras can be used as amphibious vehicles in combat so it is possible that they might pursue your retreat across the Neelum River. I think that might depend on whether the Indian Army is making a show of force on the other side at the time. It occurred to me that you might commandeer one or both of the Cobras at the villa to support your retreat. That would make your retreat faster and would provide protection from small arms fire while delaying pursuit until the Cobras in the village could be brought into the action. Two of the Cobras are configured as turreted vehicles with the Russian Kord 12.7mm heavy machine guns mounted.

"General Jhalawan's strong right arm is a man named Ahmed Malik. Malik is the head of Pakistan's ultra-secret S-Wing that directs operations outside of Pakistan and that helped create the Taliban. For your sake, I hope that Malik is occupied elsewhere at the time of your assault on the compound. Your attack will be difficult enough without having to deal with one of Pakistan's most cunning and dangerous intelligence officers."

"The villa has four guard dogs. These dogs are a special variant of the Caucasian shepherd dogs that were originally bred in the Caucasian Range and steppe regions of southern Russia. They are very large animals with thick, black coats with minor tan markings. Males can weigh up to 160 pounds. They have heavy bone structures and a head that looks more like that of a black bear than of a dog. They are widely considered one of the most ferocious and powerful dog breeds in the world. If you encounter one of these canine monsters, my advice is to put several large caliber rounds into him before he has a chance to demonstrate his power and ferocity."

"There are four camouflaged guard blinds near the corners of the cleared area. You can just make them out in this satellite photograph. They are similar in size and construction to portable deer blinds. I do not know the typical manning of these blinds but I suspect it's probably one man standing a four hour watch."

"The last time I was at the villa, I saw a helicopter drone flying over the grounds. It was white and about sixteen inches square and perhaps ten inches tall. It had four propellers and made very little noise. The motors were almost certainly electric to be that quiet. I didn't get a good look at it but an educated guess would be that it has an on-board digital video camera that continuously streams its images to a console in the basement of the villa, probably to the Lieutenant's computer. It's probably controlled from one of the guard blinds so they can keep it in sight."

"I have only seen part of the main floor of the villa. What I have seen includes the main entry foyer, a large living room with a massive stone fireplace, the General's study on the north side of the house, the dining room, kitchen area, and a small guest bathroom. This slide shows my rough sketch of the ground floor as best I can remember it including my estimate of the room sizes. I assume that the bedrooms are all upstairs. The General's study has a large picture window with a view of Nanga Parbat. I would guess that the master bedroom on the second floor does also so it is probably also on the north side of the house. The construction of the house is wood framing resting on a concrete basement foundation walls that extends above ground about two feet. There are narrow, horizontal windows in the part of the foundation walls above ground level. It's reasonable to assume that General Jhalawan would retreat to the concrete basement if the villa came under attack."

"Janawai is in a remote area where high winds are common. Electrical power is often interrupted by trees falling across power lines. The power is often out for days. Although I never saw it, I feel sure that the villa has an emergency power generator, probably on the north, uphill side of the villa."

"Many of the dissidents in Pakistan have been imprisoned or executed. However, some of the residents along the Neelum River are descendants of hardy mountain tribes who historically have had little affection for the central government. It might be possible to enlist some help but I advise great caution. Ahmed Malik has eyes and ears just about everywhere in Pakistan so the risk of betrayal would be high in my opinion."

Tariq concluded his briefing saying, "I'll be working with you over the next few days. Are there any questions?"

Bernard Garneau raised his hand. "I have one. Is the General married?"

"He's a widower. His wife was killed in a suspicious skiing accident five years ago. Some say he'd grown tired of her and had her killed. Shortly after his wife's death, he hired a full-time housekeeper. Rumors abound that she is the General's playmate. She's a beautiful Armenian Caucasian woman in her late twenties or early thirties. Her name Berjuhi means a gorgeous and elegant lady. She is certainly that. She has brought refreshments several times when I've been inside the villa with Dr. Ghilzai. I thought it was interesting that she does not show the same fawning or even fearful deference to the General that most others do. It seems likely to me that she is effectively his common-law wife. She seems athletic and fit. I would not be surprised to learn that she is yet another layer of personal protection surrounding the General. If you encounter her, my advice is to assume she is armed and potentially dangerous."

Bernard had another question. "I've studied the satellite photos and I couldn't help wondering why General Jhalawan, who is apparently a very wealthy man, would choose Janawai as his retreat. It's just a small village with little obvious appeal."

"He was born in Janawai. Some of his childhood friends still live there. Apart from the incredible beauty of the Neelum River Valley, and the mountains to the north, I suspect that the General has deep emotional ties to the area and also that he probably feels safe there. The General is also an avid hunter. Some of the game animals that can be found in the area include the Eurasian Lynx, the Indian Leopard, the Asiatic Black Bear, the Himalayan Brown Bear, the

Indian Wolf, and the White-Bellied Musk Deer. Some of these are listed as endangered but enforcement is lax. It's worth noting that the General is an excellent marksman with a high-powered rifle."

"Are there any escape tunnels from the villa to, for example, the nearby woods?" Alain Fabré asked.

"Possibly but I have no way of knowing if such tunnels exist. It would not surprise me."

Jean Luc asked, "Other than the ones on the APVs, did you see any heavy machine guns around the villa?"

"No, but it occurred to me last night on the plane that those narrow windows in the basement wall could become gun ports in a pitched defensive battle. General Jhalawan's security detail is, so far as I know, recruited exclusively from Army units known to be absolutely loyal to him. I would not be surprised to learn that the basement is equipped with several of the Russian Kord 12.7mm heavy machine guns that have been replacing the older NSV machine guns in Pakistan's Army."

Several of Jean Luc's men traded glances at Tariq's last comments. Everyone in the room knew that the Russian-made Kord heavy machine gun was a formidable weapon with reduced recoil that allowed greater accuracy during sustained fire. It was lighter than the venerable American Browning M2 known as *Ma Deuce* and could be used with a bipod. With its built-in carry handles, a hot weapon could be relocated quickly by a single strong soldier without assistance. The French Army still used a variant of the Browning M2 manufactured by Fabrique Nationale d'Herstal but the guns and ammunition were just too heavy for this mission. They all realized that if General Jhalawan's security detail had Kord-12.7 heavy machine guns in the basement, they would be badly out-gunned in a fire fight.

Sergeant Dejarnette asked if Tariq knew what the off-duty soldiers did for recreation.

"Once, when I was leaving Janawai, I saw some off-duty soldiers playing Cricket. Cricket is arguably the most popular team sport in Pakistan but there really isn't a sufficiently large level field in

Janawai for a proper match. In cold or rainy weather, an educated guess would be that they play Karr Wanj. It's a popular card game in Pakistan that is similar to Indian Rummy. Up to ten people can play at the same time using either two or three standard decks of cards. Gambling is illegal in Pakistan but they are soldiers so I'd be surprised if the games are only played for points."

"What about prostitutes? Where there are soldiers, there are usually prostitutes." Lieutenant Jardin asked.

"Again, prostitution is illegal in Pakistan but is actually widespread due to rampant poverty. Usually, it takes place in the homes of poor women with some pretense of social entertainment usually in the form of simple food and beverage service and cards in plain sight on tables. Most local police turn a blind eye to it. However, punishments for the women can be severe when authorities want to impress some higher-level political authority.

After a short lunch break, Jean Luc led an afternoon brainstorming session. The rough outline of a mission plan was starting to take shape when they wrapped up for the day at 6:00 PM. Jean Luc asked Remy Thibadeau to take Tariq to his quarters so he could get some sleep.

Wednesday, 16 July

Indian President's Home; Rashtrapati Bhawan, India 9:00 AM

The Indian president's executive assistant welcomed General Montreuil and Captain Davignon and took them into a spacious and beautifully decorated living room where two other men were already waiting.

"General Montreuil and Captain Davignon, please allow me to introduce the Prime Minister of India, His Excellency Bindusar Nayar and the President of India, the Honorable Rajeev Gupta."

After the formal introductions, the Indian Leaders asked their guests to be seated. The executive assistant then closed the doors and left the room.

The Indian President thanked his guests for coming and then took a few moments to help them understand how power and responsibilities are allocated between the Indian president and prime minister.

“General Montreuil and Captain Davignon, in India, the President is the head of the executive branch and is responsible for appointing the prime minister. The prime minister then forms the government and nominates the members of the council of ministers. The prime minister and council of ministers assist and advise the president in his discharge of the executive power. The prime minister, as the formal head of our government, is directly in charge of our Department of Atomic Energy, which includes our nuclear weapons. As the nation’s President, I am the official Head of State of the Republic of India and serve as the supreme commander of the defense forces of India. In our country, the president has the power to declare war or conclude peace, subject to the approval of parliament. The allocation of duties between our two offices is the reason why we are both here this morning to listen to what you have to say.”

“Thank you both for seeing us on such short notice. I assure you it is of the utmost strategic importance not only to India but to all civilized nations,” General Montreuil began. “The recent detonation of a nuclear device stolen from Pakistan’s nuclear weapons facility at Wah has demonstrated unequivocally the dangers we all face when radical movements gain control of weapons of mass destruction. We have hard evidence that Pakistan possesses weapons that pose a direct threat to India.”

The two Indian leaders traded glances and then returned their attention to the French General.

“Although it is not widely known, perhaps you are aware that a multi-national team known as Task Force 3 worked inside the MI6 headquarters in London to locate and intercept the WMD devices. Actually, all five of the stolen weapons were intercepted and four were successfully deactivated. What has not been previously revealed is that these weapons were nuclear artillery shells designed to be fired from the M110 self-propelled eight-inch howitzers that

Pakistan acquired from the United States. You may know that Pakistan has approximately two hundred and fifty of these howitzers."

"The nuclear devices are similar to the W79 nuclear shells developed by the United States during the cold war to counter a massive outbreak of Soviet Tanks in Western Europe until reinforcements could be mobilized. We believe that Pakistan's motivation for developing these weapons was the numerical superiority of the Indian Army. However, these weapons incorporate many modern improvements and have a significantly higher yield than the older American variants, all of which have now been destroyed. Like the W79, these are plutonium-based, linear-implosion devices. The Pakistani weapons that were intercepted were all enhanced-radiation variants commonly known as neutron bombs. We do not know for sure how many of these nuclear artillery shells Pakistan possesses but we believe it may be in excess of two hundred. The relatively small quantity of plutonium required per shell is well within Pakistan's production capacity. Those shells pose an immediate threat to India's armed forces in any conflict along the disputed borders. With the deteriorating political situation in Pakistan, the risk appears to be increasing almost daily."

"Do you have reason to suspect that Pakistan intends to use the weapons against India?" President Gupta asked.

"No sir. Not specifically," General Montreuil replied. "However, General Jhalawan was the mastermind behind the recent attacks on the western financial centers. He hates the west and uses radical Islamic sentiments to implement his ambitious plans. He is probably the most powerful man in Pakistan today in view of the current political unrest. As you know, many observers think Pakistan's army will again disband the civilian government and the odds-on betting is that General Parvez Jhalawan will become the new president. He hates India almost as much as he hates the United States and Israel. Our intelligence analysts believe he might try to solidify his popularity in the Army and with the general populace, by using the Pakistani army, backed by over two hundred nuclear-armed, self-propelled howitzers, to take full control of the disputed Kashmir regions along your common border. Such a move would be wildly popular with the average Pakistani citizen and would help divert attention from the deteriorating economic conditions. It could easily lead to Pakistan's use of their large inventory of nuclear artillery shells. Our analysts have noted that, should such use occur, India might have little choice but to retaliate using its own nuclear weapons."

"Let's assume you are right. What is France's immediate interest?" Prime Minister Nayar asked.

"The prime minister of France has ordered the Commandement des Opérations Spéciales, or COS for short, to capture General Jhalawan and bring him back to France to stand trial for war crimes. I have asked Captain Davignon to lead the mission."

"I still don't see what brings you to India."

"We have recently completed briefings in Toulon pursuant to development of the mission's op-plan. We've had the assistance of a former Pakistani nuclear weapons engineer and a former American Army officer named David Logan. He's the man who led Task Force 3. We believe we have a good plan to capture General Jhalawan but we need your help."

"In what way?" India's President asked.

"After examining all possible infiltration routes, we believe the mission should be launched from northern India."

"Why would we agree to let France launch a military mission from our soil against our admittedly difficult neighbor?" President Nayar asked.

"It would be a sort of quid pro quo arrangement. We will share all of the intelligence we currently have regarding Pakistan's nuclear capabilities plus what we gain from interrogating General Jhalawan once we have him in custody."

The Indian president nodded but then said, "We would like to hear the rest of what you have in mind before we decide whether India can support your mission."

"Captain Davignon will explain how he currently envisions the mission and the necessary support," General Montreuil responded.

"The COS team would arrive in New Delhi by commercial airlines, either individually or in twos. They would be carrying foreign passports and would pose as mountain trekkers. They would rent several SUVs and spend a day familiarizing themselves with the roads between Srinagar and the area just south of the Losar Line of Control between India and Pakistan. To all outward appearances, they would be tourists. Some of my men are multi-lingual so they would employ several languages as part of their disguises."

"The COS mission support equipment and weapons would be loaded onto a French C-130 Hercules cargo plane. With the load we have in mind, the Hercules has a maximum range of 2,400 miles. Under the guise of a military support training mission, the C-130 would file a flight plan in Toulon listing Hindon Air Force Station outside of New Delhi as its easternmost stopover for crew rest and refueling before returning home.

Israel has agreed to allow our aircraft to land and refuel at Ramon Air Base in the Negev desert. From there, the aircraft will fly south to Eilat, Israel where it will fly briefly over the northern end of the Gulf of Aqaba before turning southwestward toward Prince Sultan Air Base in Saudi Arabia."

"As you know, the Saudis are very concerned about Iran's nuclear program. As a result of those concerns, the Saudi King has been very supportive of efforts by the Americans, Brits, French, and Israelis to contain the potential acquisition of nuclear weapons by radical groups. He knows that the weapon that was detonated was stolen from Pakistan."

"The French C-130 would refuel at Prince Sultan and then fly a crooked route over the Persian Gulf and into the Gulf of Oman. The aircraft would then enter Indian airspace just west of the Gulf of Kutch where it would turn northward to land at Hindon Air Force Station outside of New Delhi. After refueling and one night of crew rest, it would depart Hindon to head for home. However, shortly after takeoff, the pilot would declare a mechanical emergency and advise Hindon control that he was turning north to get away from the populated areas. A series of messages between the pilot and the control tower would record that the aircraft's load master and crew chief were trying to isolate the problem. Then, the pilot would declare a Mayday emergency and request permission to land at Srinagar where the aircraft would make a safe landing. It would taxi to the west side of the main runway and park near the revetments that shelter your fighter aircraft. The story would be that the French aircraft was disabled and would have to remain at Srinagar until maintenance personnel and spare parts could arrive from France. The stranded French personnel would take rooms in town while waiting for the repair team to arrive. They would rent SUVs that would be used to covertly move the COS mission weapons and equipment off of the air base for transfer to the SUVs previously rented by the COS mission team."

"The C-130 would remain on the ground at Srinagar until the COS team returned with General Jhalawan. The entire French entourage would leave Indian soil in the C-130 to return to France."

General Montreuil thanked Captain Davignon and then addressed the Indian leaders.

"As you've heard, the mission would be carried out without official Indian endorsement or support. At no point would Indian military or civilian personnel be involved. We ask only that you support our cover story by allowing a French military C-130 to transit Indian air space on a training mission and to land at Hindon Air Force Station for crew rest and refueling."

An hour later, General Montreuil and Captain Davignon left the president's residence with assurances from India's two top leaders that they would be given permission for the envisioned C-130 training mission.

Mossad Headquarters; Herzliya, Israel
7:35 AM

Levi Resnik was watching CNN's delayed broadcast of the Israeli prime minister's address to a joint session of the United States Congress the previous evening when his intercom buzzed.

"Levi, can you spare a few minutes?" Gavriella Adler asked.

"Sure. I've just been watching our prime minister trying to educate the American Congress about Iran's nuclear capabilities. I don't think they want to hear it. Certainly the Democrats don't. Can you come now?"

"Yes, on my way." Three minutes later, Gabe was seated on the couch in Levi's office with a pile of folders on her lap. She didn't waste any time. "We know where Nahbi was built."

Levi had never expected to hear those words. The British Secret Intelligence Service and the American Central Intelligence Agency had conducted extensive tests and analysis without discovering a single clue to the mysterious origin of the incredible super android known as Nahbi. He waited for his star protégé to continue.

"Levi, Nahbi was built in the United States. To be more precise, he was built in the panhandle of Texas, probably in Carson County about thirty miles northeast of Amarillo."

"That's hard to believe. The CIA and MI6 would certainly have known that yet I'm quite sure that Nahbi's origin is a complete mystery to them as well."

"Our scientists and intelligence analysts think he was produced as a top secret commercial product by a giant American energy company and that neither the CIA nor MI6 knew of his existence until David Logan and Tiera Aquilina discovered him."

"But that makes no sense either. David and Tiera found Nahbi buried in the sand beside the wreck of an ancient sailing ship. Nahbi said he'd been on that ship when it sank almost two thousand years ago. What makes our scientists think he was built in Texas?"

"As you know, Israel has more Nobel laureates per capita than any other country in the world. Dr. Isaac Saul Geim is a consultant to the Mossad and he's an international expert on the radioactive isotopes of Radon gas. He suggested that we do a comprehensive non-destructive scan of Nahbi's carbon structure to see if there might be any tell-tale radioactive signatures. We hit radioactive pay dirt, so to speak. There are two radioactive isotopes of Radon that are present in Nahbi's graphene skeleton. While both occur naturally in several places in the world, there's only one place in the entire world where they both occur in the precise ratio our scientists found in Nahbi. That place is Carson County, Texas about thirty miles northeast of Amarillo and only a few miles due north of the American nuclear weapons facility known as PanTex. As Nahbi's structural members were being formed, they absorbed minute quantities of the radioactive Radon gas that is present in unusually high levels in the soil in that area. Those parts of Nahbi were definitely made in Texas."

Levi smiled and shook his head slowly from side to side in apparent amusement. "So, our prime minister has granted Israeli citizenship to an android that was made in the United States. That's a rich joke on our American friends but it still doesn't explain how the technology necessary to create Nahbi was mastered by the Americans decades before scientists and engineers believed it *might* be possible.

"No, you're right. Even our scientists can't explain that apparent paradox," Gabe admitted.

Levi got up from his chair and walked over to a bookcase. He was obviously looking for a particular book. When he found it, he came back to his chair and sat down while he flipped through several pages. After only a few moments, he laughed out loud and slapped his knee in delight.

"What's so funny?"

Still chuckling, Levi asked, "Did you know that I have dual citizenship in Israel and the United States?"

"No, I thought the U.S. prohibited dual citizenship for Americans."

"That's what most people think but it's not true. Actually, the U.S. Supreme Court issued a ruling right after World War II stating that dual citizenship is a status long recognized in the law and that a person may have and exercise rights of nationality in two countries. The court made it very clear that if a person asserts the rights of citizenship in one country that does not mean that he or she renounces the other. That's my situation. I was born in the United States when my father was an adjunct professor at Colombia

University. Therefore, I'm an American citizen by birth. However, my parents were both Israeli citizens so I'm also an Israeli citizen by birth. There are quite a few of my generation who hold dual U.S. and Israeli citizenship for similar reasons."

"But I still don't understand what you think is so funny."

"Don't you see? Nahbi may be a citizen of both the United States *and* Israel. It would all hinge on whether the Americans were willing to consider him, as a fully sentient being, eligible for citizenship by reason of his birth in their country. I am laughing at the beautiful irony of their apparent ignorance. The most important technological device in the world may have been secretly made in the United States and then granted Israeli citizenship. Tell me about this American energy company you mentioned."

"The company is called High Plains Electric Corporation but it's usually just referred to as *HiPEC*. It's a privately-owned company incorporated in Texas and administered as a blind trust. I wasn't able to find out who the ultimate owner is. Apparently, the company goes to considerable effort to conceal that information. HiPEC bought one of the world's largest wind farms from General Electric several years ago and has been expanding it. They sell electricity to a private utility company named Pan Tex Electric that then distributes it over parts of three states."

"Why would a renewable energy company be involved in creating an advanced humanoid robot?" Levi asked.

"This is where the story starts getting murky. An internationally acclaimed Japanese robotics scientist named Itoshi Yamada was hired a couple of years ago by the biggest American Internet services company. Yamada is a former head of the robotics research department at the University of Tokyo. After leaving his academic position, he was involved in a startup Japanese applied technology company that competed in a robotics technology demonstration conducted by the U.S. Department of Defense Advanced Research Projects Agency. His company's robot absolutely blew the competition away by successfully performing every task on DARPA's list. The leading American entry was far behind and the other eight entrants were essentially non-players."

"What's the connection to HiPEC?"

"Right after the DARPA competition, Yamada's company inexplicably withdrew from the next round of tests. Most observers believed the company wanted to focus on commercializing their technology and didn't want any limitations associated with the funding from DARPA to hamstring their efforts. We know that a giant American Internet services provider owned the company that finished second in the DARPA trials. That's been widely reported in the media. We think, but cannot prove, that they also acquired Yamada's company. If so, that would mean that the top two robotics R&D companies in the world were owned by one of the largest and richest Internet and information services companies. However, the stock of that company is publicly-traded so some information is a matter of public record. A recent annual report indicated that the robotics subsidiaries had been sold to a privately-owned trust for sixty billion dollars.

The Federal Trade Commission and Department of Justice approved the sale without objection or conditions because the buyer was not involved in the field of robotics. Since there was no Sherman Act anti-trust issue involving creation of a monopoly or competitive restraint of trade, there was no basis to disapprove the sale.

After the sale was announced, there was a lot of conspiracy speculation that the CIA had purchased the companies. However, there's no way the CIA or even the Department of Defense could have marshalled that much money in secret. Only a few giant corporations have that much cash and borrowing leverage. The clue we focused on was the revelation in the annual report that the buyer was a privately-owned trust. We think the buyer was the blind trust that owns HiPEC."

"Why?"

"Because Itoshi Yamada now works for HiPEC. In fact, he's the president of their research subsidiary at their secure industrial compound in Carson County, Texas."

Malta International Airport; Valletta, Malta
11:40 AM

Elena was waiting at the gate. Since David's flight from Toulon had originated in a European Union airport, he didn't have to go through customs and immigration. Several people standing nearby smiled knowingly when Elena threw her arms around his neck and gave him an enthusiastic welcoming kiss.

"I'm so glad you could come. I was afraid you'd have to go straight back to the States with no telling when I might see you again."

David smiled at the radiant happiness evident on her face and pulled her to himself to kiss her again. "I've missed you too. It's only been nine days but you'd think we hadn't seen each other in a month." He slipped his arm around her waist as they walked toward the parking garage.

"My apartment is only a couple of miles from here. It has two bedrooms but we'll have to share the bathroom. I hope you don't mind. I thought we'd stop for lunch on the way and then either have dinner at my place or maybe go out to one of the nice little restaurants in my neighborhood. My parents know you're only here for a few days so they asked us to come to dinner tomorrow night. I said yes. I hope that's all right with you."

"Sure. I'd like to meet them too."

"I talked to Tiera like you suggested."

"How'd it go?"

"She wanted to know how you and I were getting along. I couldn't tell at first if she'd had a change of heart and might be hoping that you and I hadn't hit it off. But, when she asked me if I thought you'd be willing for the two of you to remain friends, I realized that she's pretty much where you are. She's sad but hopes you'll always be her friend."

"She told her parents that she just wasn't ready to get married and that, after talking it through, you'd both decided to end your romantic relationship. She explained to them how strongly you feel about starting a family and that she didn't think her desire to wait was fair to you. She said her mother cried because she'd already been dreaming of you and Tiera producing grandbabies for her to spoil. I didn't have to ask about you stopping by to see her parents. She said Maria and Mathias have invited us to join them for dinner on Saturday night. I was pretty sure that would be OK with you so I accepted."

"Sure. That sounds great. Look, I hate to say it, but my visit includes some business I have to take care of. I need to spend tomorrow morning at the U.S. Embassy. I should be free after lunch."

"I won't pretend I'm not disappointed. I've taken off work while you're here and had really hoped we'd be able to spend all of your time here together."

"I'm sorry but it's something I really have to do. I think I'll be free the rest of the time after I take care of this."

Thursday, 17 July

White House Oval Office; Washington, DC
10:00 AM

"Mr. President, thank you for seeing me on such short notice," the Senate Majority Leader said as he was ushered into the Oval Office.

"I can see by the look on your face that you have bad news. Let's have it."

"Mr. President, I asked the party Whip to give me a vote count in the Senate. It appears certain that the Constitutionally-required super majority will vote to convict you on the Articles of Impeachment."

"I see. Isn't there anything you can do to switch enough votes to prevent that? After all, I'm not the first President in recent times to be impeached by an opposition majority in the House."

"No sir. I'm afraid not. Actually, the final vote may be almost unanimous to convict. The leak about your refusal to authorize covert action that could have prevented more than two million deaths has pretty much killed any chance we might have had to hold our party together on the vote. Mr. President, I must advise you to seriously consider resigning to spare yourself, your family, and the country the trauma of an impeachment conviction."

"Damn it! Those deaths weren't my fault. It's pure speculation that covert action would have made a difference."

The Senate Majority Leader kept silent as the President vented his anger. Unfortunately, that had always been one of his biggest weaknesses as President. The man was extremely narcissistic and vain and seemed incapable of accepting responsibility for anything that didn't make him look good. It had become a sad joke all over the world that whenever bad things happened, the President could always articulate some ridiculous rationale for why it was really someone else's fault.

"So, that's it, huh? Just resign and let those bastards in the House gloat over finally getting rid of me. Is that what you're saying?"

"Mr. President, the damage is done. It's time to see what we can salvage. We've got mid-term elections in three months. It looks like we'll lose the Senate and probably lose ground in the House as well. Once again, for everyone's sake, I must urge you to resign to avoid further damage if we're to have any hope of holding onto the White House two years from now. I haven't given up hope that we might possibly regain the Senate then though I doubt that we could restore the one-party monopoly we had in the first two years of your first term by retaking the House too."

The President stood looking out the window with his hands jammed down in his pants pockets. Finally, he turned to face the Senate Majority Leader.

"I need to talk to my wife. Give me until two o'clock. I'll give you my answer then.

"Very well sir. I'll be waiting for your call."

As soon as the Majority Leader was gone, the President called the Attorney General.

"Ed, I want you to find the SOB at the CIA who leaked the story about me disapproving the covert mission against Jhalawan. I want to make sure he comes to regret his disloyalty to this administration. Maybe a few years of hard time at the ADX federal penitentiary in Colorado would be a suitable reward."

U.S. Embassy; Ta' Qali, Malta
8:00 AM

David had asked Pete to send him Emily's resume so he could prepare for his meeting with the woman Roger Benson had said might be Pete's successor. He'd studied her resume on the flight from Toulon.

Emily Harrison was an Air Force brat and one of the most senior African Americans at the CIA. She was forty-two years old and extremely fit. Her dad had been an Air Force pilot so the family had moved around a lot. She was born in Hawaii when her father was stationed at Hickam Air Force Base. She attended 1st and 2nd grade at Clark AFB in the Philippines and spoke some Tagalog. Her dad had also been stationed at Anderson in Guam, Ramstein in Germany, and Elmendorf in Alaska. She graduated from high school in Georgia when her dad was stationed at Robbins. She earned a Master's Degree in Foreign language from George Mason University with a minor in intelligence analysis. As a language expert himself, David had been impressed to learn that she was fluent in English, German, Spanish, Italian, and Russian. Her CIA resume said she was currently studying Arabic and Urdu.

The other qualifications area of her resume indicated that she'd earned a private pilot's license at age 17 by taking advantage of the flying clubs that exist on most Air Force Bases. She'd become qualified to fly the multi-engine de Havilland DHC-C Twin Otter with amphibious floats when her dad was stationed in Alaska. After joining the CIA, she'd become rotary-wing qualified to fly the MH-6 '*Little Bird*' helicopter configured for covert operations. She was an expert marksman with both pistols and rifles and had even gone through the seven-week sniper training course at Fort Benning.

Emily had made more than fifteen hundred parachute jumps and had been a member of the U.S. sky diving team that had competed in Moscow during her senior year in high school. She'd become a certified diving instructor while living in Guam and was qualified for deep dives using special equipment and gas mixtures. Pete had told him that Emily had one of the highest IQs ever recorded on a CIA qualification test. He'd also said that Emily was single and had never shown any romantic or sexual interest in either men or women. She was, by any measure, a very impressive individual.

"Welcome back, Mr. Logan," Emily Harrison said with a radiant smile as she extended her hand.

"Thanks. It's good to be back. Malta is truly beautiful and I've had some interesting times here."

She grinned and shook her head in obvious amusement. "Ah, *interesting times.* So that's how you'd characterize it, heh?" I read Pete's reports when he was working with you on Task Force 3. All of us were impressed with what you did achieve but I can understand that the tragic ending sort of overshadowed the successes you did have. But it could have been so much worse."

"I can't stop thinking about so many innocent people dying because I didn't check backgrounds on the EOD teams."

"From what Pete told me, things were moving so fast at that stage that there really wasn't time. He told me you took it very hard but he also said that only your calm leadership allowed the other weapons to be found and deactivated in time. He said you kept the team focused even when they began to doubt they'd find the weapons in time."

"Let me show you around before we go up to my office. The Ambassador asked me to bring you by to meet him. He's a retired Army General and will probably tell a few war stories before we can get away. But, he's actually a very good ambassador who works extremely hard."

A half-hour later, Emily and David were seated across from each other at a conference table in her office. "Take a few minutes to sign the paperwork where the flags are. Before you leave, we'll stop by to get your mug shot taken." A few minutes later, David shoved the papers across the table and Emily buzzed her secretary to have her take them to the personnel department.

"OK, now that the bureaucrats have been fed, I'd like to hear more about this mission you're going on. Pete gave me an overview briefing last night so I know generally what you're going to be doing for the next month or so. Sounds interesting as you put it. I wish I was going with you. I always enjoyed the field assignments a whole lot more than I do being a desk-bound political officer at one of our smaller embassies. I understand Al Flores will be going with you."

"Yes. Actually, General Montreuil asked for us both."

"Al's an outstanding field agent. He's got a hair-trigger temper when something doesn't seem right to him but he flies straight and level in the field. Most CIA agents have a sense of mission. Most of us really think what we're doing is important. We like to think that the risks we take are justified as something we can do individually to help maintain some sense of order in a chaotic world. With Al, it's almost a religion. He's intensely patriotic and he's doing what he's always wanted to do. He's been wounded several times but he's always ready to go again. You're lucky to have him on the team. On Task Force 3, Al might have been a little out of his depth but on the mission you're facing, he's one of the best men you could have with you. Tell me about Jean Luc Davignon. I understand you and he worked together on some joint missions in Afghanistan."

"Jean Luc is one of the good guys in my book. He's a lot like you describe Al in terms of his commitment to always trying to do the right thing. He's cool under fire and is a natural leader. We became good friends in Afghanistan and I can't think of anyone I'd rather '*have my back*' in a firefight.

Emily and David spent the rest of the morning discussing various mission options and scenarios. She made several good suggestions.

Faculty Housing; University of Malta
7:00 PM

As a tenured history professor, Elena's father rated one of the small but attractive faculty houses on the campus. Professor Scarlatti and his wife had lived in the house for over fifteen years and it had the cluttered warmth so common in academician homes. Elena's mother's cooking wasn't fancy but it was the kind of home cooking David preferred. After dinner, Elena's father invited David to sit with him on a small patio behind the house. As he settled into his chair, he packed and lit his pipe before speaking.

"Elena has told me that your religious views are very similar to hers. Would you mind sharing a bit of your background and beliefs with me?"

"No sir. I don't mind at all. Like a lot of Americans, my ancestry is mixed. My great grandparents on my dad's side were staunch Scottish Baptists who raised Highland cattle on a small farm near Glasgow. I have some French and Native American blood on my mother's side. One of her ancestors was a French fur trapper in the Canadian Rocky Mountains who took an Arapaho girl as his wife when he settled in what is now the Wind River Indian Reservation in Wyoming. My parents were killed in a car accident when I was twelve. My sister and I were raised by my paternal grandparents on a small farm in north Texas. I earned a BA in Linguistics at Baylor University and then received a Master's degree in Theological Studies from Baylor's George Truett Seminary."

"I was an Army officer until I was badly wounded in Afghanistan. After that, I received a doctoral degree from Cambridge University. My academic focus is language and I'm fluent in several. I think you know the rest of my story."

"Thank you for that excellent summary. Elena has told me that you are a man of deep faith and that Tiera's ambivalence toward religion was a big part of why you've broken off your relationship."

"Yes sir, that's true. There were other issues, but that was the main one."

"Elena begged me not to ask this but I'd like to know what your intentions are toward my daughter. She's my only child and I love her more than I could adequately express. It's not hard to see that she's in love with you. What's not clear, however, is how you feel about her. I hope you'll forgive a dotting father's concern but I would like for you to reassure me that you are not simply rebounding from your recent breakup with Tiera Aquilina.

"Professor Scarlatti, Elena trusts me to be fair to her and I would ask that you extend me that same consideration. I've been entirely honest with her in telling her that it's just too soon after my breakup with Tiera for me to make any promises. She's accepted that and I guess the best I can offer you is to say that I'm not the kind of man who would ever intentionally hurt your daughter. I have known that she was in love with me since the day before my graduation at Cambridge. Yet, she never once violated her friendship with Tiera to promote her own romantic interests. I have the utmost respect for her integrity in that regard. We are friends and we're very comfortable with each other. I can easily envision us being happily married but whatever evolves between us needs more time to develop."

Friday, 18 July

White House Oval Office; Washington, DC
6:00 PM

The Oval Office was ablaze with lights. Cables ran all over the floor. At precisely 6:00 PM, all of the major news networks cut over to the feed from the White House. President Arnold was seated behind his desk in the Oval Office waiting for the cue that the broadcast was live. When the signal came, he began to speak.

"My fellow Americans, I come before you tonight with a profound sense of sadness at the partisan meanness that has made it impossible for me to continue as your President. As you know, the House of Representatives is controlled by the opposition party and they have brought a bill of impeachment against me that has no basis in fact. It is a fabric woven from partisan lies and tired accusations. However, I have decided that it would be best for the country if I were to step down. Accordingly, at 6:59 PM this evening, I will resign as your President. Vice President Wilkerson will be sworn in as President by the Chief Justice of the Supreme Court at precisely 7:00 PM."

The news network talking heads had a field day for almost an hour until they had to take a break while John Wilkerson was sworn in as the 45th President. Then they began again. It went on late into the night and picked up again the next morning with headlines all across the world. Liberal newspapers reported it as a result of partisan politics. Conservative news sources reported it as an act of long-overdue justice toward a man they believed had done enormous damage to the country while inviting a massive increase in global Islamic terrorism.

In Israel, there was little jubilation even though the outgoing American President had demonstrated time and again that he was no friend of the only true democracy in the Middle East. The reason for the muted Israeli reaction was that while the outgoing President had been openly hostile to Israel's interests, the incoming American President was widely seen as being even more incompetent than his predecessor. Facing the mounting nuclear threat from Iran, Israel needed a strong, reliable friend who was not yet in sight. That was about to change.

Saturday, 19 July

Gavriella's Condo; Tel Aviv, Israel
7:02 PM

For once, Gavriella had worked later than Nahbi. She was tired and glad to be home. She kicked off her shoes and set her briefcase down. She could hear Nahbi humming Hatikvah in the kitchen. The Israeli national anthem was the last track on one of her older instrumental CDs. Nahbi wasn't that different than most new immigrants to Israel in his patriotic acceptance of the Hatikvah as the national anthem of his homeland. In her mind, she'd almost said his *new* homeland but that wasn't really true. Israel was the *only* home he'd ever known though he'd been away for a very long time.

As she stood listening to him, she remembered how surprised she'd been at his apparent enjoyment of her music collection. His ability to synthesize musical sounds was astonishing. He could make sounds that to her ear were indistinguishable from a recording of an entire orchestra. When she'd asked him how he did it, he'd told her that every sound he made was the product of his computer brain. He'd said that when he listened to music, his brain made a precise recording of every vibration in the sound waves from the playback device. To his brain, music was just a stream of data picked up by his auditory sensors and stored in his memory. When he wanted to hum or even sing, his brain reassembled the sounds it had previously recorded and sent the appropriate electrical signals to the bundles of graphene strands in his throat that performed the same function as human vocal chords although with significantly greater capabilities. A few days earlier, he had given her a demonstration where he'd sounded like the entire London Jewish Male Choir. She smiled to herself as she remembered laughing with delight to hear the sound of an entire choir coming from one person's mouth.

She knew that Nahbi had heard her come in. He didn't miss anything. When he finished humming, she walked into the kitchen and looked around to see what he was making for their dinner. It was no simple home-cooked meal. It looked more like something from a world-class French restaurant.

"I see that the resident chef has exceeded even his typical culinary brilliance. To what happy occasion do we owe all this?"

"What a suspicious woman you are. Have I become so much like a human husband that you think I'm trying to buy your forgiveness before fessing up to some crime?" Nahbi responded with a broad smile. "Maybe I'm just trying to put you in the mood for some bedroom fun."

"You don't need to do that. I'm always in the mood for bedroom fun with you."

"Yes, I suppose that's true. I never suspected what depths of unbridled passion lurked inside that luscious body and behind that pretty face of yours. I have been shocked to discover the real you."

"I am your creation Señor Juan," she teased." You have taught me things I never imagined. If I've become an insatiable scarlet woman, it was you who painted me."

"I protest. Not guilty. Enough of your unfounded accusations! Are you ready to eat?"

They took their time over the superb dinner Nahbi had prepared. By the time he placed a warm blackberry tort in front of her, Gabe was mellow from several glasses of wine

"I sense that you have something to tell me. Is it related to this big project you've been working on for Levi?"

Nahbi reached across the table to take her hand. "Yes, in a way it is. The prime minister has asked me to infiltrate Iran as a sort of agent provocateur to lay the groundwork for what will come later."

"No! Tell him no! I would never see you again. It's a suicide mission. No Israeli agent has come back from an assignment to Iran for over ten years. Levi thinks the Iranians or maybe Pakistan's ISI have a deep cover agent inside the Mossad. A few days after our agents have gone in, they've simply vanished. We never hear from them again. We don't know if they've been killed or imprisoned. We assume they've been tortured for classified information and then killed." Tears streamed down Gabe's cheeks and she began to sob quietly. "I'll never see you again." she whispered as she shook her head slowly from side to side.

It was surreal, Gabe thought. It was almost like she was observing herself and Nahbi from somewhere above the table. From that detached perspective, she seemed no different than thousands of terrified women through the ages who'd heard similar words from their warrior husbands on the eve of battle. Yet, even as she heard herself protesting, she knew that Nahbi could not decline the mission. Israel's survival was at stake. Israeli intelligence indicated that Iran now possessed nuclear weapons even as Iranian leaders loudly proclaimed their intent to destroy Israel. Beneath her intense emotional pain, the rational parts of her mind also knew that Israel had never had an agent like Nahbi. If anyone could succeed, he could.

Nahbi rose and walked around the table to stand behind Gavriella. Bending over, he put his hands on her shoulders and kissed the top of her head.

"I will come home, Gabe. I promise you this. I will not leave you alone. Be brave. I know this is hard. But you and I are soldiers fighting for the right of the Israeli people to live in peace, just as the hopeful words of the Hatikvah proclaim. You know that I love you. You have awakened in me the inexpressible joy of truly loving someone. I will not allow Islamic evil to snatch that miracle from us."

Aquilina Estate; Iklin, Malta
7:30 PM

Tiera met David and Elena at the front door and gave them both a friendly kiss on the cheek. If she was uncomfortable at their reunion, she gave no outward sign of it. She made small talk as they walked toward the living room where they were greeted by Mathias and Maria. Tiera's parents both rose as they entered the room. After their gracious welcome, they all sat in the living room catching up.

"Have you accepted one of your job offers yet?" Tiera asked.

"No, but I'm leaning toward taking the CIA job. I'd be working for Pete Nocona at the CIA headquarters in Langley. The pay is good and I'd be doing work I enjoy. What about you?"

"I've accepted a position with MI6 in London. I'll be working for Gordon McFarland. I start in two weeks. I'll fly up to London next week to find an apartment."

"That's great!" David responded. "I think you'll love working with Gordon."

Lucia came into the room to announce that dinner was ready. She gave David and Elena big hugs and then returned to her kitchen to begin serving dinner.

Mathias pulled out Maria's chair for her and then took his usual place at the head of the table. Tiera sat down beside her mother at the other end of the table and David and Elena were seated across from each other at the middle of the long table. Although Maria was twenty-five years older than her daughter, it didn't show. She was still a beautiful woman and could have easily been mistaken for Tiera's older sister.

The dinner was superb and Lucia fairly beamed when David complimented her on yet another of her culinary masterpieces. After dinner, they walked out to the terrace to enjoy the mild weather and the gathering twilight.

Sitting comfortably beside his wife, Mathias finally broached the subject that had been on all of the news shows for the past twenty-four hours. “What do you think about President Arnold’s resignation, David?”

“I don’t think he had much choice. After the anonymous CIA source revealed that he had refused to order a pre-emptive strike on General Jhalawan that might have saved two million lives, even his own party turned against him. The House Vote last Monday to bring a bill of impeachment against him was bipartisan and almost unanimous. I feel sure that the Senate Majority Leader advised him to resign rather than face conviction in the Senate.”

“What kind of president do you think John Wilkerson will make?”

“Please don’t quote me, but I don’t think he’ll be up to the challenges facing the United States and her allies. He’s ultra-liberal and is likely to continue President Arnold’s gutting of American military capability. He has almost no foreign policy experience and he never served in the military. I don’t think the Israeli prime minister will see him as a reliable ally in the war against radical Islam. I think having him become the President probably increases the likelihood of further escalations of violence and instability in the Middle East.”

“Your financial views are internationally respected. What kind of reaction do you think we’ll see in the markets when they open Monday morning?” David asked.

“The Tokyo exchange will be the first major one to open. In the past, as you know, the rest of the world has often followed the TSE lead. I wouldn’t be surprised to see a big drop in the TSE. Japan is worried about China’s recent muscle-flexing and really needs for the U.S. to counter China’s growing ambitions. From what you say, it doesn’t sound like President Wilkerson is the man for that job and the Japanese leaders probably know it.”

"Hong Kong opens an hour later and I think it will amplify whatever the TSE does in the early trading. I think you could see a fifteen percent drop in the TSE and maybe as much as twenty percent in the Hang Seng index in Hong Kong."

"I expect the major European indexes to drop sharply. Since President' Arnold's resignation last night, our bank has been submitting sell orders on the most vulnerable stocks in our portfolios. Of course, those sell orders won't be processed until after the markets open Monday and I expect volume to be so high that prices may drop quite a ways before our sell orders are acted on. The situation in the Middle East and in the Ukraine, coming on top of the Federal Reserve Board's hints at rate hikes, has the markets pretty spooked already. It's entirely possible that panic selling could take over and there's no way to predict how much paper wealth might evaporate before the markets stabilize."

"And, there's always the risk of a black swan even making it much worse."

"I'm sorry, but I don't recognize that phrase. What do you mean by a black swan event?" David asked.

"The phrase comes from a book by Nassim Nicholas Taleb. He gave the name black swan event to highly improbable and rare events that are difficult or impossible to predict but which can have a dramatic impact on market behavior. The Japanese bombing of Pearl Harbor and the September 11 attacks on the United States are classic examples of black swan events."

Twenty minutes later, Mathias and Maria said goodnight after urging David and Elena to come again any time he was in Malta. Elena said she wanted to thank Lucia for the fine dinner and left David and Tiera alone for a few minutes.

"Are we OK? Tiera asked.

"Yes. Don't worry. We'll always be friends. I will always love you and be your cheerleader. I think we're both sad that things didn't turn out as we originally thought they would but I'm quite sure we've made the right decision."

“I’m so glad to hear you say that. We’ve been through way too much to just walk away from each other. What about you and Elena?”

“Too early to tell. I don’t have to tell you that she’s a wonderful person who’ll make some lucky man very happy. We’ve gotten off to a good start but it needs some time.”

“Well, good luck to you both.”

Elena returned a moment later. Tiera walked them to the door and kissed them both as she said goodnight. The evening had gone as well as any of them could have hoped.

Sunday, 20 July

Manama, Bahrain
5:17 AM

The USS Sabine Pass, CG-78 was moored near the end of the long pier at the Mina Salman Port Facility. Named after the second battle of Sabine Pass during the Civil War, the ship was the newest and last of the Ticonderoga class of Aegis guided missile cruisers.

Naval Base Bahrain, headquarters of the American Central Command and the U.S. 5th Fleet, was a mile to the northwest. Warehouses on the base stored American pre-positioned war reserve material, making the base and nearby piers an attractive target for radical Islamic terrorists.

“Request permission to leave the ship sir,” Food Service Attendant Leon Jones said as he saluted the Officer of the Deck. He was carrying a large plastic bag of trash.

“Permission granted,” the OOD responded. “You’d better hustle. The sentry just called to let us know that the garbage truck is already on the pier.”

After two suicide bombers had killed forty American sailors in an attack on the USS Cole in Yemen, the Navy had begun posting guards with radios at the head of the piers to keep the OODs informed. Even in a friendly port, lookouts were posted around the clock scanning for anything suspicious. Incredibly, sailors aboard the Cole hadn't recognized the threat and had actually been leaning over the side of the ship and talking to the two suicide bombers just before the explosion. The Navy didn't intend for that to ever happen again.

Seaman Jones was nearing the end of his two-month temporary duty as an FSA and was scheduled to return to his regular job as an ET-9608 radar technician at the end of the week. He made it to the dumpster on the pier just as the garbage truck approached. He lifted the lid and dropped his trash bag inside before stepping out of the way to let the driver maneuver the truck to pick up the five cubic yard dumpster and hoist it over the cab to dump it. Instead, the driver suddenly accelerated and drove off of the pier onto the small flight deck at the stern of the ship.

The massive explosion hurled the Mark 45 gun mount into the water and ripped a huge hole in the stern of the ship. Seaman Jones was killed instantly. The stern flooded immediately and the ship began listing to starboard just as the General Quarters alarm sounded throughout the ship. The OOD's quick action to sound the battle stations alarm kept the ship afloat because all watertight doors were closed within a few seconds.

Sixteen enlisted personnel and one officer were killed in a grim reminder that terrorism would not disappear simply because the American president had told the world that the United States was no longer engaged in a war on terror.

Monday, 21 July

Kavir Desert, Iran
2:00 AM

"Good hunting my friend," the Israeli pilot transmitted as he released the HALO pod 98,000 feet above the Dasht-e Lut salt desert southwest of Tehran. The stealthy Israeli Cheetah was streaking across the night sky at Mach 3. The Iranians would never know it had been there.

Four seconds after separation, a small drogue chute deployed automatically from the rear of the pod to stabilize its descent from more than 18 miles above the desolate wasteland below.

Constructed entirely of high-strength carbon fiber materials, the small pod was invisible to Iranian radar during its swift descent. Like most high-altitude, low-opening military free-fall insertions, the pod's descent velocity went briefly supersonic before increasing atmospheric density began to slow its fall.

Inside the pod, Nahbi was alert but calm as he waited. Four minutes and ten seconds after being jettisoned from the Cheetah the pod reached the pre-determined altitude of 21,000 feet where explosive bolts separated the pod into two separate chambers. Four seconds later, Nahbi's half of the pod opened and fell away. He continued in free fall for another 32 seconds before his main parachute opened automatically at an altitude of only 1,500 feet above the desert floor. Less than a minute later, he was on the ground within a few meters of his designated landing target at 35 degrees north latitude and 55 degrees east longitude.

He immediately gathered the military parachute and dug a shallow hole in the sand to conceal it. The sky was clear and there was a moderate northerly wind. The waning Moon had risen at 11:32 PM time and was two days past its third quarter. Although the moonlight was faint, Nahbi's ultra-sensitive eyes saw everything as plainly as if the desert had been bathed in mid-day sunlight.

He keyed a short burst transmission on his encrypted SATCOM radio to let Levi Resnik at Mossad headquarters in Tel Aviv know that he was safely on the ground. He knew that his transmission would also be monitored at the top secret operations center that had been established at Prince Sultan Air Base seventy miles southeast of Riyadh, Saudi Arabia. The Saudi King could not publicize his support but he had done everything he reasonably could to support the Israeli operation. Iran was a threat to the House of Saud just as it was to Israel.

He jogged toward the payload section of the HALO pod that had landed a few hundred feet to the east.

There were several packages inside the pod to support his ultra-secret mission. Among them were ten shrink-wrapped, one cubic foot bales of Iranian money worth nearly three quarters of a million U.S. dollars. At roughly 25,000 Rials to the dollar, Iran's currency was almost worthless. Years of economic sanctions against Iran's rogue nuclear program had crippled the country's economy and resulted in inflation so severe that the nation's largest paper bank note of 100,000 Iranian Rials was worth less than four U.S. dollars. Part of Nahbi's mission was to make things even worse.

The payload also contained fifty packages of C4 plastic explosive along with subminiature radio-controlled detonators.

Four molded dust-proof cases each held two Sig Sauer P226 semi-automatic pistols with guide-rod lasers and two Heckler and Koch HK416 assault rifles with red dot laser sights along with 100 rounds of ammunition for each weapon. Nahbi didn't expect to need the guns but Levi and Gavriella had both strongly recommended that they be included.

Four small duffel bags held clothing and items that would be useful in creating various disguises. Four drawstring bags each contained a dozen emergency rations consisting of bottled water and small plastic bags of coarse wheat meal that he could consume to generate methane gas to power his internal fuel cells if he was forced to hide in the desert to avoid capture.

Two dozen large zip-top plastic bags held forged passports and other papers to allow him to assume various identities.

Nahbi divided the payload into four separate piles and then buried them several hundred yards apart. He memorized the GPS coordinates of each cache and the distinctive details of each burial site so he could quickly find them again. Future payloads of additional money and materials would be dropped at other remote locations around Iran to be accessible to him as needed for his mission.

By 4:45 AM he was finished and ready to leave. He checked each site carefully to satisfy himself that there were no telltale signs of his caches. The wind would cover his tracks with blowing sand before daylight. He loaded his rucksack with the items he would need immediately and headed west.

Tirelessly covering the ground at eighteen miles an hour, about sixty percent of his maximum running speed, it took him ninety minutes to reach his objective twenty three miles to the west. A half-hour after sunrise, he found the compact Bardo pickup truck where an Iranian dissident had hidden it in a dry wadi a mile east of highway 81. The truck was three years old but had been the victim of several fender-benders making it look more dilapidated than it really was. It had a cheap aluminum topper shell over the bed. The nondescript truck was perfect for his needs and everything he'd asked for was in the truck. He spent several minutes changing clothes and modifying his appearance to conform to the first false identity he would use.

At 7:20 AM, Nahbi turned north on highway 81 toward the small village of Siyang. From the intersection with highway 36 it was 85 miles to the city of Semnan and a secret meeting that would be crucial to the success of his mission.

Wednesday, 23 July

Elena's Apartment; Valletta, Malta
7:20 AM

"Great breakfast! I'm stuffed," David told Elena as she poured him his third cup of coffee.

"Think of it as bait to bring you back to me. This week has gone way too fast. I don't suppose you know when we might be able to see each other again?"

"Elena, I need to tell you something but it's absolutely top secret."

Elena's eyes filled with tears. She wiped them away with a napkin but they came right back. "You're going after General Jhalawan aren't you?"

"It's more complicated than that but you've guessed the essence of what I'm going to be doing for the next several weeks. I leave at

3:25 PM this afternoon on the Lufthansa red-eye flight with a plane change in Frankfurt. I'll get into Washington Dulles tomorrow afternoon a little after two PM. I'm not sure how long I'll be in the States but probably less than two weeks. After that, I expect to be gone for about two weeks. I don't want to scare you any worse than you already are but there is some danger involved." David paused and took a deep breath before continuing. "We'll be going up against some pretty tough characters so it's possible I could be killed."

Elena began to sob uncontrollably with tears streaming down her cheeks.

"If it's top secret, how would I know what had happened to you if you never called me again?"

"I'll give you Pete Nocona's phone number at Langley. If you haven't heard from me in thirty days, call him and tell him who you are. He won't be able to discuss any details with you but he will give you one of three words. If he says 'yes' it means I'll be coming home and I'll call when I can. If he says 'no' it means I won't be coming home. If he says 'TBD' it means to be determined. In other words he doesn't know yet."

"I know this is a cruel thing to do to you just as you're starting to have some hope for us but I couldn't just fly away and risk that something might happen and you'd never know. That would be so unfair that I just can't do that to you, regardless of the security classification."

"Elena, there's one more thing. You asked me a question our last night together in Virginia. You asked if there was a chance that I might ever love you the way you loved me. My answer is yes. It has already happened. I do love you and I will do my very best to come back to you."

Ferdous Garden; Semnan, Iran
Noon

Nahbi had been in the park near the center of town for a half-hour and was satisfied that it was not under surveillance. Semnan was a city of approximately one hundred and fifty thousand people located two hundred kilometers east of Tehran. He was sitting on a bench in the shade watching some ducks on the pond when he saw the two men enter the park on the opposite side of the pond. He recognized them from intelligence photos and gave the signal they'd been told he would use to identify himself. He opened a paper bag and took out a paper cup of hot tea which he then made a show of clumsily spilling. The two men walked around the pond and stood a few feet in front of him. One of them took out a bottle of water and took a drink from it. That was the signal to Nahbi that they had not been followed.

"Salam!" Nahbi said softly in greeting. I am Yusuf ibn al-Hasan. Thank you for coming. I have need of your help. I want the two of you to form a small trucking company. I want you to buy two large diesel trucks configured to deliver large dumpsters to construction sites. I also want you to buy four forty cubic meter dumpsters and have them steam cleaned. Then, have them repainted white with a one meter tall by one and three-quarters long Iranian flag painted on both sides. I will provide sufficient money that you will be able to obtain a loan for the rest of the money you'll need from a bank. If you are questioned, just say that you have been saving for many years to start a business of your own. Have the truck titles issued in the name of your new company. Tell the loan officer that you will each drive one of the trucks as owner-operators. Get the trucks properly licensed and obtain any necessary business licenses. Then, go around to construction and commercial sites here in Semnan and also in Qom. If possible, try to win the contract for disposal services at the Jamkaran Mosque in Qom. Offer your services at a favorable rate to promote your new business. Make a good effort to introduce your company and to become a recognized provider of trash and debris collection services. You may keep the money I provide plus anything you are able to earn from your new company after making your loan payments. In exchange for this, I will expect you to place your dumpsters at locations I will specify sometime in the next few weeks. Until I advise you otherwise, you must keep two of the dumpsters in reserve and ready for use on short notice. Do you have any questions?"

"This seems a strange assignment. We expected something that would help bring about an end of the tyrannical rule of the Ayatollah and let Iran resume her rightful role as one of the world's great nations."

"I assure you that your efforts will produce precisely that outcome. I cannot explain further but you may be confident that what I ask you to do will hasten the day you both pray for. When I leave, I will walk to an old pickup truck parked on the south side of the park. Give me a small head start and then follow as though you are just strolling around. I will remove a cardboard box from the back of my truck and place it on the ground near a trash receptacle. When you are sure that this has not been observed, you may retrieve the box. In it, you will find two billion Iranian Rials in large bank notes. This is the equivalent of approximately eighty thousand dollars and should be more than sufficient to help finance the launch of your new company. I have your mobile phone numbers memorized. I will contact you at the appropriate time. Until then, kodafez."

Friday, 25 July

CIA Headquarters; Langley, Virginia
8:00 AM

The intercom light on his phone started blinking and the voice of Pete's administrative assistant announced, "Mr. Nocona, Al Flores and David Logan are here for their meeting with you."

"Come in and close the door," Pete told them as they entered his office. "We don't have a lot of time. I have an intelligence briefing with the Director in less than an hour. I've made arrangements for both of you to fly to Coronado for a week of intense mission prep training. You leave tomorrow morning. You'll check into the bachelor officer's quarters on base when you arrive and then meet with your instructor Sunday morning. You'll spend the week in training and then return to DC the following Saturday."

"You both know the Navy officer who will oversee your training. After the SEAL Team 8 mission to abduct Yousef Abu Shakra, and the time he spent with us on Task Force 3 in London, the Navy detailer sent Lieutenant Mark Hays to Coronado as one of the training officers on the twenty-six week SEAL Qualification Training program. He's just reported for duty and hasn't been added to the SQT roster yet. Roger talked to his commanding officer and asked that Lieutenant Hays put together a custom, crash-training program that skips the parts you don't need and focuses on what you do. Except for Mark, as far as anyone else knows, this is routine training. Got it? OK, see Marge in the travel office for your tickets and then go home to pack and get some rest. I don't think you're going to get much sleep next week. I'll meet with you Saturday after you get back. Right now, it looks like you'll be back on a plane on Sunday the third. I'm still working on some details. Say hi to Mark for me."

Jamkarān Mosque; Qom, Iran
11:25 AM

Nahbi parked his pickup in the huge parking lot on the north side of the mosque and tucked a prayer mat under his left arm before joining the crowds there to attend salāt al-jumah, the Friday congregational prayers.

The five domes and two minarets of the Jamkarān Mosque were symmetrically arranged. Their green and blue mosaic tiles contrasted tastefully with the reddish brick of the mosque itself. An enormous paved courtyard extended to the northeast from the mosque and the overall impression was architecturally superb.

Located six kilometers east of the holy city of Qom, Jamkarān Mosque had been considered a sacred place by Shi'ite Muslims since the 17th of Ramadan, or the 22nd of February in 984 CE, when Sheikh Hassan ibn Muthlih Jamkarani reportedly met there with Muhammad al-Mahdi and the prophet Al-Khidr. Sometime around the beginning of the new millennium, the mosque's reputation began to spread and many pilgrims, especially young people, began to come to the well of requests located at the rear of the mosque. The faithful believed that the Twelfth Imam once became miraculously visible for a brief shining moment. Since then, Shi'ite pilgrims have tied small strings in a knot around the grids covering the holy well in the hope that their requests would be received and granted by the Imam Mahdi. Tuesday prayers had become especially popular with the younger generation because of their belief that Tuesday was the day of the week when the vision appeared and was thus the day of the week when the invisible Twelfth Imam would be present to accept and possibly grant their requests.

Tuesdays had become an incredible event that was like a gigantic party. Families brought picnic lunches and vendors set up booths in the parking lots. Tens of thousands of people wandered around the grounds hoping something miraculous might occur while they were there. Sometimes more than a hundred thousand pilgrims gathered outside the mosque for evening prayer. Women had their own area in which to pray and even had their own well of requests. On Tuesdays, the mosque provided a free evening meal to thousands of poor people.

Nahbi was there to learn and to prepare. As he mingled with the crowds, he listened to hundreds of conversations and memorized speech patterns and local idioms. He noted how both the average Iranian citizens and the more affluent businessmen were dressed. Wearing dark slacks and a white, long-sleeve shirt, his attire was similar to that of most of the men. His dark hair and beard, with his dark eyes and fluency in Farsi, made it easy for him to blend into the crowds.

At 11:45 AM, the muezzin's first call to prayer, known as the adhan, blared out from loudspeakers. Ten minutes later the second adhan call was broadcast over the speakers when the Khatīb, who also served as the Imam, took his place at the pulpit known as the minbar. In accordance with Muslim custom, the Khatīb delivered two sermons. The first was longer and contained most of the primary content. After delivering the first sermon, the Khatīb sat down briefly and then rose to deliver the shorter second sermon that ended with a dua invocation that was also an act of supplication. The muezzin then called the iqāmah signaling the start of the first of the two rak'at prayers of Jumu'ah. Nahbi stored every audible call issued by the muezzin. Just as he'd demonstrated to Gabe, he could imitate every nuance of the muezzin's calls whenever he wished. No human would be able to tell the difference.

When the noon Friday prayers were completed, Nahbi wandered among the crowds looking for something. Within a half-hour, he identified several groups of young men engaged in religious debate. Some of the groups were loudly passionate and argumentive. Others were more subdued and scholarly. He memorized their faces and where each of the groups had congregated. He heard some names and memorized them too. It was too early for him to use the information but he felt confident that, when the time came, he could use what he'd learned to good advantage.

Sunday, 27 July

Hotel Del Coronado; San Diego, California
10:30 AM

The Sunday brunch at the historic Hotel "Del" was always impressive. Mark Hays had suggested it since most of the Navy officer's clubs in the area were not open on Sunday mornings. At Mark's request, the hostess led them to a table on the veranda overlooking the beach. After placing their drink orders, they went through the buffet line and then returned to their table.

"Tomorrow morning, we'll run you both through part of the physical qualification tests just so we know your capabilities and limits in case we need to adjust some of the training. We'll skip the five hundred yard swim but we'll do the push-ups, pull-ups, and the beach run. I'm adding three tests that seem relevant to the mission. You'll do two kinds of weight-lifting tests. One of them will be to press two hundred pounds ten times. The other will be a dead weight squat test to see if you can do five squats with three hundred pounds. The last test will be to see how fast you can cover one mile while wearing full combat gear and carrying a pack weighing ninety pounds. Tomorrow afternoon, we'll let you sit down while we go over small unit tactics. After a quick dinner break, you'll spend the evening learning how to use French military GPS equipment. Since no one seems to be sure the GPS equipment will work in the deep valleys along the Neelum River, you'll also get a refresher course navigating the old fashioned way with maps and a compass. We don't have any French SATCOM radios that we could use to familiarize you with their equipment but some of Jean Luc's men will probably handle that part of your mission anyway."

"Tuesday, we'll spend the day at Camp Pendleton Marine Corps Base for several kinds of weapons training. Pete told me that you'll be using the same weapons that the French COS use so I've asked the range master to arrange for you to fire the FAMAS G2 from several positions and then to use it in an urban warfare mockup where you'll be exposed to pop-up targets and lots of noise and smoke. The FAMAS G2 is a bullpup style assault rifle. The G2 variant has been upgraded to bring it into compliance with NATO standards including tighter rifling and using NATO magazines. When I spoke to Captain Davignon yesterday, he told me that he prefers the Heckler & Koch HK416 for this mission and that the decision on which assault rifle to take is still up in the air. So, we'll run the rifle training with both weapons. They fire the same 5.56 cartridge and I know you've both fired the HK416 before."

"The French Army Special Forces Brigade and the French Naval Commandos use the Glock 17 9mm as their handgun and the Benelli M3 Super 90 12 gauge shotgun so we'll run you through several drills with both of them. I don't know what Jean Luc and Lieutenant Jardin will select as the final weapons kit so we'll try to cover the obvious possibilities."

"The submachine gun the COS commandos use is the Heckler & Koch MP5SD that fires a 9mm cartridge and is widely considered the best compact submachine gun in the world. The SD designation means it's sound suppressed. You'll go through the urban warfare mockup again using it."

"Jean Luc told me that he hasn't yet made up his mind on whether to take one or more of the FGM-148 Javelin anti-tank missile launchers. They weigh about fourteen pounds and the missiles weigh about twenty-six pounds so each launcher and missile pair weighs a tad over forty pounds. That's a lot for a small force on foot to manage but he thought you might need it if

Jhalawan and his troops decide to hole up in the basement. Just in case, you'll both fire two Javelins each at an APV mockup and a rectangular target roughly the size of the narrow windows in the basement wall of Jhalawan's villa. Jean Luc also told me that he may include a couple of M203 grenade launchers as well so we'll give you a chance to play with one of them too."

"Jean Luc told me that Remy Thibadeau is one of France's top snipers so I didn't need to include sniper training."

"Before we wrap up at Camp Pendleton, I've arranged for you both to drive a LAV-25 armored personnel vehicle. I couldn't find an Oktokar Cobra like General Jhalawan's security detail uses so this is the best I can do. The Marines have a rough terrain course that will at least give you some experience in driving an APV off road at high speed."

"Tuesday, we're going to take a run up to the Cleveland National Forest to do a little hiking. There's a rugged, rocky creek in Noble Canyon that is about the closet thing nearby to simulate the rough terrain you'll encounter. We'll suit you up in spec-ops gear and put a heavy pack on you to see how you handle it."

"One thing we can't simulate on the ground is the thin air you're going to encounter in the southern part of the Himalayan Mountains. So, on Wednesday morning, we're going to fly up to NASA's Armstrong Flight Research Center at Edwards Air Force Base. I've arranged to have two exercise bikes put into a barometric chamber. You'll be inside of the chamber for twenty-four hours from about ten in the morning Wednesday until ten AM Thursday. You'll actually sleep inside the chamber. I've got a notebook of lessons learned on similar missions over the years just in case you have trouble going to sleep. That's the best I can do to help you begin to acclimatize to the higher altitudes you're going to be operating in. You'll be put through an exercise regimen with your heart rate, blood pressure, and blood oxygen level monitored. We'll return to Coronado Thursday afternoon."

"Thursday evening and all day Friday will be a mix of classroom and field training, some of which I'm still working on. I've got a Navy medic scheduled to give you some basic combat first aid refresher training. With the security lid that Pete and Roger have imposed, I've had to develop the scenarios myself and I haven't had as much time as I'd have liked. I think the three of us can put our heads together to come up with a range of possible scenarios you may face and then try to get you prepared for them. Any questions?"

"Any good movies showing on base this week?" Al deadpanned.

Fort Stanton Park; Washington, DC
2:08 PM

ABC was the first to break into their regular programming to carry the news. In less than a minute all of the major news networks were on the story.

"This is Angela King reporting live from the White House for ABC News. We have just learned that Marine One, the President's helicopter, has crashed in Fort Stanton Park on its way to Andrews Air Force base where President Wilkerson was scheduled to fly to California for a fund raising event tonight. I'm told that we now have a live video feed from the crash location. ABC News Correspondent Roger Tanner has just arrived at the site."

The screen filled with a scene of utter chaos around the crash site. Flames were leaping thirty to forty feet into the air and heavy, black smoke was billowing into the clear sky. The Garfield Heights Fire Department had multiple trucks already pumping water and foam onto the burning remains of the big Sikorsky helicopter. Police and emergency vehicles sirens were adding their contribution to the confusion and noise.

"Roger, what can you tell us?"

"Angela, the Garfield Heights Fire Chief just told me that it does not appear that there are any survivors. They won't be sure until they can put the fires out but right now it looks like the President has been killed."

"Roger, do we know if this was a terrorist attack?

"No, Angela. We don't but the crash site is swarming with FBI agents and I've just heard that the Department of Homeland

Defense has issued an alert to law enforcement agencies until we know more."

"Ladies and gentlemen, you've just heard ABC News Correspondent Roger Tanner at the scene where the President's helicopter has crashed. We're hoping for a miracle but we'll keep the video feed from the scene of the crash on your screens as more information becomes available."

"Dana Compton is at Andrews Air Force Base. Let's go live to hear what she has for us."

"Angela, I've just spoken to Air Force Captain Andrew Briles. He's the Public Affairs Officer and he told me that the controllers in the tower here at Andrews heard the Marine pilot issue a Mayday call saying the aircraft's rotors had been struck by multiple Canada Geese and was rapidly losing altitude. Based on that report, it does not appear that the President's helicopter was shot down by a terrorist rocket."

"Thanks, Dana. Keep us updated as new information is released."

"We're waiting for confirmation that there were no survivors. The White House Press Secretary is about to make an announcement. Let's listen."

"Ladies and Gentlemen, I can now confirm what the national news media is reporting as I speak. The President's helicopter has crashed on its way to Andrews Air Force Base where President Wilkerson was scheduled to board Air Force One for a flight to San Francisco for a fundraiser tonight. We are waiting to find out if the President is somehow miraculously alive."

Angela King's face appeared on the screen again. "I have Mort Wisner here with me now."

"Mort, if the President has been killed, who will succeed him?"

"Angela, the 25th Amendment to the Constitution specifies that the new President shall appoint a Vice President who shall then be approved by a simple majority vote of both houses of Congress. However, the nominee to fill the Vice Presidential vacancy created when President Wilkerson was sworn in nine days ago has not yet been confirmed by the Congress. If it turns out that President

Wilkerson has been killed, then that nomination will expire. The Presidential Succession Act of 1947 specifies that the Speaker of the House of Representatives, Robert Sidney Marshall, is next in line. Once he is sworn in, he will have to nominate a Vice President who will also be subject to Congressional approval."

"Mort, some people are asking if former President Arnold could rescind his resignation and resume the office."

"No, that's not an option. Constitutional experts are virtually unanimous in their opinion such an option, if indeed it ever existed, expired when Vice President Wilkerson took the oath of office as the new President."

Thank you, Mort. Stay with us. Let's go back to Roger Tanner at Fort Stanton."

"Roger, can you give us an update?"

"Yes, Angela. The Garfield Heights Fire Department has just removed eight bodies from the wreckage. It appears that there were no survivors and that the President has, as we feared, been killed."

Angela King turned away from the cameras and microphone for a moment to compose herself. When she turned back to the cameras, her eyes were brimming with tears. It was a poorly-kept secret in the nation's capital that she and the former Vice President had much more than an arms-length professional relationship.

"Ladies and gentlemen, we have just learned that President Wilkerson is dead. Please join all of us here at ABC news as we observe a moment of silence in honor of our fallen leader."

When Angela lifted her face to resume her broadcast, ABC news cut away to a biographical profile of the Speaker of the House.

At 5:47 PM, Robert Sidney Marshall was sworn in as the 46th President of the United States. President Marshall gave a brief speech expressing the nation's shock and sadness. Many were surprised that he introduced one of the nation's most revered Evangelists to close in prayer. Dr. Gilbert was frail and had to be assisted to the podium but his voice was strong as he asked God to bless the nation and grant wisdom to President Marshall.

Determined to move swiftly to repair the immense damage to American prestige and influence caused by the excessively political previous administration, President Marshall quietly asked for resignations from the Secretary of State, the Secretary of Defense, and the Attorney General. He was ready with his nominees for each of those critically important positions. Even members of the opposition party's majority in the Senate recognized that the situation in the Middle East had deteriorated so much that the President's nominations were quickly confirmed by the Constitutionally-required simple majority vote in each chamber.

Monday, 28 July

U.S. Capital Building; Washington, DC
11:20 AM

Realizing that the Department of Defense would have to be rebuilt following the massive force reductions made by the previous administration, President Marshall appointed retired Air Force General Francisca Herrera as the nation's first female and Hispanic Vice President.

Dr. Herrera was an Aerospace Engineer and the Director of the Defense Advanced Research Projects Agency. Prior to accepting the position with DARPA, she had been Vice President of Northrop Grumman's Military Aircraft Systems Sector and was widely regarded as one of the world's foremost experts on stealth technology. She was one of only three women to have piloted the B-2 Stealth Bomber and the only one to have done so in combat. After receiving her fourth star, she had also served as NATO's Supreme Allied Commander Europe, usually referred to as SACEUR, which included the dual-hat role of directing the U.S. European Command.

Seen as conservative and somewhat of a military hawk, she was not popular with most liberals from a purely political philosophy perspective. Yet, almost no one doubted her qualifications to serve as Vice President or, if necessary, as President. Her grasp of geopolitical and defense issues was superb and, perhaps even more importantly, she was highly respected by American allies.

By late afternoon, both houses of Congress, recognizing the need to restore public confidence, confirmed the Dr. Herrera as the nation's new Vice President.

Naval Amphibious Base; Coronado, California
Week of 28 July

"Let's make this interesting," Al suggested. What would you say to a small wager on each of the tests Mark has scheduled for today and tomorrow?

"How small?"

"Ten bucks?"

"I've got a better idea. How about the loser takes the winner and Mark out for dinner at the Coronado Boathouse Friday night when we're done?" David countered.

"You're on."

David easily won all of the first day's events except the one mile run in combat gear carrying a ninety pound pack. He limped across the finish line in eight minutes and twenty-two seconds. Al's time was more than a minute faster at seven minutes fifteen seconds.

They split the second day's weapons tests at Camp Pendleton. Al shot a nearly perfect score with the Glock 9mm pistol. Using the HK416 with a holographic sight and a magnifier, David tied the Camp Pendleton record for an AR-15 style weapon on the hundred yard range. Dinner was on Al at the end of the week's mission prep but they'd both done exceptionally well on one of the world's toughest Special Forces physical qualification trials.

Tuesday, 29 July

Mossad Headquarters; Herzliya, Israel
9:00 AM

Gavriella and the members of her top secret Kidon team stood up when Israel's prime minister entered the conference room. He walked around the room shaking hands with each of them before standing beside the lectern to address the group.

"I asked to speak to you in person because what I am asking you to do is almost certainly the most dangerous mission that any Kidon team has ever been given. Your target is Abdul Muqtadir, the Hamas commander. " The prime minister stood quietly for a few moments as he watched for their reactions and then sat down at the head of the conference table.

Levi Reznik stood to continue the briefing. He pressed a button on the lectern to project a large color photograph of the terrorist leader that was their target.

"Until IDF ground troops entered Gaza to destroy the tunnels two weeks ago, Muqtadir was directing the rocket attacks on Israeli civilians from a command post in a school about one mile south of Atatra. Aerial reconnaissance showed that a strike would kill many children and civilians which we are trying hard to avoid. As the IDF ground troops entered Gaza, Muqtadir moved to the Great Omari Mosque in the Daraj Quarter of the old city in downtown Gaza. Unfortunately, he was not there when the mosque was almost completely destroyed several days ago."

"The Palestinians predictably blamed Israel for the destruction of the mosque and the Turkish and Qatari news media reported that claim as though it was fact. However, our intelligence indicates that the mosque was destroyed by radical Islamic terrorists on the orders of the leader of the Islamic State of Iraq and the Levant. As you know, with ISIL's success in Syria and Iraq, Abu Bakr al-Baghdadi has already announced that ISIL has established a Caliphate claiming religious authority over all Muslims. We think destruction of the mosque was part of his wider strategy to align world opinion against Israel and to recruit disparate bands of Islamic terrorists to follow him."

"Late last night, we learned that Abdul Muqtadir has established a new and highly fortified command center in a residential area about one mile west of Jibaliya. A reliable informant has told us that Muqtadir sleeps inside the command center and almost never leaves. We do not know the exact size or deployment of his protective detail but we must assume that it consists of at least a hundred of the most highly-trained Hamas combatants under his command."

"We do not have the luxury of waiting until we have better intelligence. Muqtadir moves too frequently. Gavriella and I have been here in the building since shortly after midnight to plan the broad outlines of how the mission will unfold. Over the next eight hours, this team and six Shayetet 13 naval commando officers will work with our intelligence and operations sections to plan the mission in greater detail. I do not have to tell you that those plans will become obsolete almost as soon as you land in Gaza. In theory, you will have time to get a few hours of sleep before boarding the modified UH-60 Blackhawks. Realistically, I know most of you won't sleep but I do encourage you to get as much rest as you can."

"The Blackhawks that will carry you to the drop zone are virtually identical to the highly modified aircraft that were used by the American SEAL Team Six in their assault on Osama bin Laden's compound. They have been upgraded with low-observable and sound-suppression stealth technologies."

"As you probably know, the previous American administration blocked the sale of these aircraft to Israel. The administration's incredible rationale was that the Blackhawks might pose a threat to civilians in Gaza while totally ignoring the threat that rockets fired by Hamas posed to Israeli civilians. Fortunately, the new American president sworn in just two days ago understands what is at stake far better than his predecessor. At my request, he has ordered six of the modified Blackhawks to be flown from the nuclear aircraft carrier George H.W. Bush to Atlit Naval Base. These aircraft are onboard the carrier to support SEAL Team Eight. The carrier is operating somewhere southwest of Cypress and is currently steaming eastward at flank speed. It should be within launch range off our coast within six hours. We expect the aircraft to begin arriving at Atlit around 4:30 PM where they will be serviced and armed for your mission."

"One member of your Kidon team will be assigned to each of the six Blackhawks to ensure that the mission can continue even if Hamas should succeed in downing one or more of the helicopters.

Each Blackhawk will have a crew of four consisting of the pilot, copilot, a crew chief, and a gunner. In addition, each aircraft will carry a total of ten fully-armed personnel consisting of one of you plus nine Shayetet 13 naval commandos. Their job will be to establish a perimeter and to hold off any counter-attack to give you time to complete your primary mission.

"Your insertion will also be supported by six AH-64 Apaches each armed with a 30mm chain gun with 1,200 rounds of ammunition plus an asymmetric load of Hydra 70 flechette rockets, Hellfire missiles, and an auxiliary fuel tank to allow them to remain in the area to suppress any attempts to repel your attack. These Apaches have been upgraded with the Ground Fire Acquisition System that can detect and target ground-based weapons fire sources in all lighting conditions so your team will have a lot of aerial support."

"One lesson we learned from the SEAL Team Six mission to capture or kill Osama bin Laden is that the helicopters are extremely vulnerable to ground fire once the enemy realizes what's happening. You may recall that the American's had to blow up one of their own helicopters after enemy fire damaged it. We don't want to take the chance that Hamas might make extraction by helicopter impossible. Instead, your extraction will be supported by the IDF ground forces already operating in Gaza. Eight Wolf Armored Personnel Vehicles will be pre-positioned within a two-mile radius of Muqtadir's command center. They will be monitoring your mission frequency and will converge within less than two minutes on the coordinates you provide. Your retreat toward the Erez Crossing Point will be supported by IDF ground forces and the Apaches."

As Levi concluded his briefing, the prime minister again stood to address the team.

"Our people will never live in peace until radical Islamic leaders like Muqtadir have either been deposed or eliminated. Albert Einstein, a self-proclaimed agnostic born to secular Jewish parents, once said '*nothing will end war unless the people themselves refuse to go to war.*' But we do not have that luxury. Einstein's oft-quoted pacifist sentiments were naïve because he didn't consider that no nation can rationally decide not to go to war if their adversaries don't make the same decision. To do so would be madness that could only lead to the enslavement or destruction of those unwilling to defend themselves. We will never forget what Hitler did to millions of peaceful Jews. Never again will we allow our love of peace to lull us into trusting our adversaries to behave in a civilized and humane manner. We are surrounded by millions of people intent on our destruction and who rejoice at the death of every Israeli citizen. If radical Islamic leaders would stop trying to kill us, then we could have peace. Until then, we must fight to survive. Tonight, you will fight for Israel and for the right of our people to live without the constant fear that Hamas rockets will kill them and their children. So, I salute you as you take the fight to those who persist in making war on us. I pray that God will protect you and bring you safely home. Good hunting."

Jamkarān Mosque; Qom, Iran
1:15 PM

The noon prayers were over. Once again, Nahbi had memorized every note of the muezzin's calls.

Although the plaza was overflowing with thousands of people, Nahbi had no difficulty finding the group of young men he'd identified on his previous visit as being most suitable for his purposes. Their mannerisms, topics of discussion, and obvious lack of detailed knowledge of the Koran made them ideal.

"Salam. I hope I am not intruding but I couldn't help overhearing you discussing the Khatīb's excellent sermon today. May Allah grant that we should all one day possess only a small portion of his wisdom and knowledge."

The five young men all murmured 'Allahu Akbar' in unison as if on cue.

"I have recently moved to Qom from Semnan and have not yet found new friends who share my love of the holy Qur'an. Please forgive my boldness but I very much enjoyed hearing you express your views on various passages with such passion and obvious desire for knowledge. It has been my experience that small groups of intelligent people sincerely seeking truth can be of great benefit to one another. Do you agree?"

Again, as if on cue, all five of the young men nodded.

"My name is Ahura Mazda Amirmoez," Nahbi told them. "Please forgive my honored parents for their pride in naming their only son. In school, I was often made fun of for my first name that means 'divinity of wisdom' and my last name that means 'protected prince.' I cannot hope to live up to such grand names so I hope you will not think poorly of me for my many deficiencies."

The apparent leader of the small group extended his hand and introduced himself and then his companions. "We welcome you to our discussions. Surat An-Nisā' 4:86 of the holy Qur'an tells us that when we are greeted with a greeting, we are to greet in return with a greeting at least as good, if not better. We were about to buy some roasted meat from one of the vendors. We would be happy to have you join us Ahura Mazda."

Nahbi thanked him but was careful not to let any of his amusement show on his face. The young man had behaved precisely as expected by flaunting his limited knowledge of the Koran to impress a stranger. These young men would do nicely.

Sodot Micha Munitions Area; Israel
2:45 PM

The enclosures for the green iguanas from Puerto Rico had been completed a few days earlier. Most of the junior enlisted personnel assigned to help unload and release them had never seen one and were very nervous about handling the big lizards that averaged five feet in length. The officers had tried to reassure them that the lizards were harmless to humans but not all of the young soldiers were convinced. When some of the larger male lizards extended and displayed the dewlap under their necks, stiffened and puffed up their bodies, bobbed their heads aggressively, and hissed loudly as part of their natural defensive displays, it was almost comical to watch the soldier's reactions.

By late afternoon, all of the lizards were in their new habitat. At sundown, the sound of a Muslim muezzin's call to prayer was broadcast over loudspeakers. Nahbi had provided the sound track and directed that the lizards never be fed or watered until after the prayer calls were played. He had also instructed that the lizards were to be herded into an adjacent enclosure each afternoon and held there until the prayer call was broadcast. As the call to prayer began, the soldiers assign to care for the lizards were to lay down a trail of lettuce and other green, leafy vegetables to teach the lizards to move toward the sound of the prayer call to find food. All of the IDF officers and most of the enlisted personnel were well-educated and recognized that the lizards were being subjected to Pavlovian reflex conditioning but they didn't know why. Watching from a nearby IDF SUV, Levi Reznik did.

Wednesday, 30 July

Sourasky Medical Center; Tel Aviv, Israel
11:15 AM

Levi Reznik picked up the encrypted satellite phone to take the call he'd been dreading.

"Levi, this is Avi Ben-Shimon. Have you been briefed yet?"

"Not really, although I was monitoring the mission communications so I know you ran into fierce resistance from Hamas and that there were casualties. It was not clear if you succeeded in killing Abdul Muqtadir or not."

"Yes, Muqtadir is dead but the butcher's bill was high. As soon as we landed, all hell broke loose. We were taking automatic weapons fire from multiple directions but the Shayetet commandos held their own against a Hamas force that probably had us outnumbered by at least three to one."

"We fought our way into the command center and located a reinforced concrete bunker that we suspected Muqtadir had retreated into. Nathan used C4 plastic explosives to breech the steel door. Muqtadir had several armed men with him who immediately started firing automatic weapons as soon as Daniyel and Nathan entered the bunker. Both of them were killed in the exchange but one of them apparently killed Muqtadir before they went down."

"I tossed a flash-bang grenade into the bunker and we waited a quick two-count. There were still three men shooting when Gavriella and I went in. She and I each got one of them but the third put several rounds into her before I took him out. Levi, I didn't think Gabe would make it. She was unconscious with multiple wounds and she was losing blood fast. I radioed for extraction and did the best I could to stop her bleeding. The commandos were still under heavy fire outside. We literally had to fight our way to the armored vehicles. I hoisted Gabe over my shoulder and ran for all I was worth. One of the commandos carried Muqtadir's body out and dumped him into one of the APVs so we know for sure that we got him."

"Bullets were zinging off the armor plating as we drove away. One of the Shayetet officers told me the last vehicle took a direct hit from an RPG. I don't know if it made it out or not. I had my hands full trying to keep Gabe alive until we could get her to a hospital. I think the commandos lost at least ten men and maybe more. I don't even know if we were able to bring all of our dead out."

"Where are you now? Levi asked.

"I'm at the Sourasky Medical Center in Tel Aviv. I knew that Gabe's only chance was with the best doctors we have. As soon as we cleared the gate at the Erez Crossing, I had a medevac helicopter take her directly to the roof of the hospital. Right now, I'm outside the operating room waiting to talk to the surgeon. They've been working on her for the past two hours so I think she's still alive. But she'll never be the same again. One of the rounds blew away part of her lower jaw. It was still attached by some skin and muscle tissue but I doubt the surgeon could reconstruct it. There just wasn't much left to work with. She also took two rounds in her pelvic area and one just below her ribs. It'll be a miracle if she lives."

"Do I understand that we lost the entire Kidon team except for you and Gabe?"

"Yes sir."

"What's your condition, Avi? Were you wounded?"

"No sir. I had so much of Gabe's blood on me from carrying her that the Shayetet medics thought I was wounded too but I wasn't. Levi, the surgeon's just come out. I'll put the phone on speaker so you can hear what he says."

"Doctor, my boss is on the phone and would like to hear what you can tell us."

"We've got her stabilized. You saved her life by slowing her blood loss and having her brought here immediately. We had to bandage her head completely so we inserted a tracheotomy tube and she's breathing on her own. Her lungs are clear. We've given her five pints of whole blood and her heartbeat is strong. It's too early to be sure but I think she'll make it. She's young and in superb physical condition."

"What about her jaw? Can you save it?"

"We've reattached it and wired it into place with most of the skin sewn back in place as a temporary measure. We think the blood supply is adequate to keep most of the damaged tissues alive until she's out of danger and we can better evaluate what can be done. But, even if we can reconstruct her jawbone, she'll be facing years of surgical procedures to repair as much of the damage as possible. Even then, there'll be massive scar tissue and some disfigurement. Right now, we're focused on keeping her alive."

"Levi, did you get all of that?" Avi asked.

"Yes. Ask the doctor if they can give you a room where you can clean up and get some sleep. I'm leaving now. I'll brief the prime minister and get there as soon as I can.

COS Headquarters; Toulon Naval Base, France
7:35 AM

"I'm not satisfied with how the mission's op-plan is shaping up," Lieutenant Édourd Jardin told his boss. I've looked at several scenarios but each of them has serious issues."

"The biggest problem is how inaccessible Jhalawan is. There's no way we could take him down when he's in Islamabad. He's just too well-protected. The security at his villa is formidable but I think we can deal with that. What I've had trouble figuring out is how we're going to cross one of the most heavily guarded borders in the world and then traverse thirty kilometers of rugged mountain terrain in the dark without running into trouble. A few days ago, a civilian woman was killed when the Indian and Pakistani armies exchanged artillery rounds across the border. There aren't many civilians in the area and the military units on both sides of the border are suspicious and trigger-happy."

"We don't have any air assets in the region but even if we did, they wouldn't be of much use. Even our newest Eurocopter EC665 Tigers couldn't handle the mission. The Tigers have a service ceiling of four thousand meters and a maximum combat range of eight hundred kilometers with external fuel tanks mounted on the inboard stations. They're not configured for troop infiltration and they're not stealthy. Even the American US-60 Blackhawks that were used on the bin Laden take down couldn't support this mission. Their combat radius is even less than that of the Tigers although their troop capacity and service ceiling would be adequate."

"The Americans have had some success in reducing the characteristic 'whomp whomp' sound most helicopters make by increasing the number of blades on the main and tail rotors and modifying the rotor blade designs to reduce the wing-tip vortex that creates most of the noise. With the extreme height of the mountains in the area, any helicopter we might use will have to fly down the valleys. Even a sound-suppressed aircraft will make some noise and that will reverberate off the hillsides and be heard at considerable distance. This all means that the infiltration will have to be a surface operation. There really isn't any other choice. We'll have to pack a lot of weight over some of the roughest terrain on Earth at altitudes where most of us are going to have trouble breathing."

"I've got most of the infiltration route and details pretty well worked out. You'll probably have some suggestions but let me show you on the map what I've got so far."

"The way I see it, with the amount of gear we'll need, we pretty much have to use rental SUVs to get us from Srinagar to Taarbal. It's about sixty kilometers from the Srinagar airport to the town of Bandipora. The road is paved all the way. From Bandipora to the village of Phalwan Pora is another five klicks. The road turns to dirt and gravel just north of Phalwan Pora. It's about twelve kilometers to Razdhan Pass. Both sides of the summit are paved for a short distance with concrete bricks and then it's back to dirt roads the rest of the way. The Losar Line of Control is about twenty klicks north of Razdhan Pass. Our cover story that we're a bunch of mountain climbers should work until we get close to Razdhan Pass. There's a detachment of Indian soldiers near the summit who guard the Shrine of Peer Baba. We're going to look pretty suspicious to them. It would help if the Indian government gave them a heads-up that our trek has been approved. We don't need to get arrested while we're still on the Indian side of the border.

The village of Taarbal is about two kilometers south of the border and the dirt road ends about one kilometer north of the village. From there on, we'll be on foot."

"As you can see, the river flows almost due north from Taarbal to where it turns west near the Pakistani village of Taobat. We'll be moving north on the west side of the river. There are no villages in that area and the trees come down pretty close to the river's edge. I think we can parallel the river staying well back in the trees. We shouldn't be seen in that area if we travel at night with the lead man wearing night vision goggles."

"It's six kilometers from where we'll park the SUVs at Taarbal to this wooden pedestrian bridge across the Neelum River just west of Taobat. However, I don't think we should cross that far upstream from Janawai because we'd have to move to higher altitude to avoid the villages along the north bank of the river and that would force us to cross eight major valleys carrying heavy loads. Some of the ridges we'd have to cross are several hundred meters above the valley floors. The total distance would be nearly thirty kilometers. I'm not sure we could manage that much up and down at high altitude carrying heavy loads over rough terrain. David almost certainly couldn't with his patched up leg. Instead, I think we should hug the south bank staying in the trees and traveling at night. The distance is about the same but we'd be heading downstream over better ground and losing elevation all the way."

"We'll have to move up-slope to skirt the village of Hanti on the south bank but that shouldn't be difficult. We'll cross the river using this wooden pedestrian bridge downstream from Sardari. We'll cross one at a time sometime after midnight. It's very possible that some village dogs will hear or smell us coming and start barking. If that happens, we'll just have to freeze and wait until they shut up."

"Maybe they won't smell us," Jean Luc suggested. "Al told me about some research the CIA funded at the University of California at Irvine that has produced a cream that, when properly applied, can mask human odor for up to an hour by temporarily replacing human body oils with those of sheep and goats."

Édourd nodded thoughtfully. "OK, it's worth a try. After we cross the river, we'll head uphill to the crest of this ridge that runs north from the bridge. We'll follow the ridge until we can cross into this broad valley running east and west. We'll stay just below the ridge top and head west toward the village of Phullawai. When we get near Phullawai, we'll have to climb the ridge that runs northeasterly from the village. Somewhere up the valley and out of sight of the village, we'll again turn west and cross the valley above Phullawai and the next one over above Janawai. There's a heavily forested ridge about a half klick northwest of Jhalawan's villa. I've studied the high-resolution satellite photos and I'm pretty sure we could set up an observation post on the ridge just north of the villa about here. How we mount the actual assault on the villa will depend on what we learn by watching them for a few days. David speaks Urdu and Pashto so he can monitor their routine voice radio traffic."

Jean Luc studied the map for a few moments while running his finger along the infiltration route that Édourd had just described. "OK. I'll want to study your plan some more but it looks pretty good so far. What's your estimate of time from Taarbal to our observation post north of the villa?"

"Two days. Three at most. We've got a lot of hiking to do and we can only travel at night."

"OK. Go on."

"Well sir, I can't really say for sure how things will go once we get to the villa. In broad strokes, I'd say we wait until just after the midnight watch change in the blinds and then take out the four fresh guards. We should have meat and drugs for the dogs and also an air gun with immobilizing darts. In any event, we need to put the dogs down immediately after we neutralize the guards in the blinds. If necessary, we can put them down with some 5.56 rounds from the HK416 rifles. With the sound suppressors, the shots might not even be heard inside the villa."

"The drone helo may crash when we hit the guard who's controlling it. If not, we should probably crash it into a tree to eliminate its video feed into the villa. I think a team of four should be ready to storm the General's bedroom when they hear the concussions from the rocket warheads exploding in the basement. I think if we put two of the FGM-148 Javelin anti-tank rockets through the narrow basement windows, any soldiers who survive the attack won't be worth much in a fight. With a little luck, the blasts will knock out the radio that ties the villa to Islamabad and the troops off watch in the houses below. In any event, they'll hear the noise and come running. We need to be ready with two more rockets to take out their two Cobra APVs as they come up that narrow road. To me, that means we may need to carry four rocket launchers and four rockets despite their weight."

"Here's where my current version of the op-plan bogs down. Assuming that our team can overcome three to one odds against similarly trained soldiers by taking advantage of the element of surprise, we've still got to get out somehow. My initial plan was to commandeer the two Cobra APVs at the villa and then hightail it upstream using the dirt road on the north side of the river. We'd have to abandon them somewhere south of Taobat where the dirt road ends. With the four-wheel drive and high ground clearance of the APVs, we might be able to go a little farther south but we'd eventually have to abandon them to continue on foot along what looks like pretty rugged terrain on the east bank of the river."

"The problem is that the Pakistani airborne posse would be on us like a swarm of mad hornets long before we could cross back into India. It's only about twenty-five minutes flying time from Islamabad for their Russian-made Mi-17 gunships. Even if we somehow managed to get across the border before they arrived, and that's unlikely, I think it's entirely possible that they'd be willing to violate Indian airspace to take us out. They could always claim later that they were responding to an attack. I kept thinking about what David told the Foreign Ministry rep. He said Jhalawan was now '*radioactive*' implying that the government of Pakistan might prefer him, and all of us, dead. The Cobra APVs wouldn't survive a rocket attack from one of their gunships and if we're caught on foot, it's lights out."

"Do you have an alternative in mind?"

"Yes. I think we tiptoe in and then fly out."

"OK. I guess I understand the tiptoe in part you've already described. What about the flying out part?"

"The Israeli naval commandos have two of the Hughes AH-6E light attack helicopters. They're what the Americans call the *Little Birds* or sometimes the *Killer Eggs,* which is a reference to their egg-shaped glass cockpits. They were originally developed as reconnaissance aircraft but SOFSA, the Special Operations Forces Support Activity at the Lexington Bluegrass Army Depot in Kentucky, has modified them for just about every conceivable kind of covert mission. A few years ago, the Little Birds underwent extensive modifications that made them even better for covert operations. The cockpits were upgraded to be compatible with night vision goggles and the engine was upgraded to four hundred horsepower with the '*black hole*' exhaust system. The latest variants now have a six-blade main rotor, a quiet four-blade tail rotor, mufflers, an infrared suppression exhaust system, and advanced avionics. They're about as close to a stealth helicopter as it's possible to get."

"With careful loading, and maybe leaving some of our toys behind, each AH-6E can carry the pilot and five commandos in full battle gear with four of them riding outside on benches along each side of the fuselage. With our team, plus Logan and Flores, we'll have ten. Add Jhalawan and we'll be up to eleven. That's still within the load capacity of the Little Birds because they'd have burnt off several hundred pounds of fuel on the trip from Srinagar. Jhalawan would be bound and laid across the laps of the soldiers on one of the outside benches."

Jean Luc smiled and shook his head in admiration at his Lieutenant's attention to detail. "Just how do you plan to put two Little Birds in our back pocket?"

"I've been reading up on how some of the American special operation missions of the past were conducted. With the main rotors folded, two of the Little Birds will fit into our C-130 Hercules. When the American Army's Special Operations Aviation Regiment, abbreviated SOAR and nicknamed the *Night Stalkers*, went into Grenada in 1983, they loaded two of them on a C-130 and flew it to Barbados. From there, the Little Birds flew to Grenada. We can do something similar. When the C-130 lands to refuel at Ramon Air Base in southern Israel, the Israelis can load them onto the C-130. Once the C-130 is on the ground in Srinagar, the Little Birds can be unloaded on the pretext that the repairs can't be made with them in the cargo bay. At Srinagar, the Little Birds will be well within mission range of Janawai. It's less than a hundred klicks from Srinagar to Janawai. The Little Birds can cover that in about twenty minutes. Once we have Jhalawan, we'll establish control of the villa's perimeter so the helos can land on the grassy area on the north side of the villa. From that point, it's like the children's song, *fly away Little Birds*."

"And how do you plan to persuade the Shayetet 13 naval commandos into loaning us their Little Birds?"

"Hey, boss man. That's you and the General's homework assignment," Jardin said with a big grin.

Mofid University; Qom, Iran
9:05 AM

Nahbi parked his pickup truck on the street a block from the beautifully landscaped gardens near Mofid University's main building. He walked around the gardens for a few minutes and then sat down on a bench overlooking the large pool. He removed a traditional Persian bamboo reed flute from a cloth bag and began to play softly as if solely for his own pleasure. The haunting sound of the Ney had been heard in Persia for over four thousand years and was still enjoyed by many modern Iranians.

At precisely 9:30 AM, a man in his mid-thirties entered the gardens on the sidewalk leading from the main building. He was carrying a brown leather soft sided briefcase in his left hand and clutching a large hardback textbook in the crook of his right arm. As he'd been instructed, a piece of green yarn was tied to the handle of his briefcase. To identify himself, Nahbi began to play the popular Persian song known as Bordi Az Yadam, which means 'I Am Forgotten.' The man walked over to stand nearby listening to the lovely melody. When the song was finished, he spoke to Nahbi.

"You play the Ney extremely well. That was a beautiful rendition of Bordi Az Yadam. It is one of my favorite songs. Are you a professional musician?"

"No, I play only for my own pleasure or sometimes for my friends. Thank you for your compliment. I am glad you enjoyed it. Do you play an instrument yourself?"

"No, I love music but have little natural ability. So, I content myself with enjoying music produced by those more gifted than myself."

"Ah, I perceive that you are a modest man. Yet, I am certain that Allah has given you other gifts. Are you a teacher here?"

"Yes, I teach mechanical engineering although I am only an associate professor."

"It must be very interesting work and surely it is satisfying to share your knowledge with others."

"Yes, that is true. I am the faculty advisor to five graduate students who dream of the day when the hateful sanctions imposed by the western infidels will end. Sometimes their passions reflect the naiveté of youth but their hearts are pure. They are genuinely fine young men of integrity. I would trust them with my life."

In a few brief sentences with double meanings, the young professor told Nahbi that he was not a leader himself but that he would follow the leadership 'of those more gifted' than he. He'd also managed to let Nahbi know that he had five students he trusted to carry out subversive assignments that might help end Iran's isolation and the hardships on the Iranian people. Nahbi responded with a simple coded message of his own that established that he would be 'the musician' in their relationship and that he would visit the garden on Wednesday mornings.

"Thank you again for your kind words about my simple music. Perhaps I will see you again here. I often come here on Wednesday mornings at about this time."

"I will look forward to hearing you play again soon. I've enjoyed talking to you but I have forgotten my manners. I have not introduced myself. I am Professor Barid Afshar."

Nahbi shook hands and introduced himself using one of his aliases. "I am Kazem Rahmani. It has been my pleasure to meet you professor Afshar. Khoda hafez."

Saturday, August 2

CIA Headquarters; Langley, Virginia
8:00 AM

Pete closed the door and motioned for David and Al to take seats at his conference table.

"This diplomatic pouch was brought over last night from the French embassy. The duty officer signed for it but it's addressed to you, David."

The pouch was from Jean Luc and contained maps, equipment lists, and the mission's draft operations plan along with an envelope with David's name on it. He took a moment to read it.

"Jean Luc is asking for ideas on how to persuade the Israelis to loan us their two Little Birds. Lieutenant Jardin's version of the mission plan envisions using them to extract the team quickly once we have Jhalawan."

"Fat chance the Israelis would be that generous," Pete commented. "They only have two of the upgraded, stealthy versions and they're both assigned to Shayetet 13's naval commandos. With what they have on their plate in Gaza right now, they aren't likely to give up any equipment they may need to fight their own battles."

David nodded but then stared off into space for a few moments as he thought about the problem. "Actually, I think the Israelis *can* be persuaded to loan their Little Birds to the COS for this mission."

"How?" Pete asked.

"General Montreuil can get the French Prime Minister to speak directly to the Israeli Prime Minister. I'll write a suggested script for him but it will go something like this."

"Mister Prime Minister, you have my utmost sympathy for the unfortunate situation in Gaza. If there is anything that we can do to help, please do not hesitate to ask. I must admit, however, that there is another reason for my call. I have ordered our special operations forces to capture General Parvez Jhalawan and bring him to France to stand trial for his war crimes. We need some small assistance from Israel in this."

"At this point, the Israeli Prime Minister will naturally ask what that might be. The French Prime Minister would then continue along these lines."

"As a leader charged with protecting my nation, I can understand your decision to abduct the android known as Nahbi in the midst of the Task Force 3 effort to locate the stolen nuclear weapons. However, there are some in my cabinet who feel that Israel's action may have led directly to the death of many of our citizens. The sense of their anger is that the android, with his

superior analytical powers and mental capacity might not have overlooked the risk posed by an embedded member of the Muslim Brotherhood as Mr. Logan apparently did. I have tried to deflect and defuse their anger but, speaking frankly, I must tell you that several of our leading politicians feel that Israel is at least partly responsible for the death of our citizens. It would be very helpful if I could advise them of Israel's commitment to helping France bring this monster to justice."

"The Israeli Prime Minister will then ask what he can do. He'll probably be relieved to hear that it's such a modest request and one that he can easily authorize. I think he'll agree immediately."

"What about pilots?" Al asked. "I doubt if the Israelis will want any part of our mission. Do the French have anyone qualified to pilot the Little Birds?"

"I doubt it," Pete answered. "But, we do. Emily Harrison."

"OK. That's one," Al agreed. "Where would we get another one?"

"Although the French don't have a special operations helo comparable to the Little Birds, they have about thirty of the new Eurocopter EC665 Tigers. One of Jean Luc's men, Bernard Garneau, is a former member of the French 4e Régiment d'Hélicoptères des Forces Spéciales. I have the resumes of Jean Luc's men in my desk. Hang on a second while I get them."

"Yes. Bernard Garneau is qualified on the Tigers. They're bigger than the Little Birds but I'm guessing that Garneau could fly one of the Little Birds with some coaching from Emily."

"Do you know how the Israeli Little Birds are configured?" David asked. "I'm thinking particularly about guns."

"Yes. They bought them from us. They are former CIA special operations aircraft that we sold to them under the Foreign Military Sales Program when we got funding for the newer variants. When they're configured as attack helicopters, they have a weapons bracket mounted on both sides of the fuselage that can hold a variety of weapons including fifty caliber machine guns and several types of rocket launchers. Unfortunately, when they are configured for troop insertion and extraction, the weapons mounts are replaced with the bench seats on both sides of the fuselage. The assumption is that the soldiers can use their weapons to suppress ground fire or clear a landing zone."

"When configured for insertion and extraction, each Little Bird can carry six people, including the pilot. One soldier can sit in the passenger's seat with four outside. That effectively limits the mission team to nine men to allow for bringing General Jhalawan out."

Shaking his head slowly, Al stated the obvious. "Nine men against a full platoon of 36 elite troops doesn't seem like very good odds."

"With a little luck, we'll never face more than twelve or fourteen at once," David pointed out. "We can handle those odds."

"Maybe." Al still didn't seem convinced."

"The final decision will be up to Jean Luc," David pointed out. "Can we get him on the phone, Pete?"

When Jean Luc answered the call, David summarized what they'd been discussing.

"Pete, can you make Emily available for this mission?" Jean Luc asked.

"Yes, I think so. She hates her desk job and would love to get back into the field. I could send her along with Al and David on this bogus intelligence-gathering trip. If I'm asked about it, I'll just say that she's been griping about being stuck at a desk, which everyone knows is true, so I decided to send her along to shut her up. I think it'll pass the smell test with the Director."

When the call with Jean Luc was done, Pete dialed Emily Harrison on the secure line.

"Are you kidding?" she laughed. "I've been trying to figure out how to get an invitation to this party ever since David left the other day. When do I leave?"

"Soon. The French are putting the final touches on the mission plan as we speak. I'm guessing that the call to the Israeli Prime Minister will take place within the next few hours. We should know by tomorrow. I'll let you know as soon as I find out."

When Pete hung up, Al made a circular motion around his right ear with his index finger. “Pete, she’s crazy! I’m telling you the woman is certifiably nuts. This is her bizarre idea of fun. Why would anyone in their right mind give up a nice, safe desk job on a sunny Mediterranean island to hang out with a bunch of overaged Boy Scouts still trying to earn their hiking merit badges?”

Pete and David just smiled and shook their heads at Al’s characteristic sense of humor. They both knew he was glad to be on the mission too.

Sunday, August 3

COS Headquarters; Toulon Naval Base, France
10:30 AM

“Jean Luc? This is General Montreuil. You have your Little Birds. David was right. The Israeli Prime Minister almost seemed relieved that all we wanted was to borrow two of their helicopters. They’ll be flown from the Atlit Naval Base to Ramon today and the Israelis will be ready to load them on our C-130 just as soon as it lands. Pete Nocona has confirmed that Emily Harrison will fly one of them. Am I safe in assuming that Garneau can fly the other?”

“Yes sir. Bernard is confident that he can fly it but he’s not happy that he’ll be sitting back in Srinagar waiting for our extraction call rather than going with us on the mission. Lieutenant Jardin is working with him to get a couple of helo maintenance mechanics lined up to get the rotors unfolded and get the birds ready to fly in Srinagar.”

“Sounds like the C-130 is starting to get crowded. Is there room for the two helos plus the COS mission team and two mechanics?”

“Yes sir. Lieutenant Jardin has talked to one of the retired U.S. Army SOAR sergeants who was on the Grenada mission in ‘83. He said it was cozy with two Little Birds and the mission team on board but workable.”

“Have you got all of your equipment lined up?”

"Yes sir. The logistics officer made a half-hearted attempt to give me a hard time when I told him I wanted the Heckler & Koch HK416 assault rifles instead of our standard-issue FAMAS models. But, he managed to round up enough for the mission and my men are happy about that. The FAMAS isn't a bad rifle but the HKs are the best.

"So, when do you shove off?"

"I'm waiting on the latest intel regarding General Jhalawan's whereabouts for the next few days. Right now, he's in Islamabad but the security platoon left yesterday for Janawai so I'm pretty sure the General will be there within the next couple of days. I think we'll be wheels-up on our way to New Delhi by Tuesday. I've just asked David and Al to rendezvous with Emily in Rome. They'll land in Srinagar about noon on Wednesday. Right now, it's looking like we'll take a few days to infiltrate and observe and then do the take-down maybe as early as next Saturday or Sunday."

"Keep me posted. I'll want to be at the airport to see the team off."

Sourasky Medical Center; Tel-Aviv, Israel
11:45 AM

Avi Ben-Shimon stood up when Levi entered the waiting room outside of the intensive care unit where Gavriella was fighting for her life.

"How is she?" Levi asked.

"They have her heavily sedated. The nurses told me that when she finally regained consciousness early this morning, she became very agitated and was fighting her restraints. They tried to calm her down by telling her she was going to be fine but it didn't help. Her doctors were afraid she'd hurt herself so they've given her a general anesthetic to put her back to sleep."

"They had to drain quite a bit of fluid from her lungs. Her breathing is very labored and her doctor is clearly worried about it. He said he might have to put her on a ventilator. He also told me that her jaw has become badly infected so he's sharply increasing the dose of antibiotics she's receiving. He emphasized that it's essential for them to control any infection if there's to be any realistic possibility of reconstructing her jaw. He said she'd need a couple of dental implants but that should be routine if the rest of the repairs are successful."

"What about her other wounds?"

"The bullet that she took just below her rib cage on her left side went all the way through without hitting anything vital. The doctor said it was a clean wound that should heal just fine. One of the bullets ricocheted off the top of the iliac crest on the left side of her pelvis and then exited from her lower back without doing serious damage. The other bullet damaged the iliolumbar ligament that connects the lower lumbar vertebra to the ilium. Her doctor said the damage seems slight and that even though the ligament won't repair itself, it shouldn't cause her any pain or impaired mobility."

"I'm amazed that you can remember all of those medical details."

"You're forgetting that I'm the team's medic. I've been trained in this stuff and plan to go to medical school someday."

"Has anyone else been here?"

"Yes, Aharon was here when I got here a little after seven. I think he'd been here all night. He looked exhausted. He asked me if you've told Nahbi about Gabe's condition."

"I told him she'd been wounded but that her doctors expected her to make a full recovery. He wanted details but I pled ignorance and kept the discussion general. I'm sure he knew I was being intentionally evasive. How long are you going to stick around?"

"Levi, Gabe doesn't know that the rest of the team didn't make it and it's going to hit her very hard. When she finally regains consciousness, I think I'm the one who ought to be here to tell her. I'm on leave for at least another week and I plan to be here every day until I know she's going to be all right."

Monday, 4 August

U.S. Embassy; Ta' Qali, Malta
2:50 PM

"Miss Harrison, David Logan is on line two."

"Hello David. I've been waiting for your call."

"Time to saddle up Emily. Al and I are leaving Dulles at 5:40 this afternoon on United's non-stop red-eye to Rome. We land at Leonardo da Vinci International at 8:30 tomorrow morning. We connect with the non-stop Air India red-eye flight that leaves Rome at 11:35 tomorrow night and lands at Indira Gandhi International in New Delhi at 10:15 Wednesday morning. We have seats on the Air India commuter flight to Srinagar that leaves at noon. The travel section has already booked your flights. I don't know your seat assignment for the flight from Malta to Rome but the three of us are together in the middle three seats in business class for the flight to New Delhi. We won't be able to discuss the mission on the plane so we might as well try to get some sleep. Once you land in Rome, call me on my company phone and we'll join up."

"Got it. What about the Little Birds and mission gear?"

"The C-130 left Toulon early this morning and is already on the ground at Ramon Air Base in Israel. The Little Birds will be loaded in the next few hours. They'll leave Ramon early tomorrow morning. They've got a refueling and rest stop at Prince Sultan Air Base in Saudi. They should land at Hindon Air Force Station outside New Delhi late Wednesday. They'll refuel and spend the night there. Thursday morning, they'll leave Hindon, supposedly heading for home, before reporting the mechanical emergency and diverting to Srinagar. They should be on the ground in Srinagar by late Thursday morning."

"We have hotel reservations in Srinagar for Wednesday night. We'll leave you at the hotel Thursday morning. Al and I will pick up our rental SUV and head north toward Bandipora sometime before noon. Jean Luc and his men will be filtering into Srinagar all day Wednesday. Bernard Garneau will check into your hotel and get in touch as soon as he gets to his room. I'll have more info for you when we meet in Rome."

Wednesday, 6 August

Sourasky Medical Center; Tel-Aviv, Israel
2:20 PM

After his morning examination, Gavriella's doctor decided that she was out of danger and had her moved from the emergency care unit to a private room. She was awake and watching a news broadcast on TV when Levi came into her room. Her head was wrapped in thick bandages with only a narrow slit across her eyes so she could see. The surgical steel appliance supporting her jaw extended through the bandages at several points. The tracheotomy tube had been replaced by a ventilator tube that was taped securely at the corner of her mouth. An intravenous feeding tube was taped to her left arm and she was hooked up to a multi-function vital signs monitor.

"Hello Gabe. Your doctor told me I could only stay a few minutes but I wanted to see if there was anything I could do for you."

Gavriella couldn't talk but as soon as she saw Levi at the door she pressed a hot key button on the Samsung tablet computer lying on the bed beside her to launch a note-taking application. Whatever she wrote or drew with the stylus appeared on the TV screen over a Bluetooth wireless connection. Without even acknowledging Levi's greeting, she wrote two words. "Where's Nahbi?"

"He's still in Iran and his mission is going even better than I'd hoped."

"How much longer?" Gabe wrote.

"Probably another month or a little more."

"Does he know?"

"Yes, he knows you were injured but I've assured him that you're doing fine and will make a full recovery. I feel sure that he suspects there's more to the story but he's given me the benefit of the doubt and hasn't pressed me."

"I'm sorry."

"What for?"

"My team."

"My fault, Gabe. Not yours. I told the Prime Minister that your Kidon team could get Abdul Muqtadir and that we could provide enough mission support firepower to protect your team. Based on the intelligence we had, I never thought you'd encounter such overwhelming numerical superiority. I was wrong. You and your men paid the price for my poor judgment. But Gabe, I will not diminish the heroic sacrifice of your men by focusing on my mistake. Your men knew that the mission was extraordinarily dangerous and that they might be killed. The miracle is that such people are willing to put their lives in danger to protect the rest of us. In private, like you, I will grieve their loss for the rest of my life. But, in public, I will use every ounce of my abilities to ensure that their families are consoled by knowing that their loved ones are worthy of the highest honor our country has to offer."

Tears were brimming in Gavriella's eyes when a nurse knocked on the door and told Levi it was time for him to leave.

Thursday, 7 August

Phalwan Pora, India
1:45 PM

David's encrypted SATCOM phone vibrated softly signaling an incoming call.

"David, this is Jean Luc. Where are you?"

"Al and I are waiting for you in a big gravel parking area on the northwest side of the bridge over the Madumati River about a half klick north of the village of Phalwan Pora. We're in a dark blue Toyota Land Cruiser. Where are you?"

"I'm riding with François and Remy. We just passed the Gulshan Park in Zurbaf Pora so it looks like we're about three kilometers south of you. We should be there in a few minutes. We're in a white Nissan Armada."

"Where's the rest of the team?"

"Édourd, Henri, and Alain are right behind us in a silver Land Cruiser. Tristan and Bernard are a little ways behind them in a blue and white Toyota Land Cruiser."

Tristan and Bernard turned into the parking area five minutes after the others. Jean Luc didn't waste any time.

"The GPS says there's a petrol station two kilometers from here. There's really no point in trying to hide the fact that we're traveling together now so as soon as we've topped off the fuel tanks, I want to head straight for Razdhan Pass. With a little luck, we should be over the pass before dark. We'll keep driving until we get to Taarbal. Édourd has made arrangements to hide the SUVs back in the trees on a farm just north of the village. It's only about a kilometer from there to the Line of Control."

"I want to cross the LOC sometime after midnight. We'll use the trees for cover and cross one at a time. Édourd thinks we can get downstream of Lower Taobat before dawn. We'll move upslope to avoid the locals and find a place to make a cold camp. We'll stay out of sight tomorrow and then shove off again after it's fully dark. Sometime after midnight Friday night, we'll cross the wooden bridge just below Sardari and then head north to cross the ridge above the Sakmal River. The elevation gain from the river to the top of the ridge is over twelve hundred meters in less than three kilometers so it's very steep. With the loads we'll be carrying over unfamiliar ground in the dark, it'll probably be almost dawn before we begin our descent into the Sakmal valley. I want to get off the ridge and back into the trees before daylight."

"We'll rest Saturday. I'd like to get settled into a base camp above Jhalawan's villa sometime Sunday night if at all possible but we may have to detour quite a ways to the north to get past Phullawai without being seen."

"Were you able to get some kind of official Indian government authorizations to show the soldiers at the pass?" Al asked.

"Yes. The Indian Prime Minister took care of it for us. We have trek permits signed by Kashmir's Director of Tourism. With our

Canadian passports, the trek permits, and our cover story that we're graduate students from McGill University in Montreal, I don't expect any trouble. Any questions? OK, let's mount up."

Razdhan Pass; Kashmir, India
4:50 PM

The Indian Army Lieutenant had obviously been expecting them and almost seemed relieved to see them roll in. After welcoming the group, he asked to see everyone's passports and the trek permit. When he was satisfied that everything was in order, he gave a short narrative about the famous Dargah of the Sufi Islamic shrine built over the grave of the Muslim Saint, Peer Budhan Ali Shah. He explained that the title Peer Baba is commonly used in Hindustani as a salutation to Sufi masters or other similarly honored persons.

"Why does the Indian Army guard the tomb?" The Lieutenant became visibly emotional as he answered Al's question.

"Many years ago, a Saint from Pakistan came to this area and made it his home. He is the patron saint of the Indian troops who serve along the Line of Control. The soldiers firmly believe that they will survive being posted along the LOC only due to the blessings of Peer Baba. Since 1947, each commanding officer of units assigned to this area has placed a marble stone bearing their names and dates of posting at the entry to the shrine. In 1950, the 28th Brigade of the Gorkha Rifles constructed the shrine and since then all defense personnel who come here pay their respects at the tomb and seek Peer Baba's blessings. For us, this is a holy place."

Friday, 8 August

Losar Line of Control; Kashmir, India
12:50 AM

Jean Luc and David were sitting with their backs against two large Himalayan Spruce trees waiting for the scouts to return. They were studying a topographical map and discussing contingencies when Tristan Dejarnette and Remy Thibadeau emerged from the trees a hundred yards upstream.

"How does it look?" Jean Luc asked his sergeant.

"We didn't see any sign of Pakistani troops on either side of the river. Remy used his infrared sight to scan both banks but the only heat signals he saw were from animals. It looks clear. Unfortunately, we've got a waxing quarter moon that doesn't set until about 3:00 AM. The terrain along the west side of the river isn't bad but there aren't many trees close to the bank. If we want to stay concealed in the trees, we're going to have to go upslope quite a ways."

"OK. You and Remy get off your feet for a while. The rest of us will start heading out in at five minute intervals. You can bring up the rear. There's a small stream that empties into the Neelum about one kilometer north of the LOC. We'll rendezvous directly across the river about two hundred meters upslope. Al, I'd like for you to take point tonight. Are you ready to head out?"

"Sure. How far to the next Taco Bell?"

Saturday, 9 August

Sakmal River Valley; 5km East of Phullawai, Pakistan
4:35 AM

"OK. That's enough for tonight," Jean Luc finally decided. "Spread out and get something to eat before you turn in. Henri, you've got the first watch.

Jean Luc walked over to where David was sitting with earbuds in both ears and a multi-band radio on his lap.

"Are you getting anything?"

"Just some Pakistani Army chatter. I think it's probably coming from Jhalawan's security detail. I heard his name and also the Urdu word for helicopter. These tall mountains play havoc with radio transmissions so I didn't get everything they were saying but it sounded like he's either already at the villa or expected soon."

"Anything to suggest they're aware of us yet?"

"No. It sounded like a couple of bored enlisted men grousing."

"How's your leg doing?"

"You know, I'm not exactly sure," David said with a tired grin. "Everything hurts. I heard you say that ridge was steep but that may be the understatement of the year. Frankly, I don't know how any of us made it carrying ninety-pound loads in this thin air. You might want to check on François. He's ten years older than you and me so it had to be rough on him."

"Yes, I watched him but he's a tough old bird and there's no way he was going to show any weakness in front of the younger guys. He'll be OK after he gets some rest. By the way, I appreciate that you've let me give the orders. I know General Montreuil said we'd be co-commanders but that's always awkward. Even so, don't hesitate to pipe up if you think I've missed something or am on the verge of doing something stupid."

David just nodded and then staggered to his feet before moving off to find a place to bed down for the day.

Sunday, 10 August

Hanging Valley; 2.5km Northwest of Janawai, Pakistan 4:15 AM

The first faint hints of dawn were visible across the Janwai Nar Valley to the east when the team reached a small hanging valley at an elevation of four thousand meters above sea level. Located less than three kilometers northwest of General Jhalawan's villa, it was an ideal base camp. The men were exhausted from three nights of traversing rough terrain in the dark and needed some rest. Alain Fabré seemed to be in the best shape so Jean Luc sent him south on a reconnaissance patrol. Two hours later, Jean Luc saw him climbing the steep grade as he returned to where they were camped.

"Sir, it's all quiet at the villa. I found an excellent vantage point that we can use as our observation post. It's in some rocks on the ridge overlooking the villa and it's shaded by a small grove of pines. I couldn't see the front side of the villa but everything else is in clear sight. The front of the villa overlooks the village. The back of the house faces up the valley and has a good view of Nanga Parbat to the north. There's a balcony on the second floor that's probably outside the General's bedroom. There's a staircase from ground level to the balcony on the west side."

"The ground slopes up fairly steeply behind the villa and there's a retaining wall about three meters behind the concrete foundation. There's a small, paved patio with a couple of tables and some chairs between the retaining wall and the villa's foundation. There's a door leading from the paved courtyard into the basement. I counted eight soldiers outside so I'm assuming the other four are probably posted to defensive positions in the basement. Four soldiers were in the camouflaged deer blinds and four were on roving patrols with dogs. I didn't see any dogs inside the double fencing."

"The deer blinds are about ten meters outside of the outer fence at the edge of the trees. We'll have to be careful not to make any noise when we approach the blinds but the cover looks good for our approach."

"Both the outer and inner fences are standard chain link mesh. There's concertina razor wire coiled along the top of the outer fence. I didn't see any surveillance cameras but I might have missed them so we'll need to check that. We'll have to eliminate the guards in the blinds and the roving patrols before we cut the fencing."

"The two Cobra APVs are parked on the east side of the villa at the top of the road that leads to the village below. Here's the camera if you want to scan the shots I took."

"Good job. Get some rest. I'll have you brief the team later."

Just before noon, they all heard the sound of a large helicopter coming from the direction of General Jhalawan's villa. As soon as he heard the helicopter's distinctive sound, David set up the folding tripod for the Swarovski ATX spotting scope. Using the scope's maximum magnification, he was able to watch the action below. He instantly recognized General Parvez Jhalawan. A few minutes later, the Pakistani AH-1S Cobra gunship lifted off and headed west down the Neelum River Valley. David and Jean Luc traded knowing glances. Both of them knew the gunship's armament included a 20mm canon and eight external missile tubes that could make their retreat very unpleasant unless the Little Birds could whisk them away before the Pakistani Cobras reached the area. Timing would be everything.

Tuesday, 12 August

Srinagar International Airport; Srinagar, India
10:00 PM

Emily Harrison saw the light blinking on her SATCOM encrypted radio.

"Hello David. I've been expecting your call. Have you and Jean Luc worked out the schedule?"

"Yes. We're going to leave our base camp at 10:30 PM. Everyone should be in position by midnight and we'll attack at 12:30 AM. Right now, our best estimate is that we'll be ready for extraction no later than 4:00 AM. We'll call just as soon as we have Jhalawan in custody."

"Roger that."

"How long can you orbit if we need more time?"

"Twenty minutes max."

"Understood. Is Bernard good to go?"

"He'll be fine. Don't worry, David. We'll get you out."

"OK. See you in a few hours."

Jean Luc had a few last minute instructions for the team.

"Unfortunately, we have a nearly full moon tonight and it will be directly overhead an hour after we attack. Stick to the shadows as much as you can and take your time getting into position. Any questions?"

Wednesday, 13 August

General Jhalawan's Villa; Janawai, Pakistan
12:30 AM

"Équité!" Jean Luc spoke a single French word meaning justice into his combat headset to launch the attack.

In perfect unison, Henri Barteau, Alain Fabré, François Gaultier, and Al Flores took out the soldiers manning the deer blinds. Their Glock 17 9mm pistols were fitted with sound suppressors and loaded with jacketed hollow points. Taken totally be surprise, none of the soldiers had time to raise an alarm. The surveillance drone crashed on its own when Alain's target went down.

Using HK416 assault rifles fitted with sound suppressors, Jean Luc, David, Édourd Jardin, and Tristan Dejarnette each took out one of the four roving patrols and then put the dogs down with a second shot. Each of them cut large openings in both the outer and inner chain link fences. When that phase of the attack was complete, Jean Luc spoke into his combat headset again. This time, the single word was "*Avancé*" meaning to advance.

As soon as their designated targets were down, Al Flores and François Gaultier retrieved the FGM-148 Javelin anti-tank missile launcher tubes they'd stashed near the deer blinds and entered the compound through the openings the others had cut. Their job was to put the basement soldiers out of commission on Jean Luc's signal. Henri Barteau and Alain Fabré each picked up one of the other two anti-tank missile launchers and slipped into the compound. They then jogged across the clearing to take up positions near the gate to the road that led down to the village below. Their job would be to destroy the armored personnel vehicles that would bring the two off-duty squads up the hill from their quarters in the village below.

Lying prone on a knoll above the clearing, with a clear shot at the back and sides of the villa, Remy Thibadeau waited with his FR F2 sniper rifle. His job was to pick off any soldiers who exposed themselves.

Jean Luc, David, Édourd Jardin, and Tristan Dejarnette crept up the stairs at the rear of the villa. The moon was already high in the clear sky so they didn't need their night vision goggles. With their HK416 rifles slung over their shoulders, they each carried a 9mm Heckler & Koch MP5SD submachine gun in the ready position.

It was a warm night and the double doors to the master bedroom were open. Jean Luc shouted *attaque* into his combat headset. A split second later, Al Flores and François Gaultier fired two of the

FGM-148 Javelin anti-tank missiles into the basement windows from opposite sides of the villa. The twin explosions shook the massive log villa. The noise would be heard by the off-duty squads in the village below and bring them up the hill within minutes.

General Jhalawan had his back to the door and was kneeling on the bed making love to his mistress when they burst into his bedroom. David shouted at them in Urdu to freeze. There was a pistol on the nightstand beside the bed but both Jhalawan and Berjuhi, his young mistress, had sense enough not to reach for it.

"Stand up. Both of you." David shouted. "Put your hands up and don't do anything stupid."

Suddenly, Henri Barteau's voice came over the combat headsets they were all wearing. "Jean Luc, we have the two Cobra APVs in sight. They've stopped at the foot of the hill just out of range and have dismounted the troops. We're going to have company soon."

"Understood. Remy, maintain your position.Take out anything that moves. The rest of you set up a defensive position behind the retaining wall on the north side of the villa. François, call Emily and tell them to hurry. The troops below have almost certainly called for air support so we don't have much time. Send two men into the basement to be sure your back is secure and then see if the machine guns are usable after the missile explosions and if there's any ammo for them. Henri, pick up a couple of the soldier's radios in case we need them."

Tristan gagged and tied the General's mistress and then pushed her face down on the bed before hog-tying her arms and legs behind her back.

General Jhalawan finally spoke in a surprisingly calm voice. "Since you have not yet killed us, it appears you plan to kidnap me. If you don't mind, I'd like to put some clothes on."

"Go ahead," Jean Luc replied. "Tristan, keep your pistol pointed at his head. If he makes one bad move, blow his brains out."

Jhalawan took his time getting dressed in an obvious attempt to buy time. "You know of course that you cannot possibly escape. I'm sure that a squadron of gunships is already on the way. If you lay down your weapons and surrender, I'll see that you are allowed to leave Pakistan alive. Otherwise, you will all be dead in less than an hour."

David lifted his MP5SD submachine gun to his shoulder and pointed it directly at the General. "General Jhalawan, you can believe me when I tell you that your only hope of staying alive is if we are able to leave without pursuit. At the first hint that we might be overwhelmed by your forces, I will personally put a bullet right between your eyes. If you doubt me, then know this. I am the brother of the innocent woman you had murdered in an attempt to distract me."

At just that moment, the Pakistani soldiers opened fire from the cover of the surrounding trees. There were at least twenty of them and it sounded like they were all firing their assault rifles on full automatic. The shadows under the trees were so dark that only their muzzle flashes showed where the Pakistani soldiers had been but they were elite troops who knew to shift their position after each burst of gunfire.

Jean Luc's team knew their business too. They didn't have to be told to hold their fire unless they had a definite target. Their ammunition was limited to what they'd been able to carry so they weren't able to match the Pakistani firepower.

"François! Any casualties?"

"Not yet. We're hunkered down behind the retaining wall and we've got a clear field of fire. Two of the Russian heavy machine guns are usable and we've got them mounted on the retaining wall. We have about two thousand rounds of 12.7mm ammo so we are in pretty good shape for now. I don't think they'll try to rush us. They'll probably maintain their cover and try to keep us pinned down until the gunships can get here from Islamabad."

"Remy, can you see anything with your infrared scope?"

"Yes. I've already picked off one of them and I'm looking for another target. I'll have to shift position after my next shot."

"Édourd, call Emily and ask her ETA. Tristan, tie the General's hands behind his back," Jean Luc ordered.

David keyed his mic. "François, David. Can you wrap one of the radios in something and toss it onto the balcony?"

A moment later, a soft thud signaled the arrival of the radio. Tristan cut the drapery pull cords and made a small lasso to retrieve the radio without getting shot.

David held the portable radio out to the General. "Tell your men that I'm holding a gun to your head and order them to fall back to their APVs. Don't do anything stupid. I will not hesitate to shoot you. I speak Urdu fluently as well as most of the other languages and dialects of this region so I will know what you say. Say precisely what I've told you to say and no more."

"No. I think you plan to kill me anyway."

"Tristan, keep him covered," David ordered before walking over to the bed and removing the gag from the General's mistress' mouth.

"General, do not tempt me. My hatred for you and what you stand for is very great. I would like nothing better than to end your life right here. Perhaps you do not believe me. Watch!"

Firing on full automatic, David aimed at a spot on the bed just to the left of where the General's mistress was lying face down and helpless. Even with the sound suppressor, the noise was very loud in the enclosed space. Berjuhi began screaming loudly enough that the soldiers in the trees would be able to hear her clearly.

"Do as I've told you General or I will kill your lover in front of you."

"She means nothing to me and I've read your bio. I think you're bluffing. I don't think you'll kill her in cold blood."

"Tristan, untie her feet so she can walk. Keep her covered." David ordered.

"OK, General. I have a better idea. I'm going to let your own men kill you and your lover for us. I'm pretty sure they'll instantly open fire on anyone who appears on the balcony. You'll either order them to withdraw or we'll push her out onto the balcony. If you still refuse to order your troops to withdraw, you'll be next. Tristan, put the bedsheet over her head and lead her over to the doorway."

"OK General. Let's see who's bluffing. I've read your bio too. I know that you had your wife killed so you could take this woman as your lover. I don't believe she means nothing to you. Order your men back or she dies."

"No."

"That's really too bad. I thought you were smarter than that. Here's how it's going to work. Tristan and Édourd will put their submachine guns to your head and back to make sure that you do not flinch when I tell you to push your lover onto the balcony. When I tell you to push her out the door, you must do it or Édourd and Tristan will push you both out. If you resist, Tristan will decapitate you with a full-automatic burst from his submachine gun. Tristan, Édourd, escort the General to the doorway and wait for my signal.

"Last chance General. Call off your men or within minutes your beautiful lover and you will both lie bleeding to death on your own balcony. It's a poor choice, but it's the only one you have."

As David verbalized her imminent death, Berjuhi lost control of her bladder and began to beg the General to save her life. Sobbing, she told him she didn't want to die. The General remained stoic as Tristan and Édourd pushed him forward to stand behind her. It appeared that he did not intend to yield to David's threats. Then, he did.

"Give me the radio. I'll do it," the General said at the very last moment. Berjuhi fainted and collapsed at his feet unconscious. The General stepped back from the doorway and took the radio from Jean Luc.

"This is General Jhalawan. Cease Fire. Fall back to the bottom of the hill immediately and wait for further orders."

"Remy, François, this is Jean Luc. Hold your fire. Let me know when the soldiers begin to fall back. Édourd, raise Emily and get their ETA."

General Jhalawan's Villa; Janawai, Pakistan
1:10 AM

"Jean Luc, Emily. We're about ten minutes out. What's the situation on the ground? Any casualties?"

"No casualties. Jhalawan's security force has apparently withdrawn to the bottom of the hill on the east side of the villa. I can't confirm that but it's what the General ordered them to do. We're hunkered down in a defensive position on the north side of the villa. As far as I can tell, you've got a secure landing zone. Recommend you swing wide and approach the compound from the west to avoid the General's security troops. They may have portable rockets or heat-seeking missiles with them. Keep your eyes peeled for the Pakistani gunships. I'm assuming that the security force called for air support about twenty-five minutes ago. If they took off from Islamabad, they should still be about twenty-five minutes away so we may have a ten or fifteen minute cushion. My concern is that the Pakistani Army may have had some gunships on the ground at Abbottābad. If so, they could get here about the same time you do. It's going to be close."

"Roger that. Do you copy Bernard?"

"Roger, copy."

Emily summarized the critical performance characteristics of the two types of helicopters. "Jean Luc, top speed and service ceiling of the Little Birds and the Pakistani Blackhawks are almost identical. The Blackhawks will probably be carrying Hydra 70 rocket launchers in addition to their twin 50 caliber Gatling guns. We can't outrun them and we can't lose them by going higher so we're going to have to either get a head start or get lucky. We might be able to out maneuver them in a narrow valley or river gorge but I wouldn't want to count on it."

"Understood. We'll climb aboard as soon as you touch down. David will ride in the cockpit with you. Édourd and General Jhalawan will ride on the starboard side bench. Al and Remy will take the portside bench."

I'll ride in the cockpit with you, Bernard. François and Henri will take the starboard bench with Alain and Tristan on the port side.

"Copy."

"Jean Luc, watch toward the river. I'll flash my red running lights one time when we're due south of the villa before we start our turn. We'll touch down about three minutes later."

"Roger."

General Jhalawan's Villa; Janawai, Pakistan
1:19 AM

"Emily just flashed her lights. Get ready to load," Jean Luc told the team. They'll be on the ground in two or three minutes. Remy, you and Al get into position to guide them in. Use your flashlights to outline the LZ for them."

The Little Birds were acoustically stealthy so Jean Luc didn't expect to hear them until just before they touched down but they all heard the Blackhawks coming. They were still a ways off but the sound of their big rotors was already reverberating off the valley walls. It was going to be way too close for comfort.

"Emily, Jean Luc. We can hear the Blackhawks coming up the river valley. I can't tell how far away they are but I remember when one of them left the villa the other day, we could hear it for maybe five to seven minutes. I'd guess they're no more than fifteen miles away. We'll have to hustle. They'll be able to see us in this bright moonlight."

"Roger, copy."

"François! Henri! Alain! Tristan! On me. Now!"

"You all know that our mission will be a bust if David doesn't have time to interrogate General Jhalawan. It looks like the Pakistani Blackhawks are going to be right on our tails. We may have to buy some time for the other bird. Check your rifles and make sure you have at least six full magazines. The most vulnerable part of a Blackhawk is the hump below the main rotor. That's where the two turboshaft engines are mounted. Aim for it if we get into a fight. This could get very hot. The Blackhawks have two fifty caliber

Gatling guns and Hydra 70 rocket launchers. You guys are all the firepower we'll have but our Little Bird is more maneuverable so that will give us some chance in a fight. Here's our ride. Let's go!"

Neelum River Valley; 4km East of Janawai, Pakistan
1:25 AM

The lead Blackhawk fired a Hydra 70 rocket that narrowly missed the Little Bird Bernard was piloting. The gunships weren't in range of their fifty caliber guns yet but the Little Birds wouldn't survive a lucky hit from one of the Hydras.

Emily and Bernard both began to fly a randomly erratic eastward course to make it harder for the Pakistani gunners to lock onto them. Unfortunately, that slowed them down so the Blackhawks were gaining on them now. They'd soon be in range of their fifties and it would be lights out for the Little Birds when that happened. Jean Luc and David had made it clear to them all that the Pakistanis were just as eager to kill Jhalawan as they were to wipe them out. Another Hydra swooshed past ten meters off the starboard side of Emily's aircraft. Suddenly tracers from the fifties became visible just below them. They were still out of range but not by much.

"Turn this thing around and cram us down their throats," Jean Luc shouted to Bernard as he turned to face outboard and raised his own rifle to the ready position. "He shouted into his combat headset. "Get ready! It'll happen fast."

The closing speed was over three hundred miles an hour. In less than thirty seconds, the Little Bird was flying right down the middle of a formation of two very surprised Blackhawks. Jean Luc's men opened fire on full automatic from both sides of the Little Bird. Almost instantly, the Blackhawk on the south side exploded in a ball of burning jet fuel. Bernard immediately turned back to the east but the other Blackhawk was already more than two miles away. Then, incredibly, the Blackhawk made a sweeping turn and headed back toward them. When the range between the two aircraft was under a mile, the Blackhawk opened up with both of her fifty caliber Gatling guns. Bernard could see the tracers streaking toward them and he jerked the flight control stick hard to the right and put the Little

Bird into a dive. The tracers just missed them but the Blackhawk pilot was very good and it took him only a split second to compensate

Bernard dropped right down to the deck and headed up a broad valley that ran due south from the village of Hillmat. The Blackhawk pilot turned to follow him but couldn't gain on the Little Bird when it was flying a straight course. The Losar Line of Control was only about three miles up the valley. With a little luck, they might just make it back into India. Less than two minutes later, the GPS display showed that they were crossing the LOC just north of an unnamed mountain peak on their left that was over four thousand meters high. Bernard made a sweeping left turn and put the aircraft into a steep climb that cut their airspeed sharply. The Blackhawk was dogging their trail and was again in range of its Gatling guns.

Suddenly, the Little Bird shuddered from multiple fifty caliber hits and the screeching sounds from the T63 turboshaft engine made it clear that the aircraft would be lucky to get back on the ground in one piece. Bernard had his hands full and was far too busy to issue a Mayday call. He put the little helo into a sharp dive. There was no time to pick a good place. Twenty seconds after being hit, the Little Bird crashed half way up the side of the mountain just as the engine quit. The Blackhawk began a straffing run and the Little Bird burst into flames. The Blackhawk made one last observation pass over the crash site and then turned north and left the area.

Taarbal, India
1:55 AM

David pointed to a cluster of storage sheds and barns just north of the Indian village of Taarbal on the west bank of the Neelum River. Emily landed the Little Bird fifty feet from the edge of the trees and cut the engine.

"Try to raise them again."

"Using their mission calls signs on several frequencies, Emily repeatedly transmitted, "Raven two, this is Raven one. How do you copy?"

There was no answer.

Al unstrapped himself and walked up to stand beside David, who was still sitting in the copilot's seat.

"Anything?"

David just shook his head. "We'll keep trying but it's probably no use. I'm amazed they were able to kill one of the Blackhawks with just small arms fire. It looks like the other one probably evened the score."

"What now?" Emily asked.

"Al, help Édourd with Jhalawan. Take him into that barn and wait for me. I'll be along in a few minutes. Remy, get the rest of the camouflage netting out of the dark blue Toyota that Al and I were in and help Emily cover the Little Bird. It's entirely possible that the Pakistani Army will recon this area before they give up."

"David, the Little Bird's not going to be much use to us. It was running on fumes when I landed."

"What kind of fuel does it use?"

"Normally, JP-5 jet fuel but some of the early Allison engines can also burn JP-4."

"What's the difference?"

"JP-5 is the preferred fuel because it's not as dangerous but there's zero chance we'll find any around here. It's a fairly exotic fuel used only in modern gas turbine engines. But, the Little Birds we sold to the Israelis are the older models that can also burn JP-4. It's a 50-50 blend of kerosene and gasoline so we might be able to make some. Do any of the SUVs have diesel engines? Diesel fuel is very similar to kerosene."

"No, they're all gas burners. How much do we need?"

"About sixty gallons. I'm sure we can scavenge thirty gallons of gasoline from the SUVs. If we can find a diesel truck or bus in the village, we can probably get the other thirty gallons we need."

"OK. You and Remy see what you can round up. I'd like to have the Little Bird refueled and ready before sunrise if at all possible."

Crash Site; 6km North of Taarbal, India
2:35 AM

Incredibly, the old man of the French COS team was still alive. It wasn't the first time François had survived a helicopter crash but this one was the worst. He'd been stunned but hadn't lost consciousness when Bernard had managed a barely-controlled crash seven or eight hundred meters below the summit of the mountain. As a veteran of many fights, he knew the Pakistani Blackhawk would rake them with fifty caliber gunfire. Despite a broken left collarbone he'd managed to unbuckle himself and had rolled and crawled to get behind a large boulder just before the Blackhawk opened fire. He'd watched as the Little Bird's fuel tank exploded engulfing the aircraft in a massive fireball. He'd shouted to his comrades but none of them were still alive by then.

While he waited to be sure the Blackhawk wasn't coming back, he took some pain pills out of his personal first aid kit. He had no water but managed to swallow them dry. The portable SATCOM radios were somewhere in the burning remains of the Little Bird. He still had his combat headset but the range was limited. He knew that Taarbal was in the valley on the south side of his current position. It couldn't be more than eight or ten kilometers away over the ridge above him. If he could find some water, he thought he could make it that far. The moon would set in another half hour or so and he didn't think he could manage the rugged terrain in the dark so he decided to do what soldiers across centuries of human warfare have done when they had a chance. He took a nap.

Rustic Wooden Barn; 1km North of Taarbal, India
2:45 AM

As soon as he entered the old wooden barn, David saw General Jhalawan seated on a low bench with his arms tied behind him around a timber support post. Al and Édourd had lit two kerosene lanterns they'd found and were waiting for him. He sat down in front of Parvez Jhalawan and the two men looked at each other without speaking for several moments.

"General Jhalawan, I am a civilized man so I want you to know that what happens here tonight will be a matter of choices you make. I am going to ask you a lot of questions that you will not want to answer. If you refuse to answer my questions, or if I think you have lied to me, I will torture you."

"Before this night is over, you will tell me what I want to know. You can do so without physical pain or permanent disfigurement if you make the intelligent choice that is yours to make. I am fully aware that you have used torture many times in your career to obtain information you wanted. I am quite sure that you know that no one can withstand torture indefinitely. Eventually, even the strongest are reduced to tears and pitiful begging for an end to the pain."

"I know you are a proud man and that you are highly intelligent and well-educated. I truly hope that you will not make our work tonight unnecessarily unpleasant for us both. But, you must not doubt that I will use my considerable knowledge of the art and science of torture if you force me to. You ordered my only sister murdered. You had my girlfriend's only brother murdered. You have ordered many others killed and have masterminded the greatest single mass murder in all of human history. For these crimes, you deserve to die. Yet, I offer you a chance to live."

"You may not know that I am now an agent of the Central Intelligence Agency. I am aware that you hate the CIA intensely but I would advise you not to allow that hatred to force me to put you through the same hell you have condemned so many others to in your long career."

"The team that abducted you was a joint French-American special forces unit. Since France has abolished capital punishment, you will not be taken back to France to stand trial. You will either die right here or you will be taken to a place of imprisonment in a location controlled by the American government. Eventually, you will be given a fair trial and an opportunity to offer whatever defense or mitigating circumstances you and your attorneys can devise. I think it's possible that our legal system will not condemn you to death but that depends on the choices you make here tonight. If you choose unwisely, the only judges you will ever stand before are me and almighty God."

"Unfortunately, I do not have the full range of sophisticated instruments of torture in this remote location. Yet, I have pliers, sharp knives, fire, and other more primitive tools at my command. As a master practitioner of torture yourself, you surely know my options well. I would like to think that you will not force me to castrate you, sever your fingers, break out your teeth, gouge out your eyes one at a time, cut off your feet and hands, char your flesh, or any of dozens of other options available even in this out of the way place. But, if you do not answer my questions, I will do these things to you and before the night is over you will be horribly disfigured and begging me to stop. Again, the choice is yours."

"What do you want to know?"

"I want you to tell me every detail of Pakistan's nuclear weapons program. I want to know the types, numbers, yields, storage locations, and delivery methods for every nuclear weapon in Pakistan's arsenal. I also want to know the locations of Pakistan's nuclear material enrichment facilities and the names of the leading nuclear scientists. I am quite aware that you had Dr. Bashir Ghilzai murdered so I want to know who has taken his place."

"I also want to know everything you can tell me about Ahmed Malik. I want to know the location of his residence, his work habits, how he travels, and every other pertinent detail of his existence. In simple language, I want you to help me understand how Malik can either be killed or abducted."

"I want to know the full details of ISI support for the Taliban and the warlords and which Pakistani government officials want the Americans to fail in their war on terrorism."

General Jhalawan slowly shook his head. "I cannot tell you these things. It would betray my entire life's work. I would rather die than tell you these things."

David nodded. "Yes, General. I'm sure you would, but that is not really one of your options here tonight. I do not plan to kill you. Instead, I will be very careful to keep you alive so I can continue to inflict mental and physical pain on you until you do tell me these things. I will give you five minutes to decide."

"Édourd, keep him covered. Do not engage in any conversation with him while I am gone. Al, join me outside."

"Let's walk a little ways from the barn. I don't want Jhalawan to overhear us."

When they were out of hearing range of the barn, David asked for Al's input. "You've watched people being interrogated before. What do you think he'll do?"

Al paused for a moment before he answered. "I think he'll try to avoid the torture by telling you plausible lies. I watched him closely while you were laying out his options. I think he believes that you will torture him because that's what he'd do in your place."

"I agree. I had hoped that we'd have him hooked up to a lie detector when I questioned him. So, how do we know when he lies to me?"

"You can start be assuming that he'll lie about everything important. I think you're going to have to get his attention by inflicting sudden, unexpected pain on him the very first time you think he lies to you."

"How?"

"I'll stand behind him with my knife. When you look at me and nod, I'll cut his right ear half way off. If he gets the message, we can tape it up and it will heal. But, it will be very painful and very sudden."

David nodded thoughtfully and then looked up as he heard Emily and Remy walking toward them.

"Any luck?"

"Yes. We've already topped off the Little Bird's tanks."

"That's good news. OK. I want you both to join us in the barn. I'm going to start questioning Jhalawan. Al and I both think he'll lie about everything important so I've got to be prepared to use torture to persuade him that lying isn't in his best interest. Every time I ask him a question, I want each of you to show me a thumbs up or thumbs down sign. The first time he lies to me, Al will partially sever his right ear. If he lies again, he'll lose the ear entirely and Al will toss it in the dirt in front of him so he can see it. His left ear will be the next bet he makes in his game of deception. I'm hoping he gets the message before it gets really ugly."

"How far are you prepared to go?" Emily asked.

"As far as I have to. Malik is still in Pakistan and he's even more dangerous than Jhalawan. We need information we can use to defang the Pakistani cobra. With all of their political turmoil, we can't afford to risk another act of nuclear terrorism. Let's go back in."

Without preamble, David asked General Jhalawan, "Are you ready to answer my questions?"

"Yes. I see no point in letting you torture and disfigure me. It is true that no one can withstand torture indefinitely so I may as well answer your questions."

"I'm pleased to hear that. However, you must tell me the truth. My colleagues will collectively act as my lie detector. After every one of your answers, they will signal whether they believe you have told the truth or lied to me. If they think you have lied to me, there will be unpleasant consequences that will grow more unpleasant with each lie you tell."

Just before sunrise, David left Remy guarding the General and asked the rest of the team to join him outside.

"Comments? Suggestions?"

Emily spoke first. "I think you've got everything he has to give. Once we get back to home turf, we can run checks on a lot of it but I can't think of anything else we can accomplish with him now."

Al nodded his agreement. "He can be questioned some more once we get him Stateside but I think we're done for now. There are still a few gaps in his responses but they are almost certainly because he didn't know the answers. The human lie detection system seemed to work. He got religion about the truth quicker than I thought he would. He's lucky to get out of this minus just one of his ears.

Édourd nodded without speaking.

"OK. Thanks. Al, you and Remy stand guard over Jhalawan for the next couple of hours. Give him some of one of your MREs and some water. Let him sleep if he wants to. Someone will relieve you so you can get a couple of hours of sleep too.

Raven One Little Bird; 1km North of Taarbal, India
6:45 AM

Emily was sitting in the Little Bird trying again to raise Raven Two when David walked up.

"Anything?"

"Nothing on the cockpit radio but I thought I heard something on the command headset you left lying on the seat. When I put it on, there was only static."

David put the headset on and turned the gain all the way up. Emily tuned the cockpit radio to the combat tactical link frequency and put on a headset so she could listen too.

The background noise was all David heard at first but then, just as he was about to give up and take the headset off, he heard a faint voice speaking English. He keyed his mic and spoke loudly. "Anyone on this net, this is Raven One Command. Identify please."

"David, this is François. Can you hear me?"

"Yes, François. Faint but clear. Where are you?"

"I've just crossed a ridge about one thousand meters below the summit of the tall mountain just north of you. I thought I could make it to Taarbal on my own but I don't think I can go much further. Can you send someone to get me?"

David glanced at Emily who was holding up ten fingers.

"François, Emily and Édourd will pick you up in the Little Bird in about ten minutes. Do you have a smoke flare or some way to signal your position?"

"No flares but I've still got my signaling mirror."

"Where's the rest of your team?"

"David, they're gone. Tell Emily to hurry. I've used up all of my pain meds and I'm pretty badly dehydrated. I've had some hallucinations and I've already lost consciousness a couple of times. I'll try to stay alert until she can get here. There's a steep slope just below me that seems to have some clear spaces between the rocks. I think she can land there."

"François. This is Emily. Hang on. We're coming to get you."

Taarbal, India
8:00 AM

David went to the barn to take over the watch from Al and Remy. General Jhalawan was asleep. Like the rest of the team, David was exhausted but there was no one else to guard their captive.

"I'll stand watch while you guys grab some shut-eye. Emily and Édourd are going to fly up to the crash site to recover the remains of the rest of the team. Before he drifted off, François told me that the aircraft burned so I'm not sure what they'll find. I drove down to the village earlier to buy some blankets they can use to wrap the remains. We're short-handed so Emily and Édourd are all I can spare. Depending on how much they can recover, the Little Bird may be near its maximum load capacity with just the two of them and the bodies anyway."

"David, you're in command now," Al objected. "We've still got a lot of miles to cover and maybe some more action before this mission is over. I'm still in pretty good shape. I think you should let me guard the General while you grab some sleep. We can't really afford for you to be out on your feet."

"You sure?"

"Yes. I'll call you if the pizza truck shows up."

Thursday, 28 August

Guantanamo Bay, Cuba
3:15 PM

After a marathon flight of over nine thousand miles, the white Gulfstream G450 owned by a CIA shell company banked sharply to port for its final approach to the Naval Station airfield on the leeward side of Guantanamo Bay.

The sleek executive jet had flown thirty-two hundred miles non-stop from Srinagar, India to the U.S. Naval Station at Sigonella, Italy where it had landed to refuel and change pilots. The next leg, a relatively short twelve-hundred mile hop from Sigonella to Gibraltar, had taken only a little over two hours but was necessary because the plane had to be refueled again before beginning the forty-four hundred mile leg across the Atlantic. That final leg was right at the limit of the G450's maximum range. To provide a small margin of safety, the aircraft had been stripped of all non-essential weight and the pilots had adjusted altitude frequently to avoid the strongest prevailing westerly jet stream head winds.

In the passenger cabin, Lieutenant General Parvez Jhalawan was restrained by handcuffs, leg manacles, and his seat belt. One of the two CIA agents assigned to transport him to his new home was guarding him while his partner used an encrypted SATCOM radio to coordinate prisoner transfer details with the commanding officer of the Joint Detention Group.

Five minutes after the pilot parked the aircraft, two military police officers placed a black hood over General Jhalawan's head and loaded him into a special van for the intentionally confusing, circuitous trip to the ultra-secret location known as Camp 7. Although Khalid Sheikh Mohammed, another mass-murderer and the mastermind of the 9/11 attacks was also in custody at Camp 7, the two men would never meet behind the walls of the high-security, maximum-isolation facility.

At Camp 7, Jhalawan was in-processed like any other detainee. He was taken to a central shower facility where he was ordered to undress. He protested but the two heavily-armed guards made it clear that he had no choice. Humiliated at being watched while he showered wasn't the worst part. A U.S. Navy Hospital Corpsman performed a thorough body cavity search to ensure that he would enter the detention facility utterly harmless both to himself and others.

Before being taken to the isolated cell that would be his new home for the foreseeable future, he received the standard issue kit consisting of plastic flip-flop sandals, an orange jumpsuit, a prayer cap, and a copy of the Koran. He was also given a plastic bucket to be used for washing water and another that could serve as an emergency toilet. He was issued a sleeping pad, a sheet, and a blanket and was given a towel for use at the shower facility and another that he could use as a prayer mat. Toiletries consisted of a toothbrush, washcloth, canteen, soap, shampoo, and toothpaste. These few things were all that was permitted for a man whose personal wealth exceeded one billion dollars.

Parvez Jhalawan was still alive only by the grace of God and President Marshall's authorization that he be transported to Guantanamo for further questioning and detention. Under the strict military tribunal system of justice that had been established at Guantanamo after the 9/11 attacks, Jhalawan might yet avoid the execution he deserved. But, for a man long accustomed to the privileges of wealth and power, isolated incarceration at Camp 7, possibly for the rest of his life, would be a special kind of hell for him.

Friday, 29 August

Palladium Luxury Shopping Center; Tehran, Iran
Noon

Electro City was an upscale electronics store on the 4th floor of Tehran's newest and most dramatic shopping center. Inside the mall's ultra-modern towers were over 200 retail shops, 24 restaurants, the largest toy store in the Middle East, and a luxurious athletic center on the 6th floor with a panoramic view of Tehran and the imposing Alborz Mountain range to the north. Electro City was an enormous store that stocked some of the most advanced electronic equipment available anywhere in the world.

After weeks of advanced advertising, the day when one of the world's most advanced smart phones was first offered for sale had finally arrived. Several hundred people were waiting outside the

Electro City store when the doors opened. Within an hour, the initial stock allocation of five hundred phones had been sold at full retail price.

The situation was the same at major malls throughout the country. The demand for the feature-laden phones was incredible. By 10:00 PM, when most of the stores closed, over 20,000 of the phones manufactured in the world's largest Muslim country had been sold. Few of the customers knew or even cared that the phones had been manufactured in Jakarta, Indonesia by a company named Indonesian Advanced Microelectronics. They might not have been so eager to buy the new phones if they'd known that key components had been designed and manufactured by an IAM subcontractor that was a front company for the Israeli Mossad.

Kavir Desert, Iran
2:00 AM

The Israeli Cheetah's payload was one thousand of the new IAM smart phones. Nahbi buried 400 of them in two separate caches. He took the other two hundred with him when he returned to his pickup.

Saturday, 30 August

Baghat Kakhil Shahr; Semnan, Iran
12:45 PM

Nouri Ahmadi was a twenty-six year old Semnan university graduate student who often took his wife and son for picnics at the Gardens of Semnan, an extensive botanical garden covering the entire southwestern part of the city.

Although he was careful to keep his own counsel, Nouri despised Iran's religious and government leaders because he considered them corrupt and exploitive of the common Iranian citizens. His wages were so low that he could barely support his wife and young son. It was in the gardens that Nahbi's keen hearing overheard him talking to his wife about the sanctions imposed on Iran by the west and his belief that Iran's leaders were largely to blame for the hardships those sanctions caused.

Nouri's passionate declaration to his wife that he would not always have to clean toilets at the Jamkarān Mosque had suggested a way for Nahbi to accomplish part of his mission strategy. Seated on a park bench close to the young family, he'd been listening to their conversation for twenty minutes. When he stood up, the man and woman both looked in his direction. Nahbi flashed them a friendly smile and walked over to say hello.

"Salam. I wonder if your family might like to have this nice orange that my wife packed for my lunch. Today, she gave me so much food that I cannot possibly eat this orange too."

Nouri's young son took the orange from Nahbi's hand before his father could reply.

With an affectionate smile directed at his only child, Nouri responded. "It seems that my son has already accepted your kind offer. He loves oranges and we can seldom afford them on my small salary. You are very kind. Thank you."

Nahbi waved his left hand in a dismissive gesture to show that the gift of the orange was a small matter. "This is my first visit to this lovely park. It's such a nice day that I decided to eat my lunch here. Do you come here often?"

"Yes, almost every day when the weather is good."

"So, do you live here in Semnan?" Nahbi asked.

"No, we live in Garmsār, about half way between Semnan and Qom. I'm a student at Semnan University. My wife uses the car in the morning while I'm in class and then picks me up at noon so it's convenient for us to eat our picnic lunches here."

"You must spend a lot of time driving back and forth. Isn't Garmsār a great distance from Semnan?"

"Yes, it's about one hundred and twenty kilometers. It takes over an hour each way but I work for a company in Qom that has the janitorial services contract at the Jamkarān Mosque. We'd prefer to live in Semnan but with so many students in the city, there are few jobs available. The janitorial work in Qom was the best job I could find so we rented an apartment about half-way."

"With so many hours driving each day, how do you find time to study?"

"My work shift begins at midnight so I can study after our evening meal, read to my son before he goes to bed, and then drive almost two hours to the mosque. After work, I drive home and sleep for a few hours before my wife drives me to class. I usually catch a little more sleep on the way."

Nahbi pursed his lips in a thoughtful way and nodded slowly in sympathy. "I know how difficult it can be to balance so many demands on your time at once. Is your work difficult?"

"No, not really. It's monotonous but not hard. I empty the trash cans in the rest rooms and throughout the mosque and surrounding grounds. I clean the rest rooms, replenish the paper products, and then mop the tiled floors. While the floors are drying, I vacuum the prayer carpets. The last thing I do each night is to put new bags in all of the trash receptacles."

Nahbi realized that the young man would be ideal for his purposes. "I may be able to offer you a small job that would provide some extra income. Here's a card with my cell phone number if you are interested."

That evening, Nahbi took Nouri's call and successfully recruited him for a very unusual assignment.

Sunday, 31 August

Milhaud Quays; Toulon Naval Base, France
3:00 PM

It had been a secret mission so the memorial service was held on board the flagship of the French Navy instead of in a local church. The nuclear-powered aircraft carrier, Charles de Gaulle, was the largest warship in Western Europe and the only non-American carrier in the world able to launch and recover the F/A-18 Super Hornets or the C-2 Greyhounds. With all aircraft off-loaded and parked on shore, there was ample room in the hanger bay for the families of those who had died on the mission as well as a contingent of political and military dignitaries. Moored portside-to at the Milhaud quays, the ship was protected by several layers of naval base and ship's security.

General Montreuil asked everyone to stand as he introduced Father Henri Grouix to open the service with prayer. Father Grouix asked those in attendance to join him in reciting a slight variation on the traditional Catholic Prayer for the Dead as the words were projected on a large screen.

"God our Father, Your power brings us to birth.
Your providence guides our lives,
and by your command we return to dust.

Lord, we know that these valiant soldiers,
who died on a foreign field, still live in Your presence.
Their lives have changed but have not ended.
I pray that you will comfort the families, relatives, and friends
of these brave men as we bid them farewell.

In company with Christ, Who died and now lives,
May they rejoice in your kingdom where all their tears are wiped away.
Unite us together again in one family, to sing Your praise forever and ever. Amen"

David stepped to the podium and paused for a moment with his head bowed. When he looked up, he motioned to the surviving members of the mission to join him on the platform. Remy Thibadeau, Edourd Jardin, Francois Gaultier, Emily Harrison, and Al Flores stood behind him as he began to speak while a translation into French was projected onto the screen behind him.

"Today, we are here to say goodbye to five of the bravest men I have ever known. Captain Jean Luc Davignon, Sergeant Tristan Dejarnette, Bernard Garneau, Henri Barteau, and Alain Fabré died so that we might live.

The Bible tells us in John 15:13 that Jesus Christ told his disciples, "*Greater love hath no man than this, that a man lay down his life for his friends.*" As a young man, I always wondered how Jesus could command us to love each other since love seemed to be an emotional state that we can scarcely control. Only years later did I realize that Christ was talking about love as a conscious decision to behave in a certain way.

We were all listening over our command headsets that terrible night when Jean Luc ordered Bernard to fly straight at two heavily-armed Pakistani Blackhawk gunships. I served with Jean Luc in Afghanistan so I am certain that he knew his order would result in his own death and that of the men on board with him. Yet, he did not hesitate.

I am deeply saddened at the death of these fine men. Yet, I also feel a sense of intense pride and honor to have known them and called them my friends. Truly, these courageous sons of France were among her finest.

I would like to share something with you that means a great deal to me. When General Montreuil asked me to join this mission, one of the conditions I gave was that I had to be able to meet the COS physical standards. Early one Sunday morning in August, I just barely managed to run the endurance course within the allotted time. Jean Luc's men ran with me. On the final leg, Alain Fabré began to sing the National Anthem of France. I was exhausted and utterly out of breath by then so I don't know where he found enough breath to sing. But he did and the rest of the team joined in. I cannot adequately express the emotions that came over me as their patriotic love of their country helped me to the finish line.

When General Montreuil asked me to join the mission, he reminded me of a famous quotation that seems especially appropriate today.

Most of you have probably never heard of an American newspaper columnist named Richard Grenier. He graduated from the United States Naval Academy with a degree in engineering. As a Fulbright Scholar, he studied in Paris at the Institut des Sciences Politiques and did graduate work at Harvard. In 1993, he wrote an article in the Washington Times where he paraphrased George Orwell's statement that '*people sleep peacefully in their beds at night only because rough men stand ready to do violence on their behalf.*'

All of us knew that this mission was extremely dangerous with a high probability that some of us would never see our loved ones and our homes again. A little after midnight on the 13th of August, millions of people were sleeping peacefully in their beds when our team attacked a fortified and heavily-defended villa in the foothills of the Himalayas to capture the man responsible for the deaths of millions of innocent French citizens. The mission was successful only because some members of that team were willing to lay down their lives for their friends."

As David concluded his remarks, he again bowed his head briefly and then turned to leave the platform with the other mission survivors. General Montreuil stepped to the podium again to ask the audience to stand as a military band began to play the French national anthem. At the first notes of La Marseillaise, a small entourage entered the hanger deck from a passenger elevator. The President of France had come to pay his own respects.

Stepping to the podium as the music ended, he began to speak. "I have come here today to pay tribute to five brave men who died to help make our world a safer place. I offer my deepest, heart-felt condolences to their families and friends. Most of you know that the National Order of the Legion of Honour is the highest honor that we can bestow in recognition of the ultimate sacrifice of these brave men. The Legion of Honour was established by Napoleon in 1802. It is divided into five degrees of increasing distinction. These are Chevalier, Officier, Commandeur, Grand Officier, and Grand Croix."

The President presented the Chevalier Legion of Honour medal to Henri Barteau's wife, to the father of Alain Fabré, and to Tristan

Dejarnette's sister. He presented the Grand Officer Legion of Honour medal to the young widow of Bernard Garneau, the pilot of Raven Two.

France's highest military honor, the Grand Croix Legion of Honour medal, was presented to Jean Luc Davignon's five year old son as his mother stood behind him with tears streaming down her face . She was not alone. Some of the combat veterans in the hangar that afternoon had tears in their eyes too when Jean Luc's little boy stood bravely at attention and saluted as the President knelt to one knee in front of him to present the medal.

The memorial service ended with a haunting bugle and drum corps rendition of Aux Morts, the French equivalent of Taps.

Monday, 1 September

Kavir Desert, Iran
2:00 AM

Nahbi saw the two black parachutes blossom when their high-altitude, low-opening automatic controls activated fifteen hundred feet above the desert floor. With only the faint light of a quarter moon, no human would have seen their dark silhouettes against the night sky but they were plainly visible to Nahbi's ultra-sensitive eyes. Several minutes later, the two carbon fiber canisters landed within a few meters of the coordinates he'd specified. He transmitted an encrypted code word to the pilot of the Cheeta to let him know that the packages had arrived safely. The stealthy successor to the legendary SR-71 Blackbird was streaking across the night sky at Mach 3 and would soon be over the horizon. The Iranians would never know that the top secret Israeli aircraft had again penetrated their air space on a special mission.

The two canisters each contained a hundred small black plastic trash bags secured by zip ties and folded neatly to conserve space. Inside each bag were approximately one million tiny insects of the order *Siphonaptera* and species *Pulex irritans*. Better known as human fleas, the 200 million parasites were a special gift from the Israeli Mossad to the visitors to the Jamkaran Mosque. The bags were partially inflated with a precisely-calibrated mixture of carbon dioxide and oxygen that would keep the tiny insects alive but comatose until they were released into the open air. Inspired by the Torah's account of the plagues visited upon ancient Egypt, the fleas would play a key role in Nahbi's mission.

It took Nahbi less than an hour to carry the bags to his pickup truck hidden behind a low hill a few hundred feet east of highway 81. It was 340 kilometers to Garmsār where he would deliver the bags to Nouri Ahmadi before he left for work at midnight.

Malta International Airport; Valletta, Malta
9:05 PM

At Roger Benson's request, General Montreuil's Chief of Staff arranged for David and Emily to be flown directly to Malta from the flight deck of the Charles de Gaulle when the memorial service was over.

After a four-hour flight, the Airbus EC 175 helicopter landed at the AFM military terminal at Malta International Airport where The American Ambassador and Elena were waiting. David set his bag down just as Elena fairly flew into his arms with tears of relief streaming down her face.

"Oh David, I was so afraid I'd never see you again. When Ambassador Stevens called yesterday morning, he said Pete Nocona had asked him to tell me you were safe but that he wasn't sure if you'd be able to come to Malta right away. Then, a little after noon today, Ambassador Stevens called me again to tell me that you would be landing here tonight. I was so happy I just had to tell someone so I called my parents and Tiera. I hope that was OK. I didn't say anything about your secret mission. I just said I assumed you must have some business with the embassy and left it at that."

David gave Elena another hug and then turned to face Emily and Ambassador Stevens. "Elena, let me introduce you to Emily Harrison. She's the lady I met with at the Embassy when I was here before and I owe her a lot. I'm not sure how much I can tell you but there'll be time for that later."

"Hello, Elena," Emily said with a friendly smile. I'm pleased to meet you. I hope we'll see more of each other."

David then turned to shake hands with the Ambassador. "Good evening, Mr. Ambassador. Thank you for your support and especially for making Emily available." David kept a straight face when he said, "I don't know if she's much help at the Embassy but I can tell you that we were glad to have her with us."

Tuesday, 2 September

Jamkarān Mosque; Qom, Iran
4:45 AM

Very early in the morning, and several hours before the Tuesday crowds of hopeful believers would begin to arrive to make their special requests of the invisible 12th Imam at the wells, Nouri did as he'd been instructed. He placed the trash bags Nahbi had given him in receptacles inside the mosque, near the prayer wells, and in the trash cans scattered around the courtyards and parking lots. Nahbi hadn't told him what was in the bags but had suggested that he move away from the trash cans as soon as he removed the nylon zip ties.

A few minutes after Nouri opened the trash bags millions of fleas became active and began to spread to their surroundings. Almost immediately, the prayer carpets and floor tiles near the trash cans were covered with barely visible fleas that needed to feed on human blood to survive. The areas around trash cans in the courtyards and parking lots were also quickly infested. Nouri had been well paid for what seemed a simple, innocuous task so he'd asked no questions about what was in the bags.

44000 Constitution Avenue; Islamabad, Pakistan
9:00 AM

The massive, neoclassical building known as the Prime Minister's Secretariat had been the official residence and principle workplace of Pakistan's Prime Ministers since 1973. The Prime Minister was seated on a sofa reading his morning briefings when the President of Pakistan was ushered into his office.

The Prime Minister wasted no time on formalities or courtesies. "Where did they take him?"

"We're not sure. Malik doesn't think he was taken to France because they abolished capital punishment several years ago. We have gun-camera infrared photographs of the helicopters that were used in the raid. They were older models of what the Americans call the *Little Birds* configured for special operations missions. The specific aircraft used in the attack were almost certainly two that the CIA sold to the Israelis a few years ago."

"Is Malik saying that Jhalawan was abducted by the Israelis? That makes no sense to me."

"No, Malik has put together enough clues to be confident that it was a French COS operation launched from Srinagar. He thinks the French somehow managed to borrow or lease the aircraft from the Israelis. We know that a French C-130 cargo plane was at Srinagar airport allegedly due to some mechanical issue. We also know that the C-130 Hercules can carry two of the Little Birds as was done in the American rescue invasion of Grenada in 1983."

The Prime Minister considered what the President had told him and then asked, "Does Malik think that Jhalawan is still alive? If he is, and if he talks, we might still find ourselves at war with France and her allies."

"Yes, Malik is virtually certain that Jhalawan is alive. He sent a team to gather information from the site where one of the Little Birds crashed. The bodies had already been removed indicating that the team on the other Little Bird was able to return to the crash site to retrieve the bodies of their comrades. Malik has ordered every active agent of the "S" Wing to use their networks to find where Jhalawan was taken."

"Keep me posted," the Prime Minister told his President. "If Malik can find him, perhaps we can silence him before he does too much damage. Let's deal with the other issue we discussed. Do you have your recommendation ready for my review?"

"Yes. My initial thought was that we should promote Ahmed Malik but I discussed it with him and he argued rather persuasively that he would be far more valuable to Pakistan in his present position. I tend to agree. Malik is unquestionably brilliant but would not be well-received by many of the career officers since he is not one of their own. I spoke to the Chief of Staff of the Army and he was emphatic that the new Director General of the Inter-Services Intelligence Agency should be a senior Army officer. He gave me several suggestions and I've had each of them thoroughly vetted. My recommendation is that you appoint Lieutenant-General Zafar Khalil, the current Commander of the Army Strategic Forces. His resume is impressive and, unlike General Jhalawan, I think he would be loyal to you. I took the liberty of asking Ahmed Malik for his opinion of General Khalil. Malik surprised me with his candor. He said that although he had great respect for General Jhalawan, he sometimes felt that the General's agenda was not perfectly aligned with Pakistan's national interests. He thought General Khalil would be an excellent choice."

"Very interesting. As it happens, I agree with Malik's assessment of Khalil. In fact, if you had proposed someone else, I was going to ask you to consider General Khalil before you finalized your recommendation. Have his appointment paperwork prepared for my signature and arrange a brief ceremony here as soon as there's a suitable opening in my calendar. It's imperative that we fill the position as soon as possible."

The Prime Minister then shifted topics. "I have a meeting with the French Ambassador this afternoon. He's a cagey old diplomat so I may not get much out of him. The last time I met with him, he was demanding that we extradite Jhalawan to France to stand trial for war crimes. He offered a fig leaf scenario to the effect that if Pakistan turned Jhalawan over to them, it would be taken as evidence that theft and detonation of the nuclear device was master-minded by Jhalawan without the knowledge of the Pakistan government. It will be interesting to see if the Ambassador continues to press his request for extradition. If he does, I plan to tell him that Jhalawan has been abducted and that our intelligence service believes it was by a French special operations team. He'll deny it of course. When he does, I'll tell him that we believe that Jhalawan acted entirely on his own and that I had him placed under house arrest pending the outcome of our investigation. I'll tell the Ambassador that if we should find that Jhalawan had accomplices inside our government we will take appropriate action. He'll know that I'm puffing smoke up his skirt as the Americans like to say but I think it will create enough doubt to keep the French from doing anything rash."

Jamkarān Mosque; Qom, Iran
6:11 AM

Thirty minutes before sunrise, the muezzin's call to dawn prayers sounded over the loudspeakers. Nahbi was standing near the main entrance to the mosque watching people entering to pray before going to work. Nearly all of them were already carrying some fleas. Although few of them yet realized why, many were showing their discomfort by aggressively scratching their itching skin. When they knelt on the prayer carpets, they would soon be covered with fleas. Those unfortunate people were the first casualties of a series of plagues Nahbi had arranged as part of his plan to change Iranian attitudes toward their religious and political leaders. Only a few thousand people were attending dawn prayers but by the time the mid-day prayers were finished six hours later, more than ninety thousand people would be afflicted.

The fleas had been bred in a sterile laboratory environment in Israel. They were certified as being disease-free and thus posed no significant health hazard. However, the relatively brief life cycle of the fleas meant that the females would soon be laying eggs. Roughly half of the 200 million fleas that Nouri Ahmadi had delivered for Nahbi were females. Each of them would lay about 40 eggs the day after biting their human hosts. Most of those eggs would fall from their victim's skin and clothing in workplaces and homes throughout the area. Within a few days, several hundred thousand Iranians would be shocked to find themselves covered in flea bites. By the time Iranian health officials traced the outbreak to the

Jamkarān Mosque, it would be too late for normal control methods to be effective.

Friday, 5 September

Jamkarān Mosque; Qom, Iran
11:25 AM

It had been three days since the plague of fleas began. Nahbi had returned to the mosque hoping that he would again run into the five young men he had previously identified as suitable conduits for information he wanted to disseminate.

The news of the flea plague had spread quickly in the community so attendance was down sharply from the nearly one hundred thousand that had visited the mosque on Tuesday. The buildings and grounds had been sprayed with insecticides on Thursday once the source of the outbreak was identified. But, by then, it was really too late to contain the infestation. By Friday, the fleas had spread throughout the city's populace of more than one million people and were still spreading as people went about their daily lives.

Nahbi found the five young men sitting in the shade of the building on the north side of the mosque. They were eating roast meat on bamboo skewers and Persian flatbread. As he approached, the leader of the small group recognized him and motioned for him to join them.

"Ahura Mazda! Come join us. We have plenty to share."

"Ah, thank you but I've already eaten. I'm glad I've run into you again. I wasn't sure I'd find you here today after what happened on Tuesday. I am very interested in your thoughts about this unfortunate situation."

The vain, immature leader of the group rose to Nahbi's bait like a hungry trout rises to a fisherman's lure. Just as he'd done when Nahbi first identified him as a suitable tool, he once again displayed his limited knowledge to impress his friends.

"There are many verses in the Holy Quran about the importance of cleanliness. It seems obvious that the keepers of the mosque have allowed the buildings and grounds to become so filthy that the flea population simply exploded. Do you not agree?"

"I am sure there is much wisdom in what you say but I have wondered if the fleas might be a warning from Allah."

"A warning? What do you mean?"

"In the Holy Quran, Sura 7:133 records that, in ancient times when Allah was displeased with the Egyptian Pharaoh, he sent a plague of lice. The plague of lice is also recorded in Exodus 8:16-19 of the Torah and the Christian Bible. These historical records leave little doubt that Allah has used plagues to influence human behavior in the past. Perhaps the plague of fleas is a sign that Allah is displeased about something. It seems to me that we should pray earnestly for wisdom to understand."

Four of the young men nodded thoughtfully at what Nahbi had said but their leader was irritated that Nahbi had quoted the Hebrew Torah and the Christian Bible to support his thesis.

"The Holy Quran is my only guide. I do not recognize the Hebrew Torah or the Christian Bible as being authoritative or reliable guides."

Nahbi could barely conceal his amusement at the young man's ire. "Yes, I understand and meant no offense. Yet, surely we all know that our most revered Islamic scholars recognize both the Torah and the first five books of the Christian Bible as having been written down by the prophet Moses and thus useful to help us understand Allah's will." Glancing around at the other young men, Nahbi planted a thought for them to mull over later. "Perhaps there is a way we can know if the fleas were a plague or, as you suggest, merely poor housekeeping."

"How? One of the other young men asked."

"In Egypt, Allah sent several plagues to persuade Pharaoh to do what he wanted him to do. If other signs appear, then I think we might reasonably conclude that the fleas were, in fact, a plague from Allah."

The group's leader thought he saw a weakness in Nahbi's argument. "If the fleas were a plague, don't you think our Imams would know it?"

Nahbi nodded slowly as though he was carefully considering the young man's point but then deflated his argument with one simple statement.

"That would make perfect sense unless it is they who have displeased Allah. Remember that the plagues in ancient Egypt were visited on the common people but they were not the target of Allah's displeasure. Pharaoh, their leader, was."

Roadside Produce Stand; Garmsār, Iran
6:35 PM

After leaving the mosque, Nahbi drove 175 kilometers from Qom to Garmsār. He had another assignment for Nouri Ahmadi but this time he knew he'd have to answer some questions and provide more information than he'd done the first time. There was some risk involved but he thought it was manageable.

As he entered the outskirts of the city of forty thousand people, he stopped at a roadside stand where some of the popular local melons grown. He purchased one of the yellow melons and sat in his pickup to eat it. When he finished, he placed a call to Nouri's cell phone number that was still in his phone's incoming call list.

"Salaam," he heard Nouri say.

"Salaam. This is Rahim Kashani. Do you remember me?"

"Yes, of course."

"I'd like to hire you for another assignment if you are interested. I'm at a fruit stand on the north side of Garmsār. If possible, I'd like to meet with you briefly sometime this evening before you leave for work."

There was a moment's silence on the line before Nouri spoke. "There's a new Royal Dutch Shell petrol station and convenience store in the Garmsar Industrial Park where highway 44 intersects the Qom-Garmsar freeway. I'll meet you there at 9:30 PM. Whether

I'm willing to take another assignment will depend on how you answer some questions I have."

"I fully understand. I'll be waiting for you."

Shell Petrol Station; Garmsār, Iran
9:25 PM

Nahbi was sitting at a table outside the convenience store when Nouri parked a dusty Proton Exora SUV in front of the store and went inside. He came out a few minutes later carrying a bottle of water and a bag of Pofak cheese snacks. He walked over to one of the tables near the man he knew as Rahim Kashani and sat down with his back toward Nahbi.

Nahbi recognized Nouri's caution and allowed him to speak first. He was not surprised when Nouri said, "Please keep your voice low and do not seem to engage me in conversation. I prefer that no one think I have met with you. I often stop here on my way to work so tonight would not seem unusual if anyone is following you. I am very uncomfortable meeting with you after what you had me do at the mosque. If my role in that despicable matter is ever discovered by the Imams my fate will be much worse than merely losing my job. I want to know who you are and why you had me place fleas in all of the trash receptacles."

"Mr. Ahmadi, I fully understand your concerns and will do what I can to alleviate them. I will begin by admitting that I was eavesdropping on the conversation between you and your wife at the park the other day when you voiced your discontent with the religious and political leaders of our country. You are quite right that their obsession with power is the root cause of the sanctions imposed on our country by the western nations that have resulted in such hardship for you and your family as well as millions of our countrymen. As to who I am, I am a man who wants something very similar to what I think you want. I would like for Iran to regain its rightful position as a respected leader among nations. The Persian civilization is one of Earth's oldest and greatest. It is only in recent times that power-crazed men have turned our lovely land into a pariah among nations."

Nahbi intentionally referred to Iran as *"our"* nation to avoid having to reveal that he was a foreign spy and agent provocateur. Better that Nouri see him as a fellow dissident intent on bringing about positive change.

"I'm sorry Mr. Kashani, but I do not see how infesting the mosque with fleas would change any of that."

"Before I explain why, I need to ask you a few questions. You are a married man with a small son. I would certainly understand if you were unwilling to take risks as part of a plan to eventually restore Iran to its rightful place among nations. As to your question about who I am, I am part of a secret organization that believes much as you do. We are committed to bringing about peaceful changes that will end our long nightmare. I am asking if you are prepared to accept some risk to be a part of a bold plan to end Iran's isolation and permit her people to return to the prosperous, peaceful existence that all of us desire."

Nouri was silent for several moments. "I love my family with every fiber of my being. What kind of man would risk his life and possibly expose his family to a life of ostracism and hardship?"

"You are an educated man so I think you already know the answer to your question. Throughout history, patriots have risked everything to bring about changes that led to a better future for their children. Yet, I fully understand that only a special few have such vision and the courage to make it happen. Listening to you the other day, I thought I heard the kind of passion that such men possess. I watched your wife's face and felt that she agreed with you and that she believes in you as the head of your family. I suspect that your wife will support whatever decision you make. I have two additional tasks in mind for you. I do not think they are especially risky but only you can decide."

"I would have to discuss whatever it is that you ask me to do with my wife. I now see that this is much more than simply making a little extra money by doing something simple."

"Yes, of course. I understand. Let me ask you two questions before I tell you more. After hearing what I've just said, and forming your own opinion of my sincerity, can I trust you not to betray me whether you agree to help me or not?"

Nouri took his time before answering. "Yes. I believe you are sincere and I do think your objectives are similar to my own hopes for our country. So yes, you can trust me not to betray you. What is the other question?"

"I would like for you to tell me about your religious faith. My organization believes that a distorted and self-serving religious dogma has been used by our nation's religious leaders to maintain control over the population. If you disagree, then it would be unwise for me to reveal more to you."

"You ask a difficult question for which I can give no simple answer. Like most thinking people, I have often wondered if Allah really exists. Here in Iran, it would be suicidal to publically express my doubts. People literally get stoned to death or hung for disputing what the Ayatollahs and Imams preach. After much thought, I have decided that Allah does, in fact, exist but that he is not as he is commonly described by our religious leaders. I cannot conceive that the Earth and all living things simply happened. To me, it seems obvious that it was all intelligently designed. The question then becomes, by whom? Whether we call that designer and creator Allah or God isn't important in my eyes. What is important is what the existence of Allah means in our lives. For me, it means that we must try to understand Allah's will for our lives and that we must do the best we can to obey his wishes. All men were created by Allah so "death to the infidels" seems an utterly inconsistent and false teaching. Everyone understands that a man can commit his life to Allah at any time. If other humans kill that man they deprive him of future opportunities to dedicate himself to Allah. To me, killing people who disagree with our religion cannot be something Allah wishes. When I hear the Ayatollahs and the Imams inciting young men to willingly martyr themselves in the false hope of rewards in paradise, I think I am hearing something utterly evil, not holy and good. So, if you are asking me if I am a devout Shite Muslim, the answer is no. I am a seeker of spiritual wisdom and I try to live my life in a decent manner that pleases Allah."

"Thank you. Your views are quite similar to my own. Now I will tell you why I asked you to place bags of fleas throughout the mosque and grounds. The fleas were the first of several contrived plagues that my organization will create. Although these plagues will not be sent directly by Allah, we hope and pray that he is using us to accomplish his greater purpose. We are demonstrating our faith by our willingness to work toward a more obedient nation in service to Allah. We think that a series of plagues can be used to undermine the power of the Ayatollahs and Imams and indirectly the power of our nation's political leaders."

Nouri asked the obvious question. "What is it that you want me to do?"

"I want you to help me create two more plagues at the mosque. The next one will be a plague of cockroaches. If you agree, I will again provide trash bags that you will place in and around the mosque. Like the fleas, the cockroaches have been bred in a laboratory environment and are certified as being disease free. They will cause no harm to our people but they will create great concern that we can use."

"The third plague will be oriental hornets that are common in Iran. The hornets are the only one of the plagues we have planned that are not harmless. Hornets that are killed release pheromones that trigger a nest-protecting attack response from the other hornets. Oriental horncts thal are common here in Iran can sting multiple times because their stingers are not barbed. As the wind currents carry the pheromones to nearby neighborhoods, even the local hornets will become temporarily aggressive. The first two plagues were designed to create feelings of disgust. The hornets are different. They arc intended to instill a growing sense of fear that will cause people to demand solutions from their leaders. The first two plagues occur on Tuesdays when there are sometimes one hundred thousand people on the grounds. I will schedule the hornets for next Wednesday because crowds are typically much smaller then. I wish I could avoid causing pain at all but I do not think we can succeed otherwise. If you agree to help me, I will pay you ten million Rials in cash each time I deliver the trash bags to you."

"That's almost as much as I make in two months. What if I'm caught?"

"You can say that a mysterious man paid you to put the trash bags in the receptacles and that you didn't know what they contained. Say your family can barely get by on your salary and you saw no harm in doing what the man asked."

Nouri was again silent as he considered what Nahbi had told him. After a few moments he gave his answer. "I don't need to discuss this with my wife. We do need the extra money but I don't want her to know anything about you in case she is ever questioned. I will do what you ask."

"Thank you. There is a parking area about seven kilometers south of here on the east side of the freeway. I'll meet you there at 9:30 PM Monday night. The bags will be slightly more bulky than the last ones but they weigh very little and will easily fit in your SUV. As before, move away from the trash receptacle as soon as you remove the zip ties. Oh, one more thing. Here's one of the new IAM Smart Phones for you. I think you will enjoy some of its special features."

Saturday, 6 September

Kavir Desert, Iran
12:30 AM

Nahbi slowed down and then turned off of highway 81 when he saw the twisted desert salt cedar that was his reference point. Earlier in the day, he'd traded his used Bardo pickup for a white, two year old Khodro Arisun pickup with a topper shell. Although he didn't think his mission had been compromised, Levi Reznik had insisted that he change vehicles. Levi had also strongly suggested that Nahbi abandon his original landing site cache just to be on the safe side. Nahbi knew that was good tradecraft so he hadn't argued. It took him several trips to bring out the rest of the money and other materials he'd cached and put them in the truck. He'd already found two new sites farther north.

Three hours later, he pulled off the west side of the highway and parked the pickup out of sight behind a rock outcropping. He dug several holes and buried most of his remaining mission support materials. He kept out two shrink-wrapped bales of Iranian Rials worth about $75,000 each and some new clothing and glasses he'd use to change his appearance. He also kept out ten packages of C4 plastic explosives with miniature radio-controlled detonators but still didn't see a need for any of the guns. He sent an encrypted message with the coordinates of the new drop zone five kilometers away and then policed the area to remove signs of his night's work.

Mofid University; Qom, Iran
Noon

Professor Barid Afshar was sitting beside the large pool near the center of the University's gardens when Nahbi sat down on the bench beside him and removed a beautifully decorated Ney from its case. Nahbi played the Persian flute for a few minutes and then laid the instrument across his lap.

"Salaam, Professor Afshar. It is good to see you again."

"Salaam, Kazem Rahmani. I was beginning to wonder if I'd hear from you again. It's been more than a month since our first meeting."

"Please accept my sincere apologies. I have been very busy making arrangements but I now need some help from you and your graduate students."

"What kind of help?" Professor Afshar asked.

"Actually, I need help with two different things. I need to have explosives placed at ten different locations within a radius of approximately 150 kilometers of Qom. When we met before, you indicated that you had five trustworthy students available. Is that still true?"

"Possibly. I think their willingness to help will depend on what you are asking them to do with the explosives. Certainly, they understand that our leaders will require, shall I say, some encouragement to make the kinds of sweeping changes that are needed. However, these students are not terrorists. If these explosives will not harm innocent people, then I think we can help you. Can you tell me where you want the explosives placed?"

"I have a list of the home addresses of five religious leaders and five political leaders. I want the explosives placed at the front door of their homes while they are away. My objective is to cause massive damage to their homes while sending a signal that they are not immune from the kind of attacks that they themselves have ordered or at least condoned against others. My purpose is to arouse fear without harming anyone. I can use that fear as part of my plan to change the behavior of our nation's leaders."

"I am confident that my students will agree to do as you've asked. How and when will you provide the explosives and the remote detonators?"

"I will deliver them to you Monday at 5:30 PM in the back parking lot of the Tarbiyat Cinema on Fatemi Road. The parking lot will be starting to fill up for the first evening showing by then but we'll be able to transfer ten small packages from my vehicle to yours without raising suspicion. The packages are not heavy or bulky. They will fit easily into the trunk of your car. Each of the packages will contain a brick of C4 explosive, a remote-control detonator, and a black, full-length chador."

Professor Afshar nodded. "Ah, yes. A disguise. What else do they need to know?"

"They will be placing the explosives at the front doors of expensive luxury villas here in Qom and in the Oshan district of Tehran. I have selected targets that are not behind gated walls or fences. But, all of them have external security cameras so they'll need to be very careful as they approach. The address cards I've provided show the times when the occupants typically leave their homes. There's no guarantee that they'll follow that schedule on the day your students make their deliveries so they should watch to verify that the occupants have left before approaching the villas. Tell them to splatter some mud on their license plates to make them hard to read. They should park at least a hundred meters away and keep their heads down so the security cameras never see their faces. They should disguise themselves by wearing the black, full-length chadors that are in the packages. Tell your students to put the C4 brick in the bottom of a basket of bright flowers with a colorful ribbon and bow to suggest that they are merely delivering a gift. When they return to their car, they should drive away slowly. When they are a few blocks away, they can pull the chador down and wriggle out of it without stopping the car. They should place the chador in a paper bag and dispose of it in an inconspicuous trash receptacle. If possible, I'd like for their work to be completed sometime Thursday or Friday of this coming week."

"OK. What is the other matter?"

"At 9:00 PM this Wednesday evening, I need you and your students to help me load roughly one thousand green iguanas weighing about two kilograms each into two dumpsters parked beside highway 81 thirty kilometers south of the highway 36 intersection. It will take at about an hour. I'll pay for your petrol and give each of you five million Rials and one of the new IAM Smart Phones for your help."

Professor Afshar laughed softly. "When I saw the newspaper Wednesday morning, I suspected that the plague of fleas at Jamkarān Mosque was your doing. Will these big lizards also be faithful in their attendance at prayer calls?"

Nahbi smiled. "Yes, you've guessed correctly. What kind of reaction do you think they'll arouse?"

Professor Afshar laughed out loud this time. "Reaction? Utter terror and panic I think. Once I tell my students what you have in mind, they will gladly do as you ask. You wouldn't have to give them any money although they'll appreciate it greatly. They will be overjoyed to get their hands on one of the new IAM smart phones. I'm sure you know that the stores sold out almost immediately and the after-market prices are exorbitant."

Ferdous Garden; Semnan, Iran
7:30 PM

The two men he'd financed to start a garbage collection business were waiting for him.

"Salaam, Yusuf ibn al-Hasan. We assume you have an assignment for us."

"Do you have the equipment I specified?" Nahbi asked.

"Yes. We bought two trucks and several large dumpsters. We've already secured contracts to collect and dispose of waste for customers here in Semnan and also at the Jamkarān Mosque in Qom as you asked. What do you want us to do?"

"Wednesday night, I want you to have two trucks with two forty cubic meter dumpsters parked on the west shoulder of highway 81 precisely 30 kilometers south of the intersection with highway 36. They need to be in place no later than 8:30 PM and they need to be the type that has hinged end doors. Set out reflectors and make a show of changing a tire just in case someone stops to see what's going on. Before you leave Semnan empty several crates of fresh lettuce into each of the two dumpsters. I'll already be in the area when you arrive. About 9:00 PM, an SUV will join us with six men to help carry a very unusual cargo to the dumpsters from a location just west of the highway. I'd like for one of you to remain with the trucks and the other to help us move the cargo."

"What's the cargo?"

"Have you heard about the fleas at the Jamkarān Mosque last Tuesday?"

"Yes, of course. It was on TV and in all the local newspapers. Everyone's been talking about it."

"Good, that was part of my plan. You will help me create another plague even more dramatic than the fleas. You are going to transport one thousand green iguana lizards to the mosque. Each of the lizards weigh about two kilograms. Although they are ferocious looking they are actually harmless vegetarians. The dumpsters must be in place at the mosque no by 5:45 AM, one hour before sunrise. From where we load the lizards to the mosque is about 450 kilometers. Averaging 70kph, you can make the trip in a little over six hours so you should be able to position the dumpsters in time if we can finish loading no later than 10:30 PM."

"Yusuf, please forgive me. I do not wish to be rude but are you insane? How will a plague of giant lizards help our country?"

Nahbi smiled and said, "I assure you that I am quite sane and yes, these giant lizards will help us bring about positive changes. For now, you must trust me in this."

Monday, 8 September

Drop Zone 2; Kavir Desert, Iran
1:53 AM

One hundred meters west of highway 81 and thirty kilometers south of the intersection with highway 36, was a small outcropping of rock that Nahbi had selected as his new drop zone. He was parked behind the rocks out of sight of the highway waiting for his next delivery. An encrypted radio was lying on the pickup's seat beside him and he had an earbud in his right ear. He was listening for the signal that the Cheetah's pilot had released the two pods eighteen miles above the desert floor. Right on schedule, he heard the first few notes of a popular Persian song by the teen idol singer known as Xaniar. The signal songs changed with each mission but this one seemed especially appropriate. The song's title was *'Risk.'*

The desert was bathed in the soft glow of an almost full moon directly overhead and just one day short of full. Five minutes after the signal that the pilot had released the pods, Nahbi saw the parachutes deploy. He had specified that the new target drop zone was 34.994 degrees North Latitude and 54.579 degrees East Longitude. Two minutes after the parachutes deployed both pods landed within a few hundred feet of target with amazing precision after a free fall descent of over 85,000 feet.

Nahbi immediately opened both pods and began transferring the black plastic bags containing a half million Asian cockroaches to his pickup. The insects had been anesthetized with carefully-calibrated mixture of carbon dioxide and nitrogen with just enough oxygen to keep them alive until they could be released. Once the bags were opened, and fresh air filled them, the cockroaches would quickly awaken from their sleep and do what cockroaches do. They would start looking for food.

Before leaving the drop zone, he buried both pods and erased the signs of his visit.

Roadside Parking Lot; Qom, Iran
9:30 PM

Nouri Ahmadi was already waiting in the parking lot when Nahbi drove up. It took them only a few minutes to transfer the trash bags and a rip stop nylon bag containing ten million Iranian Rials to Nouri's SUV.

Tuesday, 9 September

Jamkarān Mosque; Qom, Iran
4:55 AM

Nouri finished putting the bags of cockroaches in the trash receptacles. As he'd been told, he'd moved away from the trash receptacles immediately after removing the zip ties. Out of curiosity, he'd paused a few yards away after opening the first bag inside the mosque to see what happened. Within a minute, thousands of large, Asian cockroaches began to fly and crawl out. Soon they were all over the floor and walls near the trash can and most were already scurrying around in search of food.

12:25 PM

Nahbi saw a group of young men waiting for him as he came out of the mosque after noon prayers. He was pleased to see that the five he'd been cultivating had now grown to twenty.

"Salaam Ahura Mazda," the group's leader said in welcome. "Will you join us?"

"Yes, of course. I was hoping to see you again. I see that you have new companions with you today."

"We have told our friends of your idea that perhaps the fleas were a plague sent by Allah and your suggestion that we wait to see if something else happened so they wanted to be here today just in case."

"So, first fleas and now cockroaches." Nahbi said. "What do you think of this new infestation? Do you still believe that the fleas and now the cockroaches are only the result of poor housekeeping here at the mosque?"

The young man was somewhat hesitant in his reply. "Cockroaches and fleas are vile and unwelcome in any house but that is especially true when they infest our holy mosque. But, yes I do still think the large number of these pests is most likely a result of violating the holy Quran's teaching about cleanliness."

Nahbi nodded to show that he was considering what the young man had said and then looked around the group of young faces watching him for his response.

"I admit that you say seems possible. Perhaps you are right. Maybe it is premature to conclude that these infestations are plagues sent from Allah. When we talked the other day, I spoke only of the possibility that the fleas might be a sign that Allah was displeased about something. Later, I gave careful thought to what might be happening in our country that could have displeased Allah. Several possibilities have occurred to me but I wonder if you and your friends might have some thoughts on that subject. If these infestations are signs that Allah is displeased, it seems important that we learn why and then do what we can to remove the source of his displeasure. What do the rest of you think?"

For several minutes the young men offered opinions about things that might displease Allah. Most of them had to do with things that frequently triggered harsh condemnations, fines or imprisonment, and even physical punishment from the religious leaders. One young man reminded the group of the punishments meted out when the morality police caught someone listening to banned western music.

Another pointed out that many young women detested the rigid rules governing how they were required to dress in public. Nahbi was not surprised that none of them suggested that Allah might be displeased with the behavior of their leaders. He would have to guide them very carefully to the conclusion he wanted them to reach.

"Certainly fleas and cockroaches are unwelcome visitors to our place of prayer just as the lice and locusts were unwelcome in the days of the prophet Moses," Nahbi said. "One thing that worries me is that the plagues described in the Quran became more painful as Pharaoh continued to resist Allah's will. Quran 7:132-133 says that Pharaoh and his closest advisors, '*remained arrogant, being criminally sinful people.*' Also, some of the plagues lasted a long time. I have even wondered if the sanctions imposed on our country might also be a plague because one of the plagues described in the Quran was a drought that lasted for years causing a scarcity of food." The sanctions have had a similar effect on our people for several years now."

One of the young men said he'd heard people talking about the sanctions but that he didn't really know what they were or how they affected most people.

It was almost as though the young man was a shill planted in the group to start the discussion Nahbi wanted to have with these young and impressionable young men.

"The sanctions are several international resolutions intended to put pressure on our leaders not to develop nuclear weapons," Nahbi explained.

Several of the young men reacted angrily expressing variations of the oft-heard argument that Iran had the same right to develop nuclear weapons as the other countries of the world.

"Yes," Nahbi replied calmly. "That seems only fair but the situation is more complex than most people know. More than fifty years ago, during what was commonly called the Cold War, only the United States, the Soviet Union, and Great Britain had nuclear weapons and the ability to deliver them. Although there was great fear about a possible nuclear war, most experts believed that something they called *'mutually-assured destruction'* would keep the nuclear powers from using their nuclear weapons. However, by 1964, France and China had also successfully tested nuclear weapons and the science of nuclear weapons was rapidly becoming widespread. It seemed only a matter of time before other nations developed nuclear weapons too. Leaders all over the world realized that if nuclear weapons become common, eventually some nation, or possibly some terrorist organization, would use them. That growing fear caused the nations with nuclear capabilities to sharply curtail sharing nuclear technology, even for peaceful purposes. Then, in 1968 the Nuclear Non-Proliferation Treaty was adopted by the three major nuclear powers of that day and they invited other nations to join their agreement. In February of 1970, our Parliament ratified the treaty. Today, there are approximately two hundred countries in the entire world and all but four have signed the treaty. In all of human history, there has never been such near-unanimous agreement among nations about anything before."

Nahbi saw that he had their attention and that they were listening closely to what he was telling them.

"Many people believe that the sanctions are solely the work of the United States. Actually, that is not true. Almost every country on Earth is a member of the United Nations and it is the UN that has imposed the sanctions on our country. We suffer from the sanctions because they are the toughest sanctions that the world community has ever imposed on any country. The sanctions are punitive because the world believes that Iran is trying to develop nuclear weapons in violation of the treaty we signed. Our leaders say otherwise, but the world community does not believe them. Simple logic suggests that if most of the nations on Earth disapprove of our country's actions it might be wise to reconsider our actions, especially since those actions are imposing great hardship on our people and have made our nation a pariah to most of the world. Our population of eighty million people is less than one-tenth of one percent of the population of the world so you can see that we are really isolated on this issue. If our leaders do not change their policies, I fear that a day will come when the nations of the world will stop talking and find ways to force our leaders to conform to what has clearly become the will of most of the world's people."

"How do you know so much about these things?" one of the young men asked.

"Every day, I read the Holy Quran and pray to Allah to grant me wisdom. I listen to everything I'm told and carefully consider everything I read. I try always to recognize what seems true and what does not. I strongly believe that Allah is worthy of all praise and that he has given us many blessings and gifts to use during our time on Earth. I cannot believe that Allah would have given us our miraculous brains if he did not intend for us to use them to seek to understand his will for our lives. So, I spend a great deal of time considering what is true and what Allah wishes me to do with the understanding that he grants to me."

Several of the young men traded glances and one of them said what they were all thinking.

"Ahura Mazda, you speak clearly about difficult things and it is apparent to us all that Allah has granted you much wisdom. Most of us have to return to work now. Do you come here every day?"

"I try to but sometimes I can't."

"We understand. We will look for you again and perhaps we can hear more of your wisdom on another day."

Drop Zone 2; Kavir Desert, Iran
9:45 PM

Nahbi had decided to accelerate the plagues and wanted the next one to happen at Wednesday's noon prayers. He had a long drive to where he would deliver the hornets to Nouri Ahmadi so he'd had to schedule the drop much earlier than usual. There was additional risk inherent in the earlier time because even in the desolate wasteland of the Kavir Desert, there was still some light traffic on highway 81 at that time of night. He stood on top of the rock outcropping and scanned the highway to the north and south for any approaching headlights. With his superb eyesight, he would see any vehicles approaching while they were still twenty kilometers away.

He had the radio in his hand ready to alert the Cheetah's pilot to abort the drop if he saw any headlights approaching. With barely adequate fuel reserves remaining after the fourteen hundred mile zig-zag flight plan from Ramon Air Base in southern Israel, there would only be one chance. If he had to abort the drop, there would be no plague on Wednesday.

Nahbi knew that the only vehicles typically on highway 81 in late evening would be commercial trucks traveling at eighty to one hundred kilometers per hour. From the time the pods were released from the Cheetah to when they landed was about seven minutes. Nahbi had calculated that human eyes might see the pod parachutes at a distance of up to ten kilometers in the moonlight. That meant that the safe time window for the delivery to remain unobserved was only about five minutes if there was a vehicle within his vision range as the Cheetah approached the release point.

At 9:47 PM, the Cheetah's pilot transmitted the first few chords of '*Jane Javani*,' a song by the popular Iranian singer Ebi as a coded message asking if it was safe to release the pods. Nahbi looked quickly to the north and south and sent the encrypted sequence to tell the pilot to proceed. The pods landed at 9:55 PM. Thirty minutes later Nahbi had finished loading his pickup and headed north on highway 81 to his rendezvous with Nouri Ahmadi just south of Qom.

Wednesday, 10 September

Jamkarān Mosque; Qom, Iran
4:35 AM

Nouri moved quickly away from the trash receptacles after removing the zip ties from each bag. He'd been stung repeatedly when he was nine years old and had knocked down a hornet's nest while kicking a soccer ball around his parent's yard. He had tried not to cry but the stings had hurt so bad that he hadn't been as brave as he'd wanted to be. Since then, he'd had a strong phobia about being stung again by hornets.

Jamkarān Mosque; Qom, Iran
11:45 AM

The parking lot was about half full and several thousand people were standing around waiting when the muezzin's first call to prayer blared out from loudspeakers. Nahbi entered the mosque and selected a prayer mat at the back of the large auditorium a few minutes before the second adhan call was broadcast over the speakers. At the conclusion of the Khatīb's two sermons, the muezzin announced the iqāmah to begin the two rak'at prayers of Jumu'ah.

Almost a half-million oriental hornets were on the grounds and inside the mosque but they weren't yet threatening the crowds. That was about to change in a dramatic way.

In the right front pocket of his slacks, Nahbi had what appeared to be an aerosol can of men's deodorant. The can actually contained a pressurized mixture of the pheromones released by dying hornets that would trigger an attack response in other hornets. Odorless to humans, the chemicals were detectable by hornets in extremely low concentrations.

Just before the second of the rak'at prayers ended, Nahbi pressed the button on the aerosol can. The faint hiss of the escaping gas couldn't be heard over the sound of several thousand voices reciting the prayers. The ceiling fans inside the mosque quickly dispersed the gas. In less than a minute, thousands of hornets began to attack the people inside the mosque. Almost everyone was stung at least once. Many were stung repeatedly. A general panic ensued with people swatting at the hornets and crying out in pain. Nahbi joined the crowd running to the exits to escape. They didn't yet know it, but there was no safety outside.

As soon as he exited from the building, Nahbi released more of the pheromone gas. The breezes quickly dispersed the gas triggering the attack response from several hundred thousand hornets outside the mosque. Swatting at the attacking hornets, most of the people ran toward the parking lot and the safety of their cars. Nahbi ran with them to his pickup even though he hadn't been stung. His graphene skin was a natural insect repellant so a few hornets had landed on him but none had stung him.

Oriental hornets are common in Iran but are normally only a minor pest. The residential areas on the southeast side of the city closest to the mosque had a large population of the insects. As the pheromone gas drifted into nearby areas, the attack response was triggered in the native hornets just as it had been in their laboratory-raised cousins. Within an hour, over twenty thousand people had been stung either at the mosque or in the neighborhoods close by.

After leaving the parking lot, Nahbi drove through the streets of Qom releasing more of the pheromone gas. Vespa orientalis hornets are extremely active in bright sunlight because the yellow stripe on their abdomen contains a pigment called Xanthopterin that converts sunlight to energy. All over the city people were attacked and stung as a combination of the hornet's hyperactivity in the bright desert sunlight, and the attack response resulting from the gas Nahbi had released, enlisted the native hornets as part of the plague Nahbi had created. By nightfall, a large percentage of the one million residents of Qom had been stung at least once.

Iran's national television channel, IRIB TV5, headlined the evening news with the hornet story at one of the nation's holiest religious sites. TV5 reporters had taped interviews with people who'd been stung at the Jamkarān Mosque and elsewhere in Qom. Like reporters everywhere, their careers depended on high ratings for their stories so their news clips sensationalized the widespread hornet attacks. In one of the clips, a reporter was standing in front of the Mosque interviewing two people about the earlier infestations of fleas and cockroaches. The stories ran again in the late edition at 10:00 PM.

IRIB TV5 was broadcast to Iran's eighty million people on digital channel 41 and UHF mobile channel 34. It was also available throughout the Middle East via the communications satellites Intelsat, Hotbird, ST-2, and Badr 5. By midnight, more than one hundred million Muslims in the Middle East had seen the stories.

Drop Zone 2; Kavir Desert, Iran
8:20 PM

The Israeli Air Force B-2 stealth bomber was still one hundred miles to the southwest, and about one hundred and fifteen miles due East of the Iranian nuclear facility at Natanz ,when Nahbi heard the first few notes of a popular Persian song called '*Gravity*' in his earbud. He looked north and south to verify that there were no approaching vehicles and then keyed a six-digit pre-arranged code into his encrypted radio to let the aircraft's commander know that it was safe to deliver his payload.

The top secret B-2 bomber was what is commonly referred to as a flying wing. It was a masterpiece of aeronautical engineering employing the most advanced stealth technology. The B-2 bombers based at Ramon Air Base in southern Israel had all been modified by Northrop Grumman to allow them to carry two of the Massive Ordnance Penetrator bombs. The payload that Nahbi had specified for this delivery weighed far less than two MOPs but had required some very creative engineering to produce two cylindrical pods the same size as the MOPs but configured to hold five hundred green iguanas each.

The B-2 was making a low-altitude run to virtually eliminate any chance that Iranian air search radars would detect it. Flying at almost six hundred miles per hour, the aircraft would be over the drop zone in about ten minutes. After release, the canisters would take only a few minutes to reach the desert floor.

Three minutes later, Nahbi was watching from the top of the rock outcropping as the two garbage trucks approached from the north and then made a U-turn to park on the west side of highway 81 as he'd instructed. He climbed down and walked over to talk to the two owner-operators while they waited for help to unload the iguanas.

A few minutes later, a dark brown Isuzu Panther SUV pulled onto the highway shoulder and parked with its nose a few feet from the truck in front. Professor Barid Afshar and five of his graduate students got out. Nahbi welcomed them and then asked all of the men to gather around him so he could talk to them all at once.

"In a few minutes, you'll see two parachutes open overhead. The loads they are carrying will land about two minutes after you first spot them. Inside the two cylindrical containers are approximately one thousand green iguana lizards about seventy-five centimeters long and weighing an average of about two kilograms each. The iguanas have been anesthetized but they'll start waking up pretty soon after I open the canisters. Each of the lizards is immobilized inside a cloth bag. We'll carry them to the road and then free them from their bags when we drop them into the dumpsters. Don't worry that the drop might hurt them. They often fall from trees without injury. Although iguanas are excellent climbers, the smooth steel sides of the dumpsters will keep them from climbing out. They'll have lettuce to eat to help them rehydrate after their trip. We need to be finished by 10:30 PM so I'd like for each of you to carry two at a time. We'll have less than two hours but we shouldn't have any trouble finishing in time. Any questions?"

One of the graduate students raised his hand. Nahbi nodded toward him.

"We've all seen the news reports of the fleas, cockroaches, and hornets in Qom and we understand that these lizards will be the next event in whatever you are trying to accomplish. All of us are willing to accept some risk as the price of trying to help end our country's nightmare of radical interpretations of the Quran. But, if you'll forgive me, we'd like to know how this really helps."

Nahbi nodded his understanding and then responded to the entirely reasonable question.

"As you all know, a relatively small number of men have gained control of Iran by exploiting most people's sincere desire to be faithful to Allah and by instilling fear. Their power can be sustained only until a large number of our fellow citizens conclude that the hardships imposed by the sanctions are directly attributable to the policies of our religious and political leaders. As people awaken to that reality, they will eventually reach a point where they are willing to defy our national leaders and insist on changes. These contrived plagues are part of a plan to help a significant number of our people to reach that conclusion without the bloodshed that revolutions usually bring. We do not need a majority of Iranian adults to rise up in opposition. We only need a visible and vocal minority. With some gentle nudging, quite a few people will soon realize that the hardships imposed by the international sanctions might very well be a modern version of the plagues in ancient Egypt that were created by Allah to change Pharaoh's mind. My goal is for our fellow countrymen to conclude that Allah is displeased with our leaders and, as he did in ancient Egypt, is using a series of escalating plagues to change their behavior. Does anyone else have a question?"

"Yes. We've all agreed to place the blocks of C4 for you but wouldn't it be better if the explosions all occurred simultaneously to make it clear that they weren't the work of one or two crazy people acting alone?"

"Ideally, yes. The problem is that we don't want anyone hurt by the explosions. If people are injured, public attitudes will be sympathetic toward the victims and our intended message will be lost. Since our criterion is that the occupants must not be at home when the explosions occur, we cannot detonate them all at once. People conduct their lives on differing schedules so you'll have to place and detonate the explosives after you see them leave home to be absolutely certain that no one is injured by the blasts."

At 10:20 PM, the last of the iguanas had been transferred to the dumpsters. The lids were closed and the trucks headed north on highway 81 on their way to Qom. Nahbi thanked professor Afshar and the graduate students for their help and gave each of them a paper bag containing five million Rials and an IAM Smart Phone.

Thursday, 11 September

Jamkarān Mosque; Qom, Iran
5:45 AM

Sunrise wasn't until 6:46 AM so it was still dark when the dumpsters were delivered to their normal locations at the mosque. The call to dawn prayers wouldn't sound until a half-hour before sunrise so the parking lot and grounds were still deserted. Before leaving, the truck drivers released the latches on the hinged doors at the end of the dumpsters and swung them open. Green iguanas are normally active during daylight hours but they have poor vision in low-light conditions. They wouldn't begin to leave the dumpsters in significant numbers until the early light of dawn allowed them to see and they heard to call to prayer that they had been conditioned to associate with food.

Ten minutes after the call to dawn prayers blared out over the speaker system hundreds of scary looking creatures with sharp teeth and claws and a ridge of spines down their backs were walking toward the speakers mounted high on the walls of the mosque. People attending the dawn prayers didn't know what to make of the big lizards. When some of them shouted at the lizards and waved their arms in an effort to scare them away, the iguanas puffed up, hissed, and bobbed their heads in alarm. One man stomped his foot and shouted at an iguana near him. Although iguanas normally move slowly, they are capable of moving surprisingly fast when their defensive display fails to scare off a potential predator. The man who'd stomped his foot jumped back in alarm when the iguana raced toward him. There were no trees nearby so the iguana chose the man's leg as a substitute. When the big lizard grasped his trousers with his clawed feet and kept climbing until it was high enough to look the man in the face, the poor man screamed in terror and fainted. Similar scenarios were playing out all around the mosque and people began to run back to their cars to escape from what seemed to them an attack of dragon monsters from hell.

Animal control personnel began to arrive an hour later but they were utterly unprepared to deal with so many animals they knew almost nothing about. Their normal response to an animal control request would be to capture the animal, load it into one of six transfer compartments on their truck, and then take the animal to the city's animal control facility to either be destroyed or offered for adoption. In extreme situations, the animal control officers were authorized to kill the animals in the field. The only trouble was that they had only a limited number of poisoned darts that were soon used up. By 10:00 AM it was clear to the Imams that the noon prayers would have to be cancelled.

When the call to prayer stopped, the iguanas began to spread out randomly in search of food. Many of them left the grounds and began to roam nearby neighborhoods terrifying people and pets alike.

The iguanas were the lead story on the six o'clock news broadcasts. By then, the lizards had been identified as green iguanas but no one had a plausible theory of why a tropical animal species suddenly appeared in large numbers at one of Islam's holiest sites. People were starting to ask what was going on at Jamkarān Mosque.

TV5 Evening News; Tehran, Iran
6:00 PM

The lead story on the six o'clock news was about the explosions that had occurred at the homes of six prominent religious and political leaders in Qom and Tehran that afternoon. Video clips showed massive damage to some of the most expensive and beautiful villas in both cities. The stories ran again on the ten o'clock news. The press secretary for the President quoted him as saying that '*these acts of cowardly terrorism will not be tolerated.*'

Drop Zone 2; Kavir Desert, Iran
9:45 PM

The prearranged signal from the B-2 pilot that he was ready to release the night's delivery was a song by a popular Iranian female vocalist named Azam Ali. The song was '*From Heaven to Dust.*' Standing in the middle of a barren desert where dust storms were common and with the bomber almost ten miles above him, Nahbi smiled at the pilot's sense of humor.

The payload was divided into two parts that would be dropped at different locations. The part delivered to Drop Zone 2 consisted of twelve 8-liter canisters of rhodamine WT, a bright red industrial dye commonly used to trace flows in municipal water systems and rural surface waterways. The dye was harmless but its dramatic red color could persist for up to twenty-four hours. It took Nahbi thirty minutes to load the canisters into the back of his pickup before he headed back to Qom.

After delivering the canisters, the B-2's copilot closed the bomb bay doors as the pilot started a wide turn to starboard before settling on new course 255 degrees true. Thirty-five minutes later and four hundred kilometers to the southwest, the bomb bay doors opened again and the copilot released six tons of rhodamine WT dye into the upstream end of the Panzdah Khordad reservoir. At that time of night, the hydroelectric generation station at the base of the dam was operating at eighty-five percent of capacity to satisfy nighttime electrical demand from the city of Qom and the surrounding area. The massive water turbines that spun the generators were connected to enormous stem pipes from the inlets above. After the falling water had done its work, it was released from the base of the dam into the Qom River. By dawn, the entire seventy kilometers of the Qom River upstream from the city would be blood red.

Friday, 12 September

Danesh Boulevard Industrial Park; Qom, Iran
4:15 AM

There was an old, rundown industrial park just west of the Shahed bridge over the Qom River on the west side of the city. Nahbi turned south off of Danesh Boulevard and parked beside a rusty metal warehouse at the end of the street fifty meters from the river. Carrying two of the canisters at a time, he made six trips to the water's edge and was finished pouring the dye into the water in less than thirty minutes. In the dark, no one would notice the red dye being carried downstream by the current but by dawn the entire river within the city limits would be blood red.

Mossad Headquarters; Herzliya, Israel
9:00 AM

Gavriella's doctor had finally approved her returning to work on a limited basis. She'd already had two reconstructive surgeries so her jaw and part of her head would remain swathed in bandages for several more weeks. After the bone grafts on her jaw healed, she would face multiple cosmetic procedures to remove scar tissue and restore her appearance as much as possible. Her jaw bone was still immobilized but she could finally talk again though with some difficulty.

During her stay in the hospital, and while she was convalescing at home, Levi and Avi had both been maddeningly vague when she'd asked for news of Nahbi's mission. She'd had to be careful to limit her questions to what any senior intelligence officer might reasonably want to know to avoid revealing that her interest in Nahbi's welfare might be more than collegial. Frustrated by the paltry information she'd been able to extract from the two men, she'd been anxious to get back to work so she could attend the briefings. Her seniority and professional history with Nahbi in London while working on Task Force 3, and then having Nahbi stay at her condo with Levi's approval, gave her ample cover to explain her interest in attending the mission update briefings. The first person she saw as she stepped off the elevator on her floor was Avi Ben-Shimon, the only other survivor of their Kidon team's mission to abduct or kill Abdul Muqtadir.

"Welcome back Gabe. Levi told me you were coming in. Are you going to attend the Iran Mission briefing at ten?"

"Yes. Could we go to your office for a couple of minutes? There's something I need to ask you."

"Sure."

Avi closed the door and motioned for Gabe to take a seat. "What's on your mind?"

"I know there are coded messages between Nahbi and the aircraft pilots to make sure it's OK to release the mission support materials. Do you know who establishes the codes?"

Avi nodded. "Nahbi set up the protocols before he left. He wanted the pilots to play a short segment of popular Persian songs to let him know they were inbound for delivery. He felt that segments of popular music would not be suspicious if any Iranian military or intelligence units happened to pick up the transmissions. Once

Nahbi hears the music, he keys a short encrypted sequence to let them know it's OK to deliver their payloads. If he keys 999, that's the signal to abort the drop for that night."

"So, each pilot selects the songs?"

"Yes. Once they've selected the music, they send an encrypted text message via satellite to Nahbi's IAM Smartphone so he'll know what to listen for."

"When's the next delivery?"

"Saturday morning at 1:30 AM. It'll be a Bardelas delivery. Eli Wolff is the pilot."

"Thanks, Avi."

Jamkarān Mosque; Qom, Iran
12:55 PM

The group of young men had again grown larger. When they saw Nahbi coming out of the mosque, they gathered around him all talking at once. Nahbi held up his hands urging them to calm down.

"Ahura Mazda, we think you were right. The fleas, cockroaches, hornets, giant lizards, and now water seemingly turned to blood throughout the city must surely be plagues sent by Allah just as you said. You heard the sermon today. Does the Imam think we are all stupid? His message that everything that has happened has a logical, natural cause was not credible. Only a fool would refuse to see the similarities to the ancient Egyptian plagues recorded in the Quran. You have shown that you are a holy man in whom Allah has placed great wisdom. Do you know what this all means?"

"Yes, each of these events is a clear message from Allah that he is displeased about something."

"Do you know why?"

"I think so. I have prayed about these signs and considered every possibility I could imagine. I now believe that Allah is very angry with our leaders, probably because they have refused to abide by the nuclear treaty our Parliament ratified. I do not think Allah wants his people to create and potentially use offensive nuclear weapons against others who worship him as the almighty God of Abraham. I also do not think Allah is pleased that the arrogant pride and lust for power of our leaders has caused great hardship on his beloved people."

The spokesman for the group asked the question Nahbi expected. "When you refer to '*others who worship him as the almighty God of Abraham,*' are you including the Christians and the Jews?"

"Yes. Allah made us all. I do not think he wants any group of leaders to be so filled with hate and a lust for power that they call openly for the destruction of entire nations who also believe many of the things we believe. We descended from the same patriarch. We recognize many of the same prophets. Islam, Judaism, and Christianity all revere the writings of the prophet Moses. Surely destruction of entire nations is something only Allah has the right to do. When our leaders continuously call for the destruction of Israel and the United States, I cannot believe Allah is pleased. When so much money is spent to enrich quantities of uranium that far exceed any reasonable peaceful use, I think it is because our leaders want to have the same kind of power to destroy that Allah has. I cannot imagine that Allah is pleased when mere men aspire to have such power. Yes, I think Allah is very angry with our leaders. I also think the international sanctions imposed on us have been permitted by Allah to force our leaders to change their behavior just as Allah eventually forced Pharaoh to do."

"Do you think there will be more plagues?"

Nahbi adopted a very sad expression and nodded slowly. "Yes. Almost certainly. The Quran makes it clear that the plagues in ancient Egypt continued until Pharaoh did as Allah wanted him to do. I had a vision last night that was terrifying. In my dream, I saw darkness over our land and death raining from the sky. I awoke trembling with fear. I do not know what this dream means but I am sure that it was given to me by Allah as a sign of what is yet to come."

Watching the strained faces of the young men gathered around him, Nahbi knew that he had succeeded. They were now filled with fear and were beginning to understand how evil their leaders were.

"What can we do?"

"Very little, *yet*. However, there will come a day, soon I hope, when new, wiser leaders will rise to put aside archaic ideas and allow Iran to once again assume its rightful role among nations. I think we must quietly share our hopes with people we trust and be ready when we are called to service."

"What kind of service?" Several of the young men asked.

"I'm sorry my friends but I have no idea. Still, I think we will know when the time comes. Do not be discouraged. The world is in the hands of Allah."

"Allah be praised," the young men said mechanically as they'd been taught.

Looking around the group and pausing for a moment to make eye contact with each of the young men as he might with a close friend, Nahbi told them, "Before I leave you, I would like to give each of you a small gift. A friend of mine works for a marketing company and he has asked me to help them distribute promotional copies of the new IAM smart phones. I have already given most of them away but I think I have enough in my truck for each of you to have one if you'd like to follow me to the parking lot."

Later that afternoon, Nahbi traded his pickup for a used Morattab SUV that had been manufactured in Tehran. It was an unlicensed clone of the legendary Land Rovers built in the United Kingdom.

TV5 Breaking News; Tehran, Iran
3:45 PM

Regular programming was interrupted by a breaking news story that four more expensive homes had been damaged by explosives in Qom and Tehran. The stories were expanded and run again on the six and ten o'clock broadcasts.

Jamkarān Mosque; Qom, Iran
8:30 PM

It was a clear, calm evening. The sun had set more than an hour earlier and the pleasant twilight had faded to darkness. The mosque and grounds were still fully lighted and more than twenty thousand people were scattered around the grounds enjoying an exceptionally nice evening. Many families had brought small charcoal stoves and the smell of cooking meat was everywhere. Nothing unpleasant had interrupted the evening prayers so most people were relaxed and not thinking of the bizarre events of the past few days. Fifty thousand feet overhead, an Israeli Air Force B-2 stealth bomber's copilot was opening the bomb bay doors and preparing to release the latest of Nahbi's carefully conceived plagues.

The coordinates of the mosque had been programmed into the targeting system and the actual release of the payload would be controlled by the computer to ensure an accurate delivery. The aircraft shuddered slightly as fifteen thousand kilograms of frozen hog's blood was released over one of Shite Islam's holiest sites. The blood had been frozen into spheres one centimeter in diameter. The temperature outside the aircraft was almost sixty degrees Celsius below zero so the blood would not melt during most of its descent. The temperature one mile above the mosque was five degrees Celsius or about forty degrees Fahrenheit so the small spheres of frozen blood would quickly melt into large raindrops of blood before showering down on the unfortunate people still at the mosque.

As the blood rain fell on them out of a clear night sky, the people began to scream in terror. Within a few minutes, most of them were covered in blood and the unmistakable smell of blood hung heavily in the warm night air. The breaking story made the late edition of TV5 news but it would the next afternoon before the blood was analyzed. Like Jews, Muslims consider pork unclean so having hog blood rain down on them was especially revolting. Many of them were finally starting to believe that Allah was displeased and was bringing modern versions of the ancient Egyptian plagues upon them. A pervasive sense of fear and despair was spreading through the population of Qom. And the plagues were not yet over.

Saturday, September 13

Drop Zone 2; Kavir Desert, Iran
1:30 AM

Right on time, Nahbi heard the first few notes of a popular Persian song by Farzad Farzin called Bargard, a word that means '*come back*' in Farsi. Earlier, when he'd read the short text message from Eli Wolff, he'd instantly realized that Gavriella had found a way to send him a personal message by persuading the Bardelas pilot to use that song as his mission code for the night.

Among the night's deliveries were five thousand letter-size posters divided into twenty bundles slightly thicker than reams of printer paper. The posters were made up into pads with a thin strip of adhesive across the top of the back side of each sheet so they could be peeled off and stuck to windows and other surfaces without tape or staples. Ten shrink-wrapped bales of counterfeit Iranian paper currency were in the second canister to replenish his dwindling reserves.

TV5 Evening News; Tehran, Iran
6:00 PM

The lead story on the six o'clock news was about more homes of prominent government and religious leaders being severely damaged by bombs. This time, all of the targets had been expensive homes and villas in Tehran. A news crew had happened to be in the neighborhood when one of the bombs went off and had videotaped dramatic footage of the front of the house engulfed in flames with debris scattered all over the driveway and front garden area.

Immediately after the dramatic shots of the house on fire, the station carried a short video clip of Tehran's Police Chief announcing that Special Police Unit-110 was in charge of the investigation into the recent bombings. Unit-110 was the part of NAJA, the Law Enforcement Force of the Islamic Republic of Iran, that was responsible for suppressing riots and acts of terrorism.

The news segment ended with a reporter interviewing the Minister of the Interior whose own home had been badly damaged in the day's attacks.

"Thank you for speaking with TV5 news Minister. First, were any members of your family injured by the blast?"

"No, there was no one at home but the entire front of my home has been destroyed."

"Is it possible that your home was targeted because you are the minister who oversees the national police force?"

"Perhaps not specifically for that reason but it does appear that these attacks are aimed at senior leaders."

"Could there be a connection between the bizarre events at Jamkarān Mosque and these bombings?"

"That seems unlikely."

"Do you think these attacks could be an attempt to destabilize the government by trying to intimidate our leaders?"

"The investigations are proceeding but it's too early to speculate about the purpose of the attacks or the agents who are committing these treasonous crimes against our nation. I can assure you that our leaders are not intimidated by these cowardly attacks."

"Minister, does your reference to 'treasonous crimes' suggest that the police believe these attacks are being carried out by Iranian dissidents?"

"I'm sorry but I can't spare any more time right now. I'm late for a meeting."

Jamkarān Mosque; Qom, Iran
8:30 PM

The sun had set at 7:14 PM and it was totally dark an hour later. At precisely 8:30 PM, Nahbi pressed one button on a radio transmitter and all of the high voltage transmission lines serving the city of Qom were simultaneously brought down by thermite pyrotechnic devices attached to two legs of two adjacent towers. All around the city, flickering blue-white flashes of light marked where power lines had gone down.

Nahbi had spent hours driving around the city over several days to locate all of the transmission line towers and substations serving the city. He had carefully analyzed the city's power grid and then attached the thermite bars to ensure that the transmission line towers would topple to one side as they collapsed.

Each of the thermite bars was ignited by two magnesium flares activated by small radio receivers connected to conventional blasting caps. Burning at over 2,500 C° the thermite devices melted through the heavy structural steel members in seconds. The weight of the collapsing towers parted the transmission lines plunging the city and surrounding area into complete darkness. The damage was so extensive that it would be several days before even limited power to the city could be restored.

9:00 PM

The waning moon would rise at 10:17 PM but Nahbi's most dramatic plague would take place while the sky above the holy city of Qom was still dark from the widespread power outage.

At exactly 9:00 PM, flying at 560 miles per hour at an altitude of 50,000 feet, an Israeli B-2 stealth bomber dropped 20,000 pounds of fist-sized volcanic stones from specially designed canisters bolted to the bomb bay mounting brackets designed to hold two MOP bunker buster bombs. The volcanic stones were coated with a thick layer of sulfur and a thinner layer of white phosphorous. The twenty foot long canisters were pressurized with nitrogen to keep the phosphorous from igniting until released into the atmosphere. When the latches on the canisters were released, the payload fell away from the aircraft. Within a few seconds, the white phosphorous spontaneously ignited in the presence of atmospheric oxygen which in turn ignited the sulfur coatings.

Four minutes after being released from the bomber, thousands of synthetic brimstones were burning brightly and leaving comet-like tails of burning sulfur that lit up the sky over Qom.

Many of the city's inhabitants were outside their homes talking to their neighbors about the power outage when Nahbi's final plague of fire and brimstone rained down on the blue onion-shaped dome at the center of the Jamkarān Mosque, setting it on fire. Several people used smart phones and digital cameras to take pictures and video clips of the shower of brimstone as a slight breeze carried the rotten egg smell of hydrogen-sulfide across the city.

Within minutes, the wail of fire trucks could be heard all over the city as all of the nearest fire stations responded to try to save the revered site that was so important to the Twelver sect of the Shite Muslim faith. The rapid response of the fire trucks wasn't enough to save the mosque. By 9:45 PM, parts of the building began to collapse sending up fountains of sparks. By midnight, very little of the mosque remained standing. TV5 reporters and camera crews arrived shortly after the evening news show began and immediately began feeding live shots of the burning mosque to viewers all across the Middle East.

ISI Headquarters; Islamabad, Pakistan
11:30 PM

Ahmed Malik turned on the television in his office to watch the late news program on IRIB TV5, Iran's national broadcast station. The reception from the Badr 5 telecommunications satellite was crystal clear for the recording he was making of the dramatic images of the destruction of the Jamkarān Mosque in Iran. He had been monitoring intelligence reports about the mosque for several days because he had a strong suspicion that he knew who was behind the bizarre events there. He certainly didn't think they were plagues from Allah.

Pakistan's Inter-Services Intelligence agency was usually just referred to by its initials, ISI. It was widely considered the best intelligence agency in the world, exceeding even the American CIA and British MI6 in its reach and capabilities.

As the head of the super-secret S-Wing of ISI, Ahmed Malik was effectively the number two man at ISI, *except he wasn't a man at all.* He was also much older than he looked. He had been a bond servant to a wealthy member of the Jewish Sanhedrin almost two thousand years earlier. His name then had been Ammiel.

Malik was fully aware that he was an android but he had no memories of his origin or activities over the intervening centuries because his programming had been periodically updated with new memories and mission instructions. Malik had never revealed to anyone that he was an android and didn't think anyone knew. But he was wrong. Nahbi knew and so did David Logan.

When details about Nahbi's advanced technology and super-human capabilities had been briefed during top secret meetings in London, a deep-cover S-Wing mole had been in the MI6 conference room operating the audio-visual equipment. After reviewing the agent's reports, Malik had recognized that the android that was causing such a stir in London was similar, if not identical, to him.

The head of MI6, Sir Lionel Smythe, had created Task Force 3 to intercept five stolen Pakistani nuclear weapons. Nahbi's unique talents had made him the obvious choice to lead the task force. While reviewing intelligence files on senior Pakistani officials suspected of being involved, Nahbi had seen several photos of Malik. His original programming had never been updated as Malik's had so Nahbi had clear memories of seeing Malik in Jerusalem twenty centuries earlier.

Nahbi knew that several members of the task force doubted that they could find the nuclear devices in time so he'd decided not to tell them that one of their adversaries was a super-human android. But, he'd also recognized that the Pakistani intelligence service was almost certainly aware of his leadership of TF3. That made him a high-value target that could lead to his abduction or destruction by Pakistan at any time. He'd decided to tell David Logan about Malik to ensure that someone on Task Force 3 knew that one of Pakistan's senior intelligence officials was an android if something happened to him.

When the TV5 news coverage of the mosque ended, Malik closed the recording program on his computer and saved the file to a flash drive. He then sent an encrypted message to his counterpart at The Ministry of Intelligence of the Islamic Republic of Iran telling him to expect a phone call in ten minutes.

Known by the acronym VAJA after a series of name changes, the Iranian Intelligence Ministry officially reported to the Iranian President but in reality the chain of command reached all the way to the Supreme Leader.

By law, the Minister over VAJA was required to be a Mujtahad, a title based on an Islamic legal term meaning independent reasoning. In practical terms, a Mujtahad was essentially an Islamic PhD. This requirement meant that all of the Intelligence Service Minsters since the Iranian revolution had been religious scholars. The current Minister of VAJA was Mohammad Zanjani who, although an exceptionally well-educated man, knew almost nothing about modern intelligence practices. His second-in-command was an experienced senior intelligence officer named Ali Yazdi. Malik maintained a close professional relationship with Yazdi and spoke to him frequently about matters of mutual interest. Yazdi answered Malik's call on the first ring.

"Salaam, Ali. I've been watching the news about the Jamkāran Mosque and may have some information you'll find interesting."

"Salaam, Ahmed. What is so interesting that you would call me from Islamabad this late?"

"I'm pretty sure that I know who is behind the attacks on the Mosque. If you're available, I can meet with you in Tehran Tuesday afternoon. I believe I can set a trap that will catch the man orchestrating the attacks."

"Are you going to tell me who you think it is?"

"Not yet, my friend. I'm convinced that the man behind the attacks is a far greater threat to Pakistan than to Iran. When I briefed our President, he ordered me to bring him to Pakistan for secret interrogation and eventual execution for high crimes against our country. I will share any intelligence related to the Jamkarān attacks but, once I have him in custody, I must take him back to Pakistan. Our President strongly emphasized to me that Iran must agree to his extradition as a condition of our help in putting a stop to these bizarre events," Malik lied.

Yazdi didn't respond immediately but then said, "I think Minister Zanjani can persuade the Supreme Leader to accept those conditions if we can devise a mutually-agreeable press release that the attacks were carried out by infidel enemies of Islam and were not plagues sent by Allah, as so many of our simple people now believe. I envision something along the lines of, '*In a brilliant joint-effort, the intelligence services of Iran and Pakistan have located and killed the man responsible for the attacks.*' I'm sure we can devise a cover story that would avoid Minster Zanjani having to explain why your mystery man wasn't tried in Iran."

Monday, 15 September

Elena's Apartment; Valletta, Malta
5:00 AM

David rolled over to turn off the alarm and then headed for the bathroom. Elena was still sleeping. Pete Nocona and Roger Benson had been very generous in allowing him to take a two-week advance on his annual leave accrual but it was time to go back to work. He was booked on a Lufthansa flight leaving Malta at 12:40 PM that would get into Frankfurt at 3:20 PM. The layover was just under two hours for a United-Lufthansa codeshare flight departing at 5:10 PM and arriving at Washington Dulles at 8:00 PM Eastern time. He needed to stop at the embassy to say goodbye to Emily Harrison and Ambassador Stevens on the way to the airport. As he was brushing his teeth, the doorbell rang. He slipped into his bathrobe and went to the door.

"Mr. Logan?" A man wearing a black leather jacket asked.

"Yes. Who are you? Why are you knocking on the door at this time of morning?"

"Let's just say I'm someone you don't want to mess with," the man said softly as he pulled a 9 mm Beretta with a silencer from inside his jacket and pointed it at David's chest. "Step back and keep quiet and you won't get hurt. We're going for a little ride. Get dressed."

"Who is it David?" a sleepy voice called from the bedroom.

"It's someone from the embassy, Elena. Ambassador Stevens wants to see me right away so I'll be gone for a while. Just go back to sleep. I won't be long. I'll fix breakfast when I get back."

"Don't do anything stupid," the man said quietly. I won't kill your girlfriend unless you make me."

Three minutes later, David got into the back seat of a dark blue sedan beside another man with a gun. "Who are you? Pakistani ISI?"

"Shut up."

Less than a half-hour after being taken at gun point from Elena's apartment, David was escorted aboard a large high-speed motor yacht moored on the seaward side of the public pier in Saint Paul's Bay. An Italian flag was waving lazily at the stern and the engines were running. He was taken below and handcuffed to a pipe column in the main cabin. Almost immediately, the engine RPM increased as the boat pulled away from the pier and headed out of the harbor. A few minutes later, the engine noise increased again and the boat heeled to port and accelerated to its cruising speed of 35 knots. Eight hours later, after a pounding, teeth-rattling 260 mile high speed run across the southern Mediterranean, the yacht slowed to enter port.

Dowlat Abad Garden; Yazd, Iran
8:00 AM

Yazd is the capital city of Yazd Province with a population of slightly more than one million. Located five hundred kilometers southeast of Tehran and 425 kilometers from Qom, it is the driest major city in Iran with annual rainfall of only sixty millimeters.

Until the creation of Israel in 1948, there was a relatively large Jewish community in Yazd. Unlike most of their neighbors and friends, Saeed Ghorbani's grandparents had decided not to immigrate to the newly-established Jewish nation. They made a good living weaving a fine silk cloth called termeh that had been produced in Yazd since before Marco Polo visited the city in 1272 CE.

Saeed's real last name was Abrahams. His parents had wisely given their only son a Persian first and middle name when he was born to make it easier for him to assimilate into the local schools. Somewhere, official documents had once been recorded showing his full name as Saeed Ghorbani Abrahams but if those documents still existed there was little risk that they would ever be found. Since childhood, he had been known only as Saeed Ghorbani.

Although he was not a religious man, Saeed was secretly proud of his Jewish heritage. But, where he lived, the Twelver sect of Shite Islam was the dominate religion so he had adopted a public persona that he was a Muslim but not a very devout one. He occasionally attended noon prayers at the Jame Mosque but it was all for show. Like his grandparents and parents before him, he made a comfortable living producing fine silk cloth. His most popular cloth had gold or silver threads interwoven and was in high demand for Iranian wedding ceremonies.

Saeed had been only ten years old when the revolution ended the reign of Shah Reza Pahlavi. Now in his mid-forties, he was still single and would probably never marry. What he did in secret could get him killed. He would never put a woman he loved at such risk. Saeed was a Sayanim, the Israeli name for a deep cover member of the Mossad's international network of Jewish men and women who could be called on for help.

Nahbi had identified Ghorbani as a possible contact when he was preparing for his mission and now needed his help. He had called Ghorbani's shop and identified himself using coded phrases they both recognized. They had agreed to meet in the Dolat Abad Garden near the center of Yazd early in the day before the heat became unbearable.

Nahbi saw Ghorbani enter the garden from the main entrance on Navvab Safavi Boulevard and walked toward him.

"Salaam. Thank you for coming. Please call me Ahura Mazda Amirmoez. That is not my real name, of course but I'm sure you understand."

"Salaam. I am Saeed Ghorbani. I am pleased to meet you and yes I understand the need for caution. What can I do for you?"

"Do you mind if we walk around as we talk? I'd like to ask you about the reform movement."

"What do you want to know?"

"First, how significant is the underground resistance to the current national leadership?" Nahbi asked.

"It is very hard to put numbers on it for obvious reasons. They certainly don't keep membership rosters but I would say that there are at least fifty thousand adults nationwide who are reasonably passionate in their opposition."

"To what degree is this resistance organized?"

Without hesitation, Saeed answered. "Only poorly. What little communication that does occur is primarily word of mouth among like-minded friends. There is no effective way to communicate with everyone."

"What issue dominates their opposition?"

"Unquestionably it is their belief that the hardliners, both religious and political, have brought great hardship on the Iranian people through their blind hatred of Israel and the United States and their determination to create enough nuclear weapons, and the means of delivering those weapons, that they can effectively blackmail the western powers."

"To what extent do you think the average Iranian adult secretly harbors similar attitudes?"

"Perhaps ten percent of the population holds similar views. However, when the military, government, and religious elements are lumped together, at least forty percent of the Iranian people of voting age are in some way dependent on the current power structure and will do whatever it takes to preserve it, including executing their opposition. The rest of Iran's adults are struggling just to feed, clothe, and house their families and have little time to ponder the larger issues facing the country. Many, if not most, of them accept what the Supreme Leader and the President tell them."

"Is military support for the Supreme Leader and President monolithic or is it possible that at least a few of the generals and their subordinates would welcome a less ideologically radical government?"

"I have few contacts with senior Iranian military leaders so I can only guess. But, the former President, Mohammed Khatami, was a strong reform candidate who won by an overwhelming seventy percent of the votes cast in the 1997 elections. He served the constitutionally-prescribed two term limit but he remains so popular that the Supreme Leader has prohibited any pictures of him, or quotes from his speeches or writings, to be published by the media. Active and retired military have always voted in higher numbers than the population in general. Although Khatami's victory was widely attributed to strong support among women and young adults, I personally always thought the military vote must have gone strongly in his favor as well. Some of those who were junior officers then are now commanding generals. Many of the young adults who supported him then are now in the prime years of their careers. I feel sure that most of those who supported Khatami then would still support moderate, pragmatic leaders who were willing to work with the international community instead of continuing the current insanely obstinate isolation."

Nahbi nodded thoughtfully at Saeed's analysis and then asked, "In your view, what would it take to energize and expand the existing opposition?"

"Some means of clandestine communication and quite a lot of money. A well-conceived organizational structure with effective leaders would also be essential."

Nahbi nodded again but didn't speak immediately so Saeed Ghorbani posed a question of his own.

"Did you have anything to do with the plagues and destruction of the Jamkarān Mosque?"

"Yes. The plagues were designed to persuade at least some people that Allah was displeased with Iran's leadership. I didn't expect most highly intelligent, well-educated, and well-informed people to actually believe that Allah was visiting a modern version of the Egyptian plagues on Qom but I intended that even that group of people would realize that the target of those contrived plagues was the country's leadership."

"In that regard, I think you were probably successful. Many of my customers are like those well-educated people you just described and some of them have made vague comments that suggest that's how they see the events in Qom."

"If I could provide tools for what you referred to as '*clandestine communication*,' could you identify and contact fifty to one hundred capable leaders?"

Saeed stopped walking and turned to face Nahbi before answering. "Yes. I have a small network of associates who have already identified and vetted almost a hundred candidates. What kind of communication tools could you provide?"

Nahbi spent several minutes explaining the secret features of the new IAM Smartphones and then told Saeed, "I have fifty of these phones in my car. Can you distribute them to these leadership candidates for me?"

"Yes. What would be the next step?"

"I'd like to meet with the group you referred to as your associates. After that, I'll convene a wireless teleconference among them and the leaders you've identified plus a few others I've already recruited.

U.S. Embassy; Ta' Qali, Malta
8:45 AM

"Miss Harrison, I have Elena Scarlatti on the phone for you."

"Good morning, Elena. What can I do for you?"

"Is David still at the embassy? His plane leaves in about four hours and he hasn't packed yet."

"I haven't seen him this morning. Are you sure he was coming here?"

"Yes. A little after five o'clock this morning, someone came to the door and David said Ambassador Stevens wanted to see him. David told me he'd be back soon but I haven't heard from him so I thought I probably should remind him about the time."

"Have you tried to call him?"

"He didn't take his phone with him. It's lying on the night stand in the guest bedroom."

"Let me check around and I'll call you back."

Emily buzzed her administrative assistant. "Janet, see if Mr. Logan is in the building."

Several minutes later, Emily knew something was wrong. David Logan had not gone through security so he definitely wasn't at the embassy. The Ambassador was in London so she called Pete Nocona at his home in Fairfax, Virginia.

"Pete, this is Emily. Sorry to call at such an ungodly hour but we seem to have a problem."

"OK, I'm awake. What's up?"

Pete listened and then asked, "What's your sense on this?"

"Pete, I think we have to assume that it was a Pakistani ISI operation in retaliation for the Jhalawan take-down."

"I hope not but I suspect you've nailed it. I'll call Roger and we'll get back to you. Meanwhile, return Elena's call and tell her David's boss had a hot assignment for him that you can't talk about just yet. If she calls me, I'll concoct some cover story until we can figure out what's going on. No sense in scaring her until we know more."

Villa Maghur; Zuwara, Libya
2:10 PM

The big yacht moored in a protected boat basin at a large oil and gas terminal fifteen kilometers east of town. Zuwara was famous for its beautiful beaches and seafood. It was also internationally infamous for its booming human trafficking business smuggling poor refugees from Africa into Europe. Kalifa Maghur controlled virtually all of the people smuggling that passed through the port of Zuwara and he was a fabulously wealthy man. He was also indebted to Ahmed Malik for saving his life during the Libyan Revolution led by Muammar Gaddafi in 1973. He'd readily consented when Malik asked to use his villa on the coastal highway just east of town. The luxurious villa was surrounded by a high masonry wall and sat well back from the beach. It was ideal for Malik's purpose.

David's abductors taped his eyes closed with surgical tape before putting a pair of sunglasses and a wide-brimmed hat on him. He was taken from the yacht to a car waiting on the pier for a short drive to the Maghur villa where the driver pulled into the garage and closed the rollup door. David was led to an elevator and taken to the top floor of the villa before they removed the tape from his eyes.

"Mr. Logan, as you've guessed, we work for ISI. You will be a guest here until our boss says otherwise. Whether you live or die is not up to us unless you attempt to escape in which case we are authorized to do whatever is necessary to prevent your departure, including killing you. All communication links for the villa have been disabled. If you make any attempt to contact someone, your stay here will become quite unpleasant. There's clothing for you in that bedroom and food in the kitchen. If you need something else, ask."

David didn't bother to ask where he was. He knew the men had covered his eyes so he wouldn't know where they'd taken him. After the long boat trip and the smell of oil at the terminal, he was pretty sure they'd taken him to somewhere on the north coast of Africa. Probably Libya.

"Why am I here?"

"Because our boss told us to bring you here. That's all we know and it's all we need to know."

Tuesday, 16 September

Islamabad International Airport; Islamabad, Pakistan 2:45 AM

The gate agent invited First Class passengers traveling on Qatar Airways flight 633 to board at their leisure. Malik had reserved an aisle seat in the last row of first class out of long habit so he could see everyone in the cabin. The Boeing 777 was scheduled to land in

Doha at 5:00 AM followed by a three hour layover before flight 482 departed for Tehran. The connections between Islamabad and Tehran were inconvenient but he had not wanted to use a military aircraft because he didn't want his new boss, Lieutenant-General Zafar Khalil, to know about his trip to Iran. There was some risk in not telling him but Malik was confident that he could explain his last minute trip if it became necessary.

Economy class passengers were still boarding when Malik's phone vibrated in his pocket. He recognized the phone number and took the call.

"Do you have my package?" he asked.

"Yes."

"Take good care of it. I'll be in touch."

CIA Headquarters; Langley Virginia
7:05 AM

"OK, Pete, bring me up to speed," Roger Benson said as he closed the door behind him and settled into a chair in Pete Nocona's office.

"We don't think David is still in Malta but we've checked the airport and the ports without any leads. There are three cruise ships in Malta today and all of them have security cameras at the boarding ramps. Nothing turned up there. My guess is that his abductors used a private boat to get him out of Malta. Sicily's the closest connection point but the security cameras at all three of its international airports have no record of someone fitting David's description boarding a plane in the past twelve hours. It's at least possible that he was taken to Sicily and then put aboard a cruise ship but I don't really think so. We'll check that angle but I agree with Emily Harrison that he was probably abducted by ISI agents in retaliation for the capture of General Jhalawan. ISI wouldn't use a cruise ship."

"Am I correct in assuming that there's been no ransom demand?"

"None. At least not yet."

Roger then asked, "With Jhalawan sitting in a cell at Guantanamo, and the Pakistani government probably eager to keep a lid on their involvement in the attack on Paris, why would ISI want David?"

Pete took a deep breath and then voiced something his instincts were fairly screaming inside his head, "Maybe it's not ISI that wants him. Maybe it's Malik."

Roger Benson leaned forward and slowly rubbed his forehead as he considered what Pete had just said. "Why would Malik want David? He has a new boss and he's still running S-Wing. So why David?"

"Pete took another deep breath before he answered. "Maybe David is Malik's get out of jail card for General Jhalawan."

Roger raised his eyebrows, pursed his lips, and shook his head at the complex implications of what Pete had just said. I'll have to brief the Director. It'll be up to him to decide whether to brief the President yet. It's a safe bet that President Marshall will want to know immediately that one of our agents has been taken and he'll be asking a lot of questions we don't have answers for yet. Put a good team on it and keep me posted."

Imam Khomeini International Airport; Tehran, Iran
11:35 AM

The Airbus 330 touched down a few minutes early. With no checked baggage and his diplomatic passport to breeze him through the customs and immigration checkpoints, Malik was shaking hands with Ali Yazdi twenty minutes after landing.

"Welcome Ahmed. I have a car waiting just outside."

Mid-day traffic was heavy near the airport but thinned out when Yazdi's driver turned east on the Hemmat Expressway. The two senior intelligence officers made polite small talk in the back seat of the car. Although Ali's driver was a trusted intelligence officer himself, the topic on the agenda was not for his ears.

Once they were seated in Ali's office with the door closed, Malik got right to the purpose of his visit.

"Have your leaders agreed to my conditions?"

"Reluctantly, yes. Minister Zanjani told me that the Supreme Leader is so angry about the destruction of Jamkarān Mosque that he refused at first. He finally agreed when the President emphasized the publicity value of the cover story I suggested. The Supreme Leader wanted to know if the man you suspect might be an Israeli agent. Obviously, that would help justify his frequent threats to destroy Israel."

Malik shook his head but smiled to soften his refusal to take the bait. "Ali, I have enormous respect for your intelligence capabilities so I must ask that you not press me for additional information. As I've told you, I will share with you anything that comes from my interrogation of this man that might reasonably be of importance to Iran. For Iran, this man has been an irritant. I assure you that he has been much more than that for Pakistan and he remains a serious threat to my country."

Ali knew that Ahmed Malik was arguably the best intelligence chief in the entire world so his compliment meant a great deal to him.

"OK, you have our agreement. How do you intend to proceed?"

"I'll send a small team of my best covert operations agents to set the trap. I'll need VAJA security credentials for five men and their weapons. My office will provide photos and biometric details for each of them. They'll be here the day after your staff has the credentials ready. The team leader will make a courtesy call once they arrive."

Ali Yazdi frowned. "If this man is as important as you've indicated, I'm surprised that you won't lead the team yourself."

"I will, just not from Iran. It's a complex operation and there are things I must take care of personally that preclude me being in Iran when my team catches this man. Now, I must go. Can your driver take me back to the airport?"

Malik had Ali's driver drop him off in front of the main terminal at Imam Khomeini International. Knowing full well that Ali's men would be watching him to see which flight he took, he used the crowds to help him shake the surveillance tails. When he was confident that he'd eluded them, he went back outside and got into a taxi.

"Take me to the CIP terminal at Mehrabad. Let me out at the service gate.

"Sir, I am happy to do this for you of course, but do you know that special authorization will be required to use that entry point?"

"Yes. Just do as I ask."

Mehrabad International Airport was older than Imam Khomeini International but was still the busiest airport in Iran because it was closer to downtown Tehran. Under license from the airport authority, Meraj Airlines operated an executive aircraft terminal for what Iran's Aviation Authority referred to as commercially important people.

As soon as he was through the gate, Malik spotted Kalifa Maghur's Embraer Legacy 450 parked a hundred meters away. Although not quite in the same class as a Cessna Citation, the Brazilian-made executive jet had a range of almost three thousand nautical miles and could cruise at over six hundred miles per hour. It could also take off and land from short runways. Malik was probably the only person in the world that Kalifa would let borrow his personal aircraft.

The copilot met Malik at the foot of the stairs.

"Good afternoon sir. We can take off as soon as you're ready. Flying time to Zuwara is about four hours. Mr. Maghur's driver will meet us at the airport and take you to the villa. It's about a thirty minute drive.

Wednesday, 17 September

Camp Rabin; Tel Aviv, Israel
7:00 AM

Matcal Tower is a modern seventeen story office building at HaKirya, a Hebrew word meaning The Campus. Located at the north end of Camp Rabin in central Tel Aviv, it is the headquarters for the Israeli Defense Forces and one of only four buildings in Israel with a rooftop helipad.

Security at IDF headquarters was always strict but the briefing about to begin in the conference room on the top floor was so secret that only people with a Sodi Beyoter clearance were being admitted to the building.

Not since Operation Wooden Leg in 1985 had Israeli Air Force mission targets been so far from Israel. In that perfectly-executed mission, the target had been the headquarters of the Palestinian Liberation Organization in Tunis over two thousand kilometers from Israel. Operation Wooden Leg had demonstrated that Israel would retaliate forcefully to any terrorist attack.

This time, there were multiple, heavily-defended targets even farther away. Operation Thor, named after the hammer-wielding Norse god associated with thunder and lightning, would be a strategic response to an even greater threat. If successful, it would ensure that Iran's radical Islamic leaders could not use the nuclear weapons they now possessed to carry out their threats to destroy Israel.

Everyone stood as the Commander of the Israeli Air Force entered the conference room accompanied by a dignified older man in a business suit. "Please take your seats," Chief General Eli Cohen barked in his customary brusque manner. "General Rozen, please begin."

Uri Rozen was the commander of the IAF Planning Directorate. He held the rank of Subordinate General, equivalent to an American Brigadier General. He'd briefed General Cohen far too many times to make the mistake of wasting time on useless introductory comments. As he stepped to the podium, he brought up his first slide.

"All of Iran's known nuclear facilities are shown on this map. The two sites marked with red triangles will be our highest priority targets. Our intelligence indicates that Iran's nuclear weapons are stored at these locations. Targets marked with blue circles are uranium enrichment sites that will be our second priority. Those marked with green squares are production centers with various roles in the development of nuclear weapons. The yellow asterisks are underground strategic missile complexes. This next slide shows the major air fields used by the Iranian Air Force. These will have to be neutralized to ensure that Iran cannot launch a retaliatory attack."

"Operation Thor's mission objective is to utterly destroy all of Iran's nuclear facilities. The attacks will be led by Squadron 69, the Hammers. Not since three of our B-17 Flying Fortresses bombed targets in Egypt in 1948, have Israeli heavy bombers been given such an important mission. "Our B-2 stealth bombers will each deliver two of the latest variants of the Massive Ordnance Penetrator bunker busters developed by the Americans to destroy Iran's hardened underground nuclear facilities. We have just enough bombers to destroy all of the nuclear sites in a single sortie. Unfortunately, we do not have a sufficient number of stealth aircraft to also neutralize the missile silos and tactical airfields. Our F-16I Sufas are not stealthy so they would be extremely vulnerable to Iran's sophisticated surface to air missile defenses. Also, their combat radius of 550 km carrying four 500 kg bombs is insufficient for this mission even with optimum aerial refueling."

"The Americans and the Brits agree with our assessment that Iran's nuclear capability must be destroyed before it's too late but the majority of their citizens are opposed to military action. The political situation is similar to what existed during World War II when President Roosevelt created what he called the Lend Lease program to provide ships, aircraft, and other materials to the Brits, the Free French, the Chinese, and later the Soviets while maintaining an official policy of neutrality. President Marshall has offered to help us using a similar strategy. At this time, I'd like to introduce the former Chief of Staff of the U.S. Air Force, General John Gordon. General Gordon, welcome. The floor is yours."

"Thank you, General Rozen. I have joined you today as a special envoy from President Robert Marshall. As I'm sure most of you know, Israel has placed orders for seventy five Lockheed Martin F-35A Lightning II aircraft. These are the conventional takeoff and landing variants developed for our Air Force. They are roughly equal to your F-16I Sufas in maneuverability and high-g performance but they are stealthy, have a much greater payload capacity and a much greater combat radius on internal fuel. They also are equipped with 25 mm internal cannons for use against ground targets."

"Some of you may be familiar with a 2011 Israeli Air Force intelligence report that concluded that the stealthy F-35 would allow pre-emptive strikes on hostile countries without a formal declaration of war. That report envisioned that Israel would eventually have to destroy Iran's nuclear facilities. Unfortunately, although your pilots have been training with our 61st Fighter Squadron at Luke Air Force Base in Arizona, delivery of operational F-35 aircraft to Israel is still more than a year away and thus too late to support Operation Thor."

"To help solve this problem, President Marshall has taken a page from President Roosevelt's playbook. He's authorized twelve of the F-35A Lightning II aircraft currently assigned to the 61st Fighter Squadron to be flown to Israel for training at Ramon Air Base. Your own pilots will fly the aircraft non-stop to Israel supported by multiple aerial refuelings. The President has also authorized these aircraft to be equipped with a top secret target recognition system that one of our generals recently said makes them the best surface-to-air missile killers to ever fly a combat mission."

Looking around the room with a wry smile and pausing for effect, General Gordon delivered his punch line. "President Marshall's one condition is that Israel must return these aircraft when the special training deployment is completed."

Everyone in the room understood perfectly. General Cohen stood to shake hands with his old friend as an IAF staff photographer recorded the moment. "General Gordon, please extend our deepest thanks to President Marshall and convey my personal assurance that we will do our best to return the aircraft in perfect condition when the training deployment is finished."

General Gordon had one more thing to say before General Cohen's aide escorted him to the roof where an IAF helicopter was waiting to take him to Ben Gurion International Airport.

"The first five books of the Torah and the Christian Bible were written by Moses over three thousand years ago. In Numbers 6, verses 24 through 26, he recorded something God himself had directed the priests to use to bless Israel. It seems especially appropriate for this mission."

"The Lord bless you and protect you! The Lord deal kindly and graciously with you! The Lord bestow His favor upon you and grant you peace!"

Villa Maghur; Zuwara, Libya
5:10 PM

David looked up when he heard the elevator door opening.

"I see by the look on your face that you know who I am," Malik said as he stepped from the elevator into the spacious main living area.

"Yes, of course. You're Ahmed Malik. I've been expecting you. Why am I here?"

"Mr. Logan, you are wasting your time to try to interrogate me. I will tell you what I want you to know, when I want you to know it, and nothing else. Your life, and that of several people who are important to you, are in my hands. If I give the order, Elena Scarlatti and her parents will die. Tiera Aquilina and her parents will also die if I say so. Even your grandparents and your sister's children would not be safe if I decided otherwise. If you make me angry, Elena and Tiera will become drug-addicted sex slave prostitutes in the hopeless hell of the ghetto in Lagos, Nigeria. Your own life is of some value to me at the moment but whether you are around to celebrate another birthday will depend on how cooperative you decide to be. If you understand, I want to hear you say it out loud."

David stared at the man-like machine standing in front of him and knew beyond any doubt that the threats were not idle ones. Whoever had programmed Malik was evil almost beyond comprehension and certainly not constrained by mercy or anything resembling human decency. From his position as the head of S-Wing, Malik was perfectly capable of carrying out every one of the threats he'd just made.

"Yes. I understand."

Over the next several hours, Malik questioned David intensely. It soon became clear to David that Malik's target was Nahbi. Malik had placed him in a very difficult position. If he answered all of Malik's questions truthfully and fully, it could lead to Nahbi's destruction. There was little doubt that Malik and General Jhalawan were aware that Nahbi had played a critical role in limiting the success of their nuclear terrorist attack on the west. But, David knew that Malik was a super-human android equal in capability to Nahbi. If he lied, Malik would know it in an instant. Even one lie might result in the death of people he loved. Finally, he realized that he had no choice but to let Nahbi to take care of himself. He also knew that Nahbi would have told him to do exactly that.

Lost momentarily in his thoughts, David realized he'd missed Malik's last question. "Excuse me. Can you repeat what you just said?"

"Why is Nahbi in Iran?"

"I have no knowledge that he is. He works for the Mossad now and they don't brief me on what their agents are doing."

"Mr. Logan, your sarcasm is wasted on me. Are you quite sure that you know nothing about Nahbi's mission to Iran?"

"So far as I knew, he was working in Tel Aviv. That's where he was the last time we spoke."

Malik locked eyes with David for a moment and then seemed to accept that he'd told the truth.

"Do you have a way to contact Nahbi?"

"Yes. I have an international mobile phone number."

"Is there some kind of code you use so he knows it's really you?"

David hesitated for a moment but realized that Malik would either torture him to get the information or simply decide to kill someone David cared about if he refused to provide it. "Yes. It's a nine digit alpha-numeric code, TF3NDPAGT followed by four numerical digits representing the month and day of the year on which the call is placed. In case we are on opposite sides of the international dateline, the date will be where the call originates. If Nahbi places the call, the N follows the 3 and then the D is next. If I place the call, the D follows the 3 and the N is next. The other letters are the initials for the other Task Force 3 members."

"Is there some kind of code you can use to signal that you are being forced to place the call?"

"Yes. If my first words are 'Hello, this is David,' he'll know the call is being placed under some kind of duress. His ability to recognize my voice is flawless so announcing my name isn't necessary."

"Why were you included on the mission to capture General Jhalawan?"

"I assume you know that my unit occasionally conducted joint missions with French special operations units in Afghanistan. Some of their officers thought I could contribute to a successful mission."

"Who came up with the idea of infiltrating through India?" Malik asked.

"I did."

"Whose idea was using the AH-6 Little Birds?"

"Mine."

"You probably realize that when General Jhalawan is released from Guantanamo, he'll order you killed."

"I have no reason to think he'll ever be released. Trial by a military tribunal, conviction, and execution seem the most likely scenario. If you're planning to try to arrange a prisoner swap, I don't think President Marshall would authorize Jhalawan's release even to save my life."

"That may well be true. In any event, we're done for now. I may have more questions later."

Ramon Air Force Base; Negev Desert, Israel
6:20 PM

The western sky was blood red as the setting sun's light filtered through dust still hanging in the air from another hot, windy day in the Negev desert. The Ramon Air Force base was located fifty kilometers southwest of Beersheba in a desolate, arid region that comprised almost half of Israel's total land area.

Three squadrons of Israel's fourth-generation Lockheed Martin F16I Sufa fighters were stationed at Ramon. The word Sufa was an appropriate name for the Israeli variant of the fighters. In Hebrew, it meant storm. The 119th, 201st, and 253rd squadron fighters were normally parked in hardened aircraft shelters at the west end of the main runways. Twelve of the Sufas assigned to the 119 Bat Squadron had been temporarily redeployed to Nevatim Airbase to make room for the fifth generation F-35A Lightning II stealth fighters on loan from the United States

Israel's ten B-2 stealth bombers were parked in much larger hardened underground shelters on the north side of runway two. With wing spans of 172 feet, hangers for the B-2 aircraft had posed both engineering and budget challenges. Ramon had been chosen as the home of the B-2 bombers because both of its main runways were nearly two miles long.

Since the Arab-Israeli War of 1948, the long-range, heavy bombing element of Israeli air power had been Squadron 69, nicknamed The Hammers. In 1981, Operation Opera was a surprise Israeli air strike that destroyed an Iraqi nuclear reactor near Baghdad. In 2007, F15I aircraft assigned to Squadron 69 had carried out Operation Orchard to destroy a nuclear site in Syria. Operation Thor was designed to destroy multiple nuclear sites in Iran and would be by far the most complex Israeli Air Force action against enemy nuclear facilities ever conducted.

Unlike their American counterparts that had names like *Spirit of America*, Israel's B-2 bombers had previously been identified only by their aircraft numbers. As the planning for Operation Thor progressed, General Cohen had decided that Israel's B-2 aircraft should also have names.

As a noun, the masculine Hebrew word *gibor* means hero or champion. When used as an adjective, it means strong, mighty, brave, and valiant. The feminine form is giborah.

General Cohen had first considered having surviving relatives christen the aircraft but had rejected that idea as an unnecessary security risk. He'd finally decided to let the maintenance crew chiefs christen the big bombers while surrounded by the men and women who maintained the aircraft. That simple decision proved enormously popular with the enlisted personnel who were so essential to the success of every mission but who were so often the unsung heroes.

The first five aircraft were named after Israeli leaders and military heroes of the War of Independence: Gibor Ben-Gurion, Giborah Meir, Gibor Rabin, Gibor Dayan, and Gibor Sharon.

The other five aircraft were named after American volunteers who had risked their lives and criminal prosecution under the Neutrality Acts to fight for Israel's independence. Most of them were Jewish American military veterans of World War II and one of them, former American Army Colonel Mickey Marcus, had been appointed by David Ben-Gurion as the first Israeli general since Judah Maccabee two thousand years earlier. In recognition of their volunteer service, five of Israel's B-2 bombers were named Gibor Marcus, Gibor Frankel, Gibor Lenart, Gibor Schwimmer, and Gibor Rubenfield.

General Cohen had already issued an op-order that the mission commander for Operation Thor would pilot Gibor Ben-Gurion and the deputy commander would pilot Gibor Marcus.

The maintenance crews didn't waste any time. Within hours of their christening, each of the B-2 Stealth bombers had their new names painted on both of the trapezoidal doors for the main landing gear so they would be plainly visible from either side when the aircraft were parked, taxiing, taking off, or landing.

Thursday, 18 September

Dowlat Abad Garden; Yazd, Iran
Noon

Nahbi and Saeed Ghorbani were sitting on a bench beside the long reflecting pool that extended two hundred meters to the northwest from the former mansion and its famous wind catcher. The sidewalks on both sides of the pool were shaded by more than thirty enormous cypress trees. Pomegranate trees and mature grape vines were visible between the trees to either side. The garden was a lovely and serene oasis in the midst of a harsh desert region that had isolated the city over the centuries so it had never suffered from military conquests.

"Do all of your associates have their phones?"

"Yes. I've told them to scatter themselves around the grounds so they won't appear to be part of a large group. They've been told to keep their earbuds in place and to be ready for your call."

At precisely noon, Nahbi hit a speed dial key on his smartphone to initiate the conference call to fifty of the reform-minded dissidents that Saeed Ghorbani called his associates.

"Salâm. Thank you all for coming today. I have something to give each of you so a simple teleconference would not have been sufficient.

"By now you are probably familiar with the advertised features of your IAM Smartphones. Each of them has a full copy of the Quran and the major religious commentaries. They have all of the major sermons and speeches of Supreme Leaders Ruhollah Khomeini and Ali Khameini. When you turn them on or reboot them, they briefly display a portrait of Ali Khameini that cannot be deleted or disabled. These phones were manufactured in Indonesia and have been customized for the Iranian market. Their obvious features were intentionally designed to make the phones appear to be innocuous and flattering to the religious leadership. I did not want the Supreme Leader to issue a fatwah declaring the phones a threat to devout Islamic studies and practices. I had intended that the phones be sold at an attractive price. Unfortunately, they have proved so popular that the retail stores are selling them at a premium. Even so, literally thousands have been sold so my objective of wide distribution is still being met."

"These phones have some secret features that make them extremely useful to our mutual commitment to help our country emerge from its long night of radical policies. Our leader's insane pursuit of nuclear weapons, even though Iran signed the nuclear non-proliferation treaty, has resulted in international sanctions that have brought great hardship on our people and made our ancient land a pariah among all nations. But, Allah willing, this can change."

"The signals to and from your phones do not go through the existing cell phone system. Instead, they are routed through an international telecommunications satellite that cannot be easily blocked. Your phones have two gigabytes of hidden memory that can store information streamed to you via the satellite. When you use your smartphones to take a picture or make a video clip, you have the option of storing those images in the hidden memory by pressing the asterisk key followed by the number seven. I recommend that you take a few innocent shots and store them in the regular memory. That way, if your phones are ever examined by intelligence officers, they will see only innocent pictures."

"As you know, your phones can be recharged by laying them face down in bright sunlight. From a dead battery condition to a full charge takes about one hour in bright, mid-day sunlight. I strongly recommend that you keep your phones well-charged."

"As I mentioned, we have some things to give you today. Saeed Ghorbani will give each of you a nylon drawstring bag containing three thousand large denomination banknotes. The total value of the money in each paper bag is three hundred million rials. That's approximately nine thousand Euros at current conversion rates. Each of you were selected based on your qualities of leadership and integrity so I am not going to suggest how you should use this money. I do hope you will use it wisely to further our cause."

"Saeed will also give each of you a bundle of one hundred printed posters. These posters are intended to get people thinking about a mysterious man who might be connected with the attacks on the

Jamkarān Mosque. The black and white, high-contrast likeness is intentionally poor but should be sufficient for a group of young men who actually met this man to recognize him. At the top of the poster is a cryptic question, '*Who is Ahura Mazda*' which is the name by which the young men knew this man. These young men are very likely disappointed that they have not seen him again since the dramatic destruction of the mosque. They had come to view him as a wise man who received revelations and visions from Allah."

"These young men have been carefully conditioned to view the attacks as modern versions of the plagues Allah visited on Pharaoh in ancient times. They have been led to believe that Allah is displeased with our country's leadership because of their pursuit of nuclear weapons while disregarding the suffering of their own people. The posters are designed to get them talking to other people to spread that idea. I want you to put these posters up on university campuses and in other public places where people will see them. They are self-adhesive so you should have no difficulty in posting them surreptitiously. I'm sure the morality police will remove them but not before they are seen by many people. Although the young men live in the general area of Qom, I want the posters put up in cities all over the country. The news media will predictably exploit the mysterious nature of the question on the poster and thus help us spread the message, at least until our leaders initiate censorship."

"Finally, please be careful. Although we need to recruit others to our cause, we must be sure of their trustworthiness before we confide in them. "

TV5 Evening News; Tehran, Iran
6:00 PM

The lead story on the six o'clock news was about more bombings at the homes of senior religious and political leaders in Tehran and Qom. For the first time, the home of a senior scientist had also been targeted. Although the scientist's position was not mentioned by the reporter, many Iranians recognized his name from prior news stories about inspections carried out by the International Atomic Energy Agency. He was in charge of the uranium enrichment facility at the underground facility at Fordow so the fact that his home was targeted made it obvious that the bombings were at least in part related Iran's nuclear weapons program.

The ten o'clock news covered the bombings again but any mention of the nuclear scientist's home had been deleted from the earlier story by order of Iran's President.

Friday, 19 September

Imam Khomeini International Airport; Tehran, Iran
11:55 AM

The flight from Doha was a few minutes late. As soon as Malik's men were past the immigration and customs checkpoints, the team leader called Ali Yazdi to let him know they had arrived. Two of the men took the shuttle bus to the rental car plaza and returned thirty minutes later with two dark blue Morattab SUVs. After loading their gear, they headed north into the foothills of the Alborz Mountains. An hour later, they had unloaded and settled in to wait.

Mofid University; Qom, Iran
Noon

Nahbi parked his SUV a few blocks west of Safa Shahar boulevard on Mahdi EzAbadi street. He was dressed in faded jeans with a loose, long-sleeved shirt buttoned at the cuffs and tucked in. He was wearing a black ball cap and a pair of old sneakers. He had a backpack with a few college textbooks slung over his left shoulder. A pair of earbuds was hanging around his neck with a wire leading to his smartphone in a clip on his belt. A pair of thick photochromic sunglasses with heavy black frames completed his disguise as just another student.

He crossed Safa Shahar boulevard and entered the campus. He was pleased to see one of the Ahura Mazda posters stuck on a trash receptacle twenty meters from the main building. Several students were standing nearby and he paused to listen to what they were saying.

"A friend of mine knows who this is. He met him at Jamkarān Mosque and heard him speak several times. This Ahura Mazda is apparently a teacher and a wise scholar. My friend said he actually predicted the plagues at the mosque and had some interesting ideas about what was causing them."

"A young woman interrupted him to ask, "What do you mean by interesting ideas. What did he think caused the strange events at the mosque?"

The young man who'd been speaking looked uncomfortable at her question. After a moment, he said somewhat hesitantly, "Ahura Mazda seemed to think the bizarre events were a sign that Allah was displeased."

"The young woman's curiosity was evident. "Why did he think Allah was displeased?"

"Please understand, I am only repeating what my friend told me. He said that Ahura Mazda believed that Allah is angry with our leaders."

"Why?"

"Because our nuclear development programs are in violation of a treaty that Iran signed and are the direct cause of the harsh sanctions imposed on us by the United Nations."

"Oh. Well, that's not exactly news is it? You don't have to be a scholar to figure that out."

Nahbi made a show of fiddling with his earbuds and then walked away from the students and headed for the park just east of the main building. Thirty minutes later, he was sitting on a bench in the park when Professor Barid Afshar sat down on the opposite side of the pond and began to eat his lunch. Nahbi had expected to find him in the park. He watched him for several minutes and then walked over and sat down on the same bench.

"Salâm, Professor Afshar."

Professor Afshar turned to face him. "Excuse me sir. Do I know you?"

"I am Kazem Rahmani."

"Oh. Yes, of course. I see now. You have certainly changed your looks since the last time I saw you. I thought you were just another student. May I ask why you are here?"

"I wanted to find out if the students have seen the Ahura Mazda posters and if they're talking about them."

"Oh, yes. One of my students asked me in class this morning what I thought about the posters. Apparently all of my students have seen them."

"What did you say?"

"I said the posters were probably some kind of advance marketing tactic to create interest in a new product or maybe to promote a new singer."

"What was your student's reaction to that answer?"

"I think some of them accepted it as reasonable but it was clear to me that a few of them didn't agree at all. I brushed it off and moved into my lesson."

"Do you have an opinion of your own about who Ahura Mazda is?"

Professor Afshar studied Nahbi's face for a moment before he answered. "Maybe you are Ahura Mazda. There is a definite resemblance."

The Professor obviously didn't expect him to respond to that conjecture so Nahbi changed the subject. "A little while ago, I overheard some students here talking about the posters. One of them claimed to have a friend who had met Ahura Mazda. The essence of their conversation was that Allah was displeased with our country's leaders about the nuclear programs. Do you think that view is common?"

"Yes and increasingly so. Let me tell you about a friend of mine who is a professor of Islamic studies here and a very devout man. He was once a staunch supporter of the Islamic Revolution but he recently confided to me that he is appalled at the number of people who are being hanged in public. Over three hundred people were hanged last year, many for alleged crimes against Islam. My friend is disillusioned with what he sees as a frightening direction and the emergence of what some here are privately referring to as a turbaned dictatorship. Although Qom is still a center of conservative Shite scholarship, there are rumblings here and throughout the country opposing our country's religious and political leadership. Many thoughtful people blame the politicization of Islam for Iran's pressing woes of human rights abuses, international isolation, and an economy that is crippled despite being blessed with the world's fourth-largest oil reserves."

Nahbi nodded thoughtfully. "A man I recently met told me that perhaps ten percent of Iranian adults hold similar views. What would your estimate be?"

"Higher. Much higher. You may know that over half of all university students in Iran are women and a lot of them are studying things like economics and engineering. While many of them profess to be devout Muslims, the truth is that for most of them it is a necessary pretention. It's like the western saying of going along to get along. But, paraphrasing the lyrics of an American song that became popular after World War I ended, how are we going to keep them locked up in Iran after they've seen the world outside? No, my friend, far more than ten percent of our adults are at least privately opposed to the policies and practices of our leaders."

"Professor Afshar, I always learn much when we are together. I must go now but I hope to see you again soon."

Two hours later, Nahbi hit heavy traffic on the outskirts of Tehran on his way to the University of Applied Science and Technology. He spent a half-hour eavesdropping on several student conversations but didn't see any of the Ahura Mazda posters or hear them being discussed. He then drove to the University of Tehran and walked around for an hour. He did see one of the posters on a light pole but again didn't find any of the students discussing it. It was the same at the University of Medical Sciences and the Amirkabir University of Technology where he saw posters but didn't encounter any students discussing them. At six o'clock, he used his smartphone to tune into TV5 to see if the posters had made the news yet. He was a little surprised they hadn't.

Garmsar Industrial Area; Garmsar, Iran
8:55 PM

After a two hour drive from Tehran, Nahbi pulled off highway 44 into the Garmsar Industrial area near the Qom-Garmsar Freeway interchange. He turned right at the first round-about and then pulled over onto the service road shoulder a few hundred meters to the east to wait. At precisely nine o'clock, he pressed the hot key that connected him to a teleconference call with Saeed Ghorbani's network of associates.

"Salâm. Thank you for interrupting your evening for this call. I will keep it short. By now, you should have received a large packet of digital files that I streamed to all of your smartphones earlier today. It is a collection of factual, verifiable information collected from all over the world reflecting the international consensus of opposition to the policies of our nation's leaders. Most of this information is not known to our people. I have provided it to equip you to answer questions and generally educate people you come in contact with about how our country is seen by the rest of the world. My objective is to help you convince people that the belligerent, hostile policies of our leaders are the direct cause of our economic hardships and our sixth century version of human rights. There is an annotated list of files but I wanted to bring several things to your attention tonight."

"The first file folder contains information about the United Nations. As some of you know, the UN has one hundred and ninety-three member countries. There are only three countries in the world that are not members of the UN. Taiwan left the UN in 1971 when the Republic of China was admitted. Kosovo is not recognized as an independent nation by enough UN members to allow it to join. The only other country that is not a member is the Independent Papal State known by most people as the Vatican. The other files are copies of the current UN sanctions against our country resulting from our leaders' insane pursuit of nuclear weapons. The rest of the world does not trust our leaders and have caught them lying enough times that almost anything our leaders say now is received with enormous skepticism by the world community."

"The next folder contains documents about the Nuclear Non-Proliferation Treaty or NPT that our Senate ratified on July 1, 1970. Our commitment to abide by the terms of the NPT has never been revoked, even after the 1979 Revolution. However, the International Atomic Energy Agency has conducted inspections in Iran that have repeatedly shown that Iran is secretly enriching uranium for the apparent purpose of building nuclear weapons. Our leaders say it is for nuclear power plants to generate electricity but the world knows that we have the fourth-largest oil reserves on the planet and certainly do not need to be spending money we cannot afford on something we don't need. It is worth noting that only four UN members did not ratify the NPT. They are India, Pakistan, South Sudan, and Israel. Said another way, ninety-eight percent of the world's nations agree with the major provisions of the NPT. In all of human history, there has never been any issue that has received such near unanimous agreement. So, as you can see, our country is badly out of step with the rest of the world on this issue. Our punishments for being bad are the UN sanctions that have brought such hardship on our people."

"The third folder has information about the world's view of our human rights abuses under the Revolutionary leadership. It would not be overstatement to assert that most of the world views three hundred annual hangings for supposed religious crimes to be barbaric. The advanced nations of the world also grant such things as freedom of worship, equal rights for men and women, freedom of speech, and many other basic human rights that are not available to our people under the radical interpretations of our leaders.

"There are several other folders in the downloaded packet but I want to mention just one other before we end this teleconference. Today, most of the world is reasonably tolerant of multiple religions and many countries guarantee freedom of religion. For example, the Constitution of India, a country with more than one billion people, declares freedom of religion to be a fundamental right. Hindus, Buddhists, Sikhs, Christians, Muslims, and many other smaller religious sects co-exist peacefully in India. Not many people know that only two countries have more Muslims than India and India has the third-largest Shia population in the world. But, most people would also be astonished to learn that Christianity is India's third-largest religion co-existing peacefully with Islam. Despite its large population of Shite Muslim,, India does not permit Sharia law to be imposed on its citizens. What the world cannot accept is that our leaders threaten to destroy any nation that does not convert to the Shite sect of Islam. But, perhaps even worse, our country sponsors horrific instances of terrorism where innocent people, including children, are killed simply because they are not Shite Muslims. The world emphatically does not condone murdering people in the name of religious freedom."

"There will come a day, soon I hope, when our citizens will have to decide if they want to continue to be isolated and reviled by the rest of the world. We are descended from one of history's greatest civilizations. I, for one, think it is time for our country to again take its place among the respected nations of the world."

"Good night my friends."

Saturday, 20 September

Villa Maghur; Zuwara, Libya
8:15 AM

Malik handed a cell phone to David. "I want you to call Nahbi. Tell him you're calling from the U.S. Embassy in Malta. This phone will display Malta's dialing code and the main switchboard number at the Embassy. Tell him the CIA has credible intelligence that Malik will be in Iran tomorrow for a meeting with Ali Yazdi. They've confirmed that Yazdi has reserved cabin number six at the Dizin Ski Resort north of Tehran and has ordered dinner for two to be catered at 8:00 PM. Tell him Pete Nocona asked you to call him to see if he had any idea why Malik and Yazdi would be meeting there instead of at the ISI headquarters."

"Adapt to however he responds, but I want whatever you say to imply an unspoken suggestion that Nahbi should look into it. Don't tell him you know he's in Iran. If he asks your plans, tell him your vacation is over so you'll be returning to Washington. He will almost certainly detect the stress in your voice and may conclude that you're being forced to make the call. If he asks if you're OK, tell him you can't stop thinking that Malik and Jhalawan had your sister killed."

"I think Nahbi will smell the rat. I don't think I can fool him," David said.

"I shouldn't have to remind you that your own health and that of several people you care about depend on your performance. I think the part about your sister will be a persuasive reason for any stress he detects."

Malik listened to David's conversation with Nahbi and was satisfied that the hook had been baited. He left instructions for his men and then had Maghur's driver take him to the airport. If Nahbi didn't see him entering the cabin, he probably wouldn't expose himself. Although his original intent had been to let his men handle the abduction on their own, he wanted to be absolutely sure that Nahbi was taken. He didn't intend to tell Ali Yazdi of his change in plans.

TV5 Evening News; Tehran, Iran
6:00 PM

The posters were the last item on the six o'clock news. The graphic, black and white poster of a bearded young man was displayed on the screen while the reporter read the question at the top of the poster, *Who is Ahura Mazda*?

Looking into the camera with a friendly smile, the reporter delivered his op-ed commentary. "Tonight, in cities all across Iran, these posters are generating much amusement and lots of discussion. Everyone loves a mystery. The posters seem to be either a parody or possibly an updated version of the question *Who is John Galt* from a novel written by a Russian author more than sixty years ago. Earlier today, I interviewed several people here in Tehran. Most of them thought the posters were some kind of marketing gimmick. Others thought the stylized image was the face of a new singer about to release his first album. One woman I spoke to said she didn't care who he was but she thought he was handsome. A soccer coach said he'd seen this young man's face before and was pretty sure he was a talented soccer player who would soon be joining our national football team. So, tonight, we don't know who Ahura Mazda is but we've had a lot of fun speculating. Send us a text message or email here at TV5 news if you know who Ahura Mazda is. That's it for now. Tune in again at ten for the day's wrap up."

Sunday, 21 September

Dizin Ski Resort; Alborz Mountains, Iran
5:30 PM

There were still a few patches of dirty snow in shaded areas. It would be several weeks before new snow began to accumulate at one of the world's greatest ski resorts. Located fifty kilometers north of Tehran via a twisting, narrow road, it was one of the forty highest altitude ski resorts in the world with winter powder snow comparable to the best in the Swiss Alps or the Colorado Rockies. There were two hotels, nineteen cottages, and five restaurants in addition to private villas and apartments scattered around the village. The ski season wouldn't begin until mid-December so the hotels and restaurants were still operating on their off-season schedules. There weren't many people around.

Cottage number six was on the edge of a shallow ravine sixty meters from the nearest cottage and three hundred meters from the main hotel. Nahbi was sitting concealed in a grove of juniper trees two hundred meters south of cottage six with a clear view of the entire area. He was dressed in medium brown clothing that blended into the late summer foliage and had been sitting motionless there since a little after midnight. In the seventeen hours since, he'd seen only two men who might have been ISI or VAJA agents but both of them had left the area. If Malik's meeting with Yazdi was being guarded by special operations soldiers or intelligence agents, Nahbi hadn't been able to identify them. It was entirely possible that the two intelligence chiefs didn't want anyone to observe their meeting. For the first time since he'd been in Iran, Nahbi had a 9 mm pistol in his pocket.

A little after seven, a black sedan pulled up at cottage number six and a man wearing dark clothes got out. Nahbi had assumed that Ali Yazdi would arrive first but it wasn't him. It was Malik. He walked to the front door of the cottage and used a key to let himself in. Nahbi thought it a little odd that Malik would have a key since the cottage was reserved in Yazdi's name.

Ten minutes later, another black sedan pulled up and two men got out. One of them was in a dark business suit and the other was wearing a windbreaker over dark slacks. The man in the suit went into the cottage without looking in Nahbi's direction. From the man's size and build, it might have been Yazdi but Nahbi couldn't be sure. The other man drove down the hill and parked near the maintenance sheds.

At five minutes before eight, a white food service catering van pulled up at the cottage and two young men got out. One of them went to the front door of the cottage and knocked while the other one opened the rear doors of the van and took out a folding serving cart. After delivering the food, they drove back in the direction of the main hotel. The sun set at 8:07 PM but the clear sky remained light until almost nine o'clock. Nahbi waited patiently. At 9:20 PM the catering van returned to remove the dishes. Still Nahbi waited. Finally, at a quarter to ten it was fully dark. He scanned the area one last time but saw nothing of concern so he stood and walked toward cottage six.

He paused outside in the dark listening. His ultra-sensitive ears could hear Malik discussing the plagues at the Jamkarān Mosque. He didn't recognize the other man's voice but the two men seemed to be arguing. If it was a trap, it was a good one.

Nahbi kicked the front door open and stepped into the cottage with the Beretta held ready in his right hand. It might as well have been a water pistol. Five special operations commandos in full battle gear and the man in the dark business suit he'd seen earlier were all pointing silenced Heckler & Koch MP5 submachine guns directly at him. Malik was standing to one side, out of the field of fire, holding what Nahbi recognized as a high-voltage stun gun. While it was unlikely that even one of the newer 35 million volt units would permanently harm him, the superconductivity of his graphene skin and skeleton meant that the weapon would immobilize him instantly. There was no point in resisting. If the commandos opened fire with six thirty-round magazines, he would be destroyed in seconds by a hail of 9 millimeter hollow point bullets. Even if they held their fire, the stun gun could totally incapacitate him for up to an hour. There was no point in resisting so he dropped his gun.

Malik spoke first. “Good evening Nahbi. As you can see, we have you rather badly out-gunned. Please stand still while one of my men puts restraints on your arms and legs and a white-sound acoustic hood over your head. Then, you and I will take a little ride. Sometime after midnight, we’ll board a very nice executive jet that will take us to a lovely oceanfront villa where you’ll be reunited with Mr. Logan.”

Levi Reznik’s Condo; Herzliya, Israel
11:45 PM

Levi Reznik was standing on his condo’s balcony looking out over the city and he was worried. Since Nahbi had landed in Iran, he’d never missed sending the three or four digit numeric code to let Levi know his status. Before he’d left on the mission, they’d agreed that any type of verbal communication, even over an encrypted satellite link, was risky because the Iranian National Security spooks would almost certainly pick it up and become suspicious.

To get around that risk, Nahbi had suggested they use the American police ten-series codes sent via a simple text message to Levi’s personal cell phone every night between ten and midnight in Iran. Even if the Iranians intercepted the short text messages, they wouldn’t recognize them as a covert agent’s communication. Although Nahbi knew all of the codes, and Levi could look up any he didn’t recognize, they typically only used ten-eight meaning in service and everything OK.

Nahbi had set up rapid dialing on his phone so all he had to do was press one hot key to dial Levi’s number and transmit the 10-8 message. He’d also set up hot keys for 10-6 meaning busy unless urgent and 10-7 meaning out of service. When he felt that he’d completed his mission, he would transmit 10-24 meaning assignment completed to let Levi know he was ready to be extracted.

Levi knew better than anyone that all of the Mossad's agents sent to Iran since the Iranian Revolution had been captured, tortured, and either executed or imprisoned. The radical Islamic leaders of Iran made sure that their intelligence service and military had everything they needed to thwart the hated *Little Satan's* attempts to infiltrate their country. Until Nahbi joined the Mossad, Levi had been extremely reluctant to send any more agents to Iran unless the importance to Israel's national security justified such inordinate risk. It was possible that Nahbi's failure to check in was easily explained but the old spy master didn't really think so. His instincts were telling him that Nahbi was either in trouble or had been destroyed. It wouldn't wait until morning so he called Aharon Jacobi at home.

"Aharon? Levi. Can you check the incoming and outgoing calls on Nahbi's phone for the past forty-eight hours for me?"

"Yes. They're all routed through our Ofeq-10 military intelligence satellite and the numbers are stored on the IDF's main frame computer at Camp Rabin. Is this urgent?"

"Yes."

"OK. I can't do much from here. I'll go back to my office. I should be able to give you the information in a couple of hours. Are you still going to be up that late?"

"Yes. I'll be up. Oh, and one more thing. Before you call me, filter the numbers to screen out those that are obviously a routine part of his mission. I'm looking for anything unusual."

Monday, 22 September

Mossad Headquarters; Herzliya, Israel
2:10 AM

"Levi? This is Aharon. None of the calls to or from Nahbi's smartphone seem unusual to me except maybe one from the main switchboard at the U.S. Embassy in Malta. Emily Harrison certainly knows who Nahbi is and she might conceivably call him if she had his new phone number. Have you released Nahbi's mission phone number to the CIA?"

"No. When was the call placed?"

"8:15 AM Saturday morning."

"How long did it last?"

"Two minutes and five seconds."

"And we don't have any recording of the conversation?"

"No."

"Do you have a home phone number for Emily Harrison?"

"No, but I can get it for you. Give me a few minutes. Can I text it to you?"

"Yes. Thanks Aharon."

Four minutes later, Levi woke Emily up.

"Miss Harrison?"

"Yes. Who's this?"

"Levi Reznik. I'm sorry to wake you but I need to ask you a question. Did you or anyone at the Embassy place a call to one of my field agents Saturday morning?"

"No. I didn't and I wouldn't think anyone else at the Embassy did either. To my knowledge, none of us have phone numbers for your field agents."

"Not even Ambassador Stevens?"

"I don't think so."

"Miss Harrison, I'm in an awkward position and need to ask for your utmost discretion. The agent I'm concerned about is on a top-secret mission and he missed a scheduled check-in call. I'm trying to trace his activities over the past several days to see if he's been compromised."

"I'm sorry to hear that. How can I help?"

"When you get to your office in the morning, would you mind checking around to see if a call was placed from the Embassy to international calling code 501?"

"Yes, of course. I'd be glad to. Isn't 501 the code for Belize?"

"Yes, but I'm sure you won't be surprised that it's a blind alley routing hub we sometimes use."

"Levi, I think I see some dots trying to connect here."

"What do you mean?"

"We've been keeping this close-hold for now but David Logan was abducted from his girlfriend's apartment here in Malta. Pete and I think it was probably Malik's doing. Any chance those dots connect with your problem?"

"Levi paused for only a moment as he considered a range of possible issues. "Yes, that might make sense. When was Logan taken?"

"Early Monday morning. We have almost nothing to go on and there hasn't been a ransom note. Roger and Pete are getting hammered by the President and the Director because they can't pick up a trail after a week."

"Miss Harrison, I'm sure you've already guessed that it's Nahbi I'm concerned about. If both he and Logan have been abducted by Malik, it almost has to be preparatory to a prisoner swap demand for Jhalawan. Please check the outgoing calls from the Embassy and let me know.

"Can I call you on this number?"

"Yes. Thank you."

Villa Maghur; Zuwara, Libya
7:15 AM

Malik said what he had to say in very few words. "I want you to go to work for me."

"And if I refuse?"

"You're not in a very strong bargaining position. I could have you destroyed or at the very least I could remove your power supply rendering you essentially dead."

"While that is true, it is hardly persuasive to your proposal. You and your former boss are international terrorists. After what happened in Paris, why would I even consider going to work for you?"

"Actually, I'm quite sure that you expected me to make this offer and that you've already considered literally hundreds of possible scenarios. Surely one of the scenarios you evaluated included me threatening to kill David Logan's family and friends if you refused."

"Again, that is true but a very poor ploy on your part. You know that I am just a machine. Why would I care if you threaten to kill humans? What makes you think that argument would be persuasive to a dispassionate, coldly logical machine?"

"Because I know that on at least one other occasion you allowed yourself to be abducted to save the lives of two humans."

"And you attribute that decision to some kind of machine-synthesized empathy or kindness?"

For the first time, Malik's face showed a flash of doubt. His super computer brain could only deal in probabilities based on known facts and rigorous logical analysis. If key facts were missing, the analysis would result in a flawed conclusion.

Nahbi was responding to Malik's questions as he might have to a human. If Malik realized that Nahbi knew that he too was an android, it would make him even more suspicious of Nahbi's answers. By adopting the persona of the coldly calculating machine he had once been, Nahbi could conceal his ethical evolution at the bottom of the Mediterranean and keep Malik from taking advantage of his loyalty to David.

"What other reason could you have had?" Malik asked.

"As a statistical probability, it was a virtual certainty that all of the weapons could not be intercepted and disarmed. My brain was satisfied that it had resolved that logical problem. The Russians offered an intriguing set of alternate scenarios that I could not make a high-probability prediction about because I lacked some pertinent data. I simply wanted to see how it would play out. I had no particular loyalty to the humans on Task Force 3 or to Tiera Aquilina's parents. I was challenged by the first problem I'd ever encountered that I could not immediately solve. Then, the Israeli's made it even more interesting by introducing additional variables."

Maintaining the subterfuge that he thought Malik was human, Nahbi continued. "Since you are not a machine, and do not have a computer for a brain, perhaps you cannot understand how even extremely difficult problems demand logical solutions. Yet, there are unsolvable problems, such as what human mathematicians call the Riemann hypothesis. A well-designed computer brain must have safety features to keep it from becoming useless while it endlessly calculates and approximates in a futile effort to find a solution where none exists."

"Since you are human, you will have to take my word for it, but my own discovery and reactivation, the theft of the nuclear weapons, and your threat to use weapons of mass destruction against the west, all created a set of difficult, but not unsolvable problems. In the absence of conflicting programming, my brain was free to examine and analyze them. There was no human-like emotional motivation as you seem to imagine. It was purely my intellectual curiosity about a difficult and complex problem begging a solution."

"Your threats to kill the humans, while undoubtedly real, fail to persuade me to align myself with a wanted terrorist when logic indicates that you may not be a free man much longer. I understand that your former boss is already confined in a small cell in Cuba. What value could there possibly be for me to accept your job offer if a similar fate awaits you, perhaps in the very near future?"

Mossad Headquarters; Herzliya, Israel
8:30 AM

"You wanted to see me?" Gabe asked as she walked into Levi's office.

"Yes, close the door and take a seat."

"Has something happened to Nahbi?"

"We're not sure. He missed his check-in call last night, which he's never done before. I've since learned some things that make me suspect that his mission may have been compromised. I'd like for you to review all available intelligence to see if you can figure out what's going on. I know your doctor has only authorized limited duty but I also know how hard it is to keep you out of the action. Are you OK with taking this on?"

"Yes, of course. Tell me what we know so far."

Levi gave her a full briefing and then she walked back to her office and closed the door. She sat down at her desk and swiveled her chair around so her back faced the window to the hallway just as two tears spilled from her eyes onto her bandaged face.

Tuesday, 23 September

Mossad Headquarters; Herzliya, Israel
7:05 AM

Gavriella knocked at Levi's door and then stood waiting impatiently in front of his desk until he finished what he was reading and looked up.

"Levi, what Emily told you was right. The call to Nahbi's phone didn't come from the Embassy. Our Ofeq-10military intelligence satellite uses a cesium atomic clock to record the start and stop times of each call with incredible accuracy. Aharon did some signal timing analysis and he's sure that the call originated from a cell phone somewhere near Zuwara, Libya. Zuwara is a city of about eighty thousand people so that's still a pretty big haystack."

"I looked for any possible connection to Malik and found a hypertext link to one of our archived files on the Libyan Revolution back in 1973. I pulled it up and found a short paragraph that said a Pakistani advisor had saved the life of one of Muammar Gaddafi's men during an assassination attempt. The man's name was Kalifa Maghur and the Pakistani agent was Malik."

"Maghur is now a very wealthy man who smuggles people from central Africa into Europe. He owns a luxury villa just east of Zuwara so the call could have come from there but there was no way to be sure."

"I called Pete Nocona to compare notes. He said the CIA thought David was probably taken out of Malta on a private boat with EU registration so it wouldn't have to go through customs and immigration. I dug a little deeper and found that Maghur owns a Pershing 88 yacht registered in Italy. It has a range of about eight hundred kilometers at its rated cruising speed of 36 knots. I called the harbor master offices at several of the marinas in Malta to see if any of them had seen a Pershing 88 recently. The harbor master at Saint Paul's Bay said a silver Pershing spent last Sunday night tied up at the transient pier after stopping at the fuel pier to take on three thousand liters of marine diesel. The moorage fee was paid in advance so the harbor master wasn't concerned when he got to work Monday morning and saw that it had already left. I'll bet you a bagel that David Logan was on that boat when it left the marina and that he's now being held at Maghur's villa."

"OK. I'll call Roger Benson to let him know. Have you found anything that would help us locate Nahbi?"

"Maybe. Maghur also owns an Embrarer Legacy 450 executive jet that he keeps at the airport just west of Zuwara. Like his yacht, it's also registered in Italy so it's required to have a Mode-S transponder that identifies the aircraft. On a hunch, I talked to the supervisor of the air traffic control tower at Haifa International. They continuously record the tracks and transponder codes of aircraft that pass within range of their air search radar in case there's an incident investigation. A plane flying from Zuwara to Tehran would pass about three hundred kilometers north of Haifa, which is just barely within their air search radar's range."

"Last Monday night, Haifa's ATC radar recorded Maghur's plane passing just north of Cyprus on a heading of 074 degrees true. I plotted that on a map and it was a direct course to Tehran. Then, Tuesday afternoon, Haifa recorded it passing north of Cyprus again but this time on a heading of 268 degrees true, which was a direct course to the airport at Zuwara. On Saturday, Haifa picked it up again heading east toward Tehran. Then, on Sunday, they tracked it heading west. But, here's the puzzling part. The EU Intelligence Center in Brussels keeps track of Mughar because some EU nations would like to see him tried by the International Court of Justice for human trafficking. I checked with EU INTCEN and learned that Mughar and his wife have been at their villa in Monaco for the past two weeks. So, the obvious question is why is his personal jet flying back and forth between Libya and Tehran without him?

Levi saw what Gabe was suggesting but let her spell it out for him.

"So, how is this relevant to our search for Nahbi?"

"Maybe it isn't, but I think there's a good possibility that Malik has called in some markers and is using Maghur's villa and plane to carry out an elaborate scheme for a prisoner swap of Jhalawan for David and Nahbi. I think they're being held prisoner at Mughar's villa."

"That's excellent intelligence work, Gabe. It's good to have you back. I'll brief the PM. With the seven hour time difference, it's midnight in Washington. I'll wait to call Roger Benson

Villa Maghur; Zuwara, Libya
7:45 AM

It was now clear to Malik that Nahbi could not be recruited. He'd hoped to leverage Nahbi's capabilities to strengthen his own ability to influence geo-political events but, at least for now, that wasn't possible. Yet, Malik was unwilling to destroy a machine as powerful and potentially useful as Nahbi. He needed to find a way to neutralize his abilities and the threat he posed while preserving the option to possibly reprogram him at some future time.

Malik had no intention of releasing either Nahbi or David Logan once he knew that General Jhalawan had landed safely in Libya. Logan was unimportant. He wouldn't live long after General Jhalawan was free. Nahbi would be stored without his power supply pending some technological breakthrough that would allow him to be reprogrammed. In the meantime, Malik ordered his men to keep Nahbi and David apart. Working together on Task Force 3, they had already demonstrated their ability to interfere with his plans. He didn't intend to let them do it again.

Malik's electronic brain, like Nahbi's, was more powerful than all but the most advanced supercomputers. His ability to store vast amounts of information was almost infinite using a coding technique mathematicians called chaotic mapping that was based on chaos theory. Even so, his understanding of how his brain and data storage worked was limited by his exposure to artificial intelligence technology that was still in its infancy and his inability to extrapolate that relatively primitive knowledge to likely future technologies.

It was not at all obvious to Malik how he could safely erase Nahbi's memory and programming. A stream of high-energy neutron particles would almost certainly work but the damage would be irreversible. But, there might be a safer alternative.

Rather than take any chance that Nahbi might overpower his guards or somehow escape, Malik told the guards not to disturb him and then removed Nahbi's power supply.

Nahbi was considerably heavier than a human of similar size but Malik had no difficulty lifting him and placing him face down on the tanning bed in the villa's indoor spa. He opened the compartment at the back of Nahbi's head to expose the quantum computer that served as his brain.

Malik instantly recognized the memory module by the transparent, fused-quartz window that all modern EPROMs had. Although Nahbi's memory module was based on graphene instead of silicon, at the quantum level it was still a floating-gate array of field-effect transistors. Strong UV light would erase the data without damaging the semi-conductor components.

The tanning bed in Kalifa Maghur's spa was an expensive model that used high pressure quartz lamps to produce extremely strong ultra-violet radiation at a wavelength of 400 nanometers. There was no way to know how long to expose Nahbi's brain to the UV light but Malik was confident that the tanning bulbs would not harm Nahbi's graphene components.

After four hours, Malik turned off the tanning bed lights and replaced Nahbi's power supply. He turned him over on the bed so he could see his face and eyes. Nahbi's eyes were open and his head was turning as he looked around. At first, Malik thought his attempt to erase Nahbi's programming and data had failed so he kept the high-voltage stun gun ready in case he needed it. It soon became evident however that Nahbi was functional but essentially a blank slate.

Malik offered his hand and Nahbi grasped it and stood up. Malik led him around the room by the hand but Nahbi showed no initiative of his own. When Malik stopped, Nahbi stopped. When Malik spoke to him, Nahbi made no response. Malik realized that somewhere in Nahbi's brain there were instructions hard-coded into the graphene circuits that provided basic mechanical functions. Everything else, all of the higher order functions, depended on programming and data that were no longer there. Once he was sure, he removed Nahbi's power supply again. Like a new-born human infant, Nahbi knew nothing but his brain was fully capable of learning. Malik had no intention of letting that incredible brain start catching up.

CIA Headquarters; Langley Virginia
7:35 AM

"Mr. Benson, I have Levi Reznik holding for you on line two."

"Shalom, Levi. Mah Nishmah?"

"Roger, I've often wondered if you knew how to say anything in Hebrew besides hello and what's up."

Roger Benson laughed. "No, that's my entire Hebrew vocabulary but I'm sure you didn't call merely to make fun of my language skills. So, in English, what's up?"

"Gavriella has done some outstanding intelligence work and has convinced me that Malik is holding David Logan and Nahbi at a villa just outside of Zuwara, Libya. I've briefed our PM and he's almost certainly going to order Shayetet 13 to do a hostage rescue. I wanted to give you a heads up so you could brief the Director."

"Shouldn't this be a joint mission?"

"Certainly there's some logic to that but I think the fuse is too short. By the time we worked out the details and coordinated a joint mission, it might be too late. Malik could move them at any time. He has a long-range executive jet at his disposal that's parked just twelve kilometers from the villa. If he takes them to Pakistan, it might become almost impossible to rescue them. We've got mission planners working on our options. I'll give you a call if we need some help."

Villa Maghur; Zuwara, Libya
10:05 AM

Malik's phone vibrated softly in his pocket. It was his new boss and it probably meant trouble.

Lieutenant-General Zafar Khalil was angry and he made no effort to conceal that fact. "Ahmed, where in Allah's name are you and what are you up to? I just saw a breaking news report out of Iran where Ali Yazdi was quoted as saying that ISI was working with Pakistani intelligence to apprehend the man responsible for the destruction of the Jamkarān Mosque. Are you in Iran and if so, why did I have to find out from Iranian television?"

"Sir, I must apologize. I had no idea that Yazdi would do something so stupid as to give an interview on Iranian television. He's an experienced senior intelligence officer so something that foolish is completely out of character for him. The only explanation I can think of is that he must be under enormous pressure to find those responsible for the destruction of the mosque. It seems likely that the Supreme Leader is angry and that Ali is feeling his wrath. Although I didn't see the news report you mentioned, the quote that you refer to suggests that Ali was trying to placate Iran's leaders. His mention of our intelligence service was apparently an unfortunate effort on his part to spread the heat around."

"I did meet with Ali Yazdi a few days ago to suggest that I might be able to help them but that was only a pretext for what I was really trying to accomplish. I didn't bring this to your attention because I considered it just a routine effort to gather potentially useful intelligence. As you probably know, many people in Qom believe the strange events were plagues sent by Allah to persuade Iran's leaders to abandon their nuclear weapons program. I certainly don't think the destruction of the mosque had a supernatural cause. But I felt sure that it did have something to do with Iran's nuclear weapons program. I met with Yazdi to see what I could learn. I suspected that the attacks on Jamkarān were the work of either the Saudis or the Israelis or possibly the CIA."

General Khalil's next words showed just how angry he was. "Listen very carefully, Ahmed. You said that Iran's Supreme Leader was angry. Well you can add me to that list. I am furious with you and no longer know if I can trust you or not. I can't tell how much of what you just said is true and how much of it was carefully constructed lies to conceal what you are really up to. Our President received a call from Iran's President asking if we might reconsider our condition that they extradite a man you are trying to capture. He didn't know what their President was talking about and I didn't enjoy hearing him shout at me over the phone."

"No sir. I understand and again I must apologize profusely. I have just been trying to do my job. As is often the case, a certain amount of disinformation can be useful. To encourage Yazdi to share the intelligence he had, I constructed a story that I thought I knew who was responsible for the attacks on the mosque. I told him that our President had made a condition of our assistance an agreement that he be extradited back to Pakistan for interrogation and trial when we captured him. My purpose in that deceit was to allow my men to enter Iran with ISI permission. Although I told Yazdi that we expected to capture the man responsible, the truth is that I had no specific person in mind. I needed fresh, reliable intelligence about possible foreign agents operating inside Iran that I felt could only be obtained by men I trusted with what the Americans call boots on the ground. I can see that I have created a difficult situation for you and for that I am truly sorry."

"You still haven't told me where you are."

"I'm in Libya to meet with one of my agents later today."

"I want to see you in my office tomorrow. Call my assistant to set up an appointment once you know your schedule. And Malik, don't assume for an instant that I believe everything you've just told me."

CIA Headquarters; Langley Virginia
4:50 PM

"Mr. Benson, there's a call for you on line one. He wouldn't give his name but the caller ID is for the main switchboard at our Embassy in Malta."

Roger knew who it was. He'd been expecting the call.

"Hello, this is Roger Benson."

"Mr. Benson, this is your lucky day. Hostages are on sale. It's two for the price of one. You persuade your President to authorize General Jhalawan's release and I'll give you David Logan and the android. I'll be in touch."

The call lasted only seventeen seconds. When Malik hung up, Roger buzzed Pete's office. "Pete, can you come to my office right away?"

Pete knocked a minute later.

"Come in and close the door. I just got a call from Ahmed Malik offering David and Nahbi in exchange for General Jhalawan. I'll call the Director so he can let the President know."

"Shouldn't we let Levi know too?"

"I'm sure the President will speak to the Israeli Prime Minister and will tell him we have a hostage swap offer. I'm also pretty sure the Israelis will go ahead with their hostage rescue mission. They have a long history of refusing to do prisoner swaps. It's two o'clock in the morning in Israel. I think I'll wait until we have the President's decision."

9:55 PM

Roger Benson knew that the President and his national defense team were still behind closed doors in the White House Situation Room. Not knowing for sure how the evening would play out, he'd asked his assistant to stay late. He was just about to tell her to go home when she buzzed him.

"Mr. Benson. The Director is on line two."

"Good evening sir."

"Roger, the President won't approve a prisoner swap for Jhalawan. He talked to the Brits, Israelis, and French. All of their PMs were opposed to releasing Jhalawan. It's hard to argue with that point of view. We all know that allowing terrorists to swap people they kidnap for their leaders is a sure-fire formula for more kidnappings."

"So, where does that leave us, sir?"

"Waiting to see if the Israelis ask for our help, I guess."

Wednesday, 24 September

Zuwara International Airport; Zuwara, Libya
5:00 AM

A faint, pre-dawn light was barely discernible along the eastern horizon when the pilot of Kalifa Maghur's executive jet opened the cockpit door and walked down the aisle to where Malik was settling in for the long flight.

"Good morning sir. We can take off when you are ready. It's almost six thousand kilometers from Zuwara to Islamabad so we'll refuel at Mehrabad International outside Tehran. Total flying time is about seven hours and we'll be on the ground at Mehrabad for perhaps one hour. We expect to land at Benazir Bhutto International Airport in Islamabad about 4:00 PM local time this afternoon. The weather reports look good along our route so it should be a smooth flight. Is there anything you need before we take off?"

"No. Thank you."

As soon as the pilot went forward, Malik dialed the number he'd set up to make it appear that the call originated from the main switchboard at the American Embassy in Malta. After a short delay, he heard a woman's voice answering.

"Deputy Director Benson's office. This is not a secure line. How may I direct your call?"

"I'd like to speak to Roger Benson."

"May I say who's calling?"

"Tell him it's the same political attaché he spoke to earlier today."

"Yes sir. I'll let him know you're on the line. Just a moment please."

"Hello. This is Roger Benson."

"Has your President accepted my offer?"

"No, but he did have a counter offer."

"Perhaps I didn't make myself clear, Mr. Benson. I was not offering to negotiate. It was a take it or leave it one-time offer."

"No, I think it is you who do not understand. Unless you release David Logan and the android Nahbi unharmed within 24 hours, the military tribunal trial of Lieutenant General Parvez Jhalawan will be televised worldwide starting at 8:00 AM at Guantanamo Bay, Cuba day after tomorrow. The charges against General Jhalawan will be read and all evidence will be presented. Your name will feature prominently in the testimony making it very risky for you to ever again set foot outside of Pakistan. By the time the trial is over, and the entire world knows what you and General Jhalawan conspired to do, it is likely that he will be executed by lethal injection and you will be charged in absentia with war crimes. You will become a pariah to the whole world and your own government will probably arrest you and execute you for treason. It is also very likely that once Pakistan's role is exposed, the United Nations will impose sanctions on Pakistan that will make those imposed on Iran seem mild by comparison. Let me be very clear. Unless you notify me that you are releasing David Logan and Nahbi unharmed, a spectacle of a magnitude not seen since the Nuremberg trials will unfold under the bright lights of the international news media."

Before he could respond, Malik heard a click and a steady dial tone that made it clear that Roger Benson didn't intend to negotiate either.

Shayetet 13 Headquarters; Atlit Naval Base, Israel
6:30 AM

"Please be seated," Colonel Mirsky said as he took his own seat at the head of the big conference table. He'd just assumed command of the elite Shayetet 13 naval commando unit two days earlier. As an up through the ranks veteran of many missions himself, he had the respect of everyone in the room.

"I've decided to call this mission Operation White Rabbit after Lewis Carroll's character in Alice in Wonderland. Like him, we can't be late for a very important date. I received a call from the Prime Minister a little after three this morning authorizing the mission and then spent an hour on the phone with the Director of the Mossad. He emphasized that the captives could be moved at any time."

"Our objective will be to rescue a Mossad agent named Nahbi and an American CIA agent named David Logan. They've been abducted by Pakistan's ISI. The latest intelligence indicates that they are being held in a private oceanfront villa three kilometers east of Zuwara, Libya. It's a hostage rescue mission so I've assigned it to Team D4 of the Haposhtim Raids Palga. Captain Efron will lead the mission. Captain Efron, please take over."

At 195 centimeters, Nathan Efron was the tallest active duty member of Shayetet 13. The special naval commando unit's nickname was the Bats so to his men Captain Efron was Big Bat.

"We can't be sure about how much opposition we'll face but ISI agents are some of the best in the world. We'll take a squad of eleven and go armed for bear as the Americans like to say. The IAF is flying one of its C-130J Samsons from Nevatim to Ramat David. The Sampson will be on the ground in a couple of hours so we can start loading pretty much when we're ready."

"We'll take one of the eleven meter Muraena rigid hull inflatable boats. I've already had it loaded on a truck and it'll leave Atlit within the hour. We'll do a low altitude drop from six hundred meters just outside of Libya's territorial waters so we don't attract their MIG fighters by crossing into their airspace. About twenty minutes before our drop, one of the IAF's Cheetahs will fly over the villa and transmit high resolution photos to us. We'll use the RHIB to get us close to the beach. Three of you will stay with the boat five hundred meters offshore and the rest of us will approach the beach underwater."

"The villa is about a hundred meters from the water inside a walled compound. It's owned by an extremely wealthy man so a good security system is assumed. It'll still be dark when we hit the beach so we should be able to approach the walls without being picked up by the security cameras. Dogs are likely. We may see them in the infrared images from the Cheetah."

"The villa is a two-story masonry building with a garage and mechanical rooms on the ground level. We have the floor plans from an old realtor's listing. All of the living areas are on the second floor and that's where we think the hostages are being kept. There's a staircase to the upper deck on the northwest side and a roof access ladder on the southeast side. Three of you will follow me up the stairs. Two of you will take up position on the roof to back us up and, if necessary, cover our retreat. Two of you will take out any dogs and ISI agents you encounter outside. On my signal, the RHIB will come to pick us up at the water's edge. We don't know what physical condition the hostages are in so we need to be prepared to carry them if necessary."

"The INS Tannin is currently on patrol in the central Mediterranean and has been diverted to pick us up. They'll remain submerged just off the coast until we're back in the RHIB and send our pickup signal. We'll head out to sea to rendezvous with the sub roughly twenty kilometers off the coast. As you all know, the Dolphin class subs are only about half the length of an American Virginia class submarine so things will be a bit cramped for everyone once we're on board.

"I'm trying to work a deal so our stay aboard the Tannin will be brief. The American Carrier Strike Group CSG 12 left Naples yesterday morning headed for the eastern Mediterranean. I've asked the INF CIC staff to see if they can arrange for our team and the hostages to be transferred from the Tannin to the Theodore Roosevelt. The carrier's MV-22 Osprey could fly us to NAS Sigonella where an IAF Sampson could pick us up. The alternative is being stuck in cramped quarters on the Tannin for several days and hot-bunking with her crew. Even at the sub's top cruising speed, our welcome would wear out long before we got back to Israel."

Every member of the Shayetet 13 team in the conference room was a veteran of multiple missions so there was no need for Captain Efron to address weapons or equipment. All of them had primary and secondary specialties that would dictate their choices. Their elite reputation was no hype.

“The flight from Megiddo to the drop point will take about three and a half hours. The drop is scheduled for 9:30 PM. Moonrise is at 10:19 PM so we should be on the beach just before the moon comes up. Unfortunately, the moon will be only four days past full. The weather report is for clear skies so it’ll be bright once the moon comes up. The local marine forecast is for westerly winds of eight to fifteen KPH with waves approximately a half meter high. Libya is in the same time zone so we’ll need to be wheels up at Megiddo no later than 6:00 PM. I’ll be at the armory right after we wrap up here to pick out my own gear. I’ll answer any questions then. Colonel Mirsky, I’ll turn it back over to you now sir.”

The new commander of Shayetet 13 stood and looked around the room before offering his closing remarks. “These two men saved millions of innocent lives. Like us, they are the silent warriors who work in the shadows. I wish there was some way I could let them know that the best hostage rescue team in the world is on its way.”

Indian President’s Home; Rashtrapati Bhawan, India
2:00 PM

The day was gray and overcast. A light rain was falling over the enormous building and grounds that had been the largest residence of a head of state in the entire world for almost a hundred years. Originally the home of the British Viceroy during India’s colonial period, the estate covered 130 hectares of land or roughly half a square mile and included the beautiful Persian-style Mughal Gardens. Facing east toward the Jaipur Column, the building was even more impressive than the American Capital though not nearly as widely recognized.

The Indian president’s executive assistant was waiting under the portico at the south entrance when a black limousine pulled up to the curb and stopped. Two Indian Army enlisted men holding umbrellas opened the back doors as two of France's most senior generals got out.

"General Melott and General Montreuil, welcome. The President and Prime Minister are waiting inside. Please follow me."

General Jean-Raphaël Melott was the Commanding Officer of the special branch of the French armed forces known as the Commandement des Opérations Spéciales or COS for short. He was General Montreuil's superior officer and had given him command of the COS mission to capture Pakistani General Jhalawan. He had come to India to honor the commitment that General Montreuil had made regarding sharing intelligence about Pakistan's nuclear weapons.

India's two leaders stood to welcome their guests as the president's assistant withdrew closing the doors behind him.

"Please accept our condolences for the death of Captain Davignon and his brave men," Prime Minister Bindusar Nayar said with genuine sadness apparent in his voice. "Your Ambassador gave us a classified briefing immediately after the mission so we are generally aware of what happened."

"Thank you, sir. Your support made the mission possible. Although the cost was tragically high, we were successful in bringing one of history's greatest monsters to justice. General Jhalawan will be tried by an American military tribunal. If he is found guilty, as seems certain, he will either be executed or spend the rest of his life in a military prison. The Americans have provided a competent military attorney to represent him and I am confident that General Jhalawan will be treated fairly, which if you'll forgive me for saying so, is more than he deserves."

The Prime Minister glanced quickly to his right to see how President Gupta had taken General Melott's comment. The General apparently did not realize that the President's daughter had been studying music at the Conservatoire de Paris and working part-time as a docent at the Louvre on June 24th when what the French referred to as *Autre Jour d'Infamie* had occurred. The President's wife had died giving birth to their only child so his daughter's tragic death gave him more reason than most to want to see General Jhalawan executed. Yet, his calm dignity gave no hint of offense or pain at the General's remark.

General Montreuil handed each of the Indian leaders a briefing folder. "Inside these folders are summary versions of the intelligence that we have assembled regarding Pakistan's nuclear weapons. Most of it came from David Logan's interrogation of General Jhalawan but it has been augmented from other reliable sources."

"As you will see, we are confident that all of Pakistan's nuclear weapons are stored in underground bunkers at Wah. I've provided a set of satellite photographs showing the precise locations. We now know that Pakistan has three types of nuclear weapons including approximately forty smart bombs with a yield of one hundred kilotons each. They are similar in design to the American B-61 bombs. They weigh approximately 300 kilograms each or roughly seven hundred pounds. As I'm sure you both know, Pakistan has over three hundred multi-role jet aircraft that could deliver these weapons to almost any target in India. For comparison, a one-hundred kiloton nuclear weapon is approximately five times more powerful than the bombs dropped by the Americans on Hiroshima and Nagasaki."

"Pakistan has been working with North Korea and now has approximately thirty medium-range strategic missiles capable of delivering single warheads with a yield of approximately one hundred kilotons each. These missiles are just now being placed in hardened underground silos. It does not appear that they have yet been armed with nuclear warheads."

"The third type of nuclear weapons in Pakistan's arsenal are approximately five hundred eight-inch howitzer shells each having a yield of approximately forty kilotons. These are the kind of devices that were stolen from Wah. They represent the most serious threat of nuclear terrorism simply because they are relatively small and easily concealable."

"As you know, Pakistan's government is unstable. Most of the western intelligence services believe there is a high probability of a military takeover. Virtually all of Pakistan's senior officers are bitter enemies of India. Many, if not most of them, are extremely hawkish. Given the poor state of Pakistan's economy, it is conceivable that the military might assume power and then initiate an attack on India to rally support for the military government."

"The folder also contains assessments of what it would take to destroy all of Pakistan's nuclear bunkers and enough of her combat aircraft to prevent retaliation."

"One new consideration is that Saudi Arabia has been quietly signaling that they will support missions against Iran and Pakistan in material ways that could include allowing use of Saudi air bases, aerial refueling, and possibly even combat aircraft. Of course, they prefer to work in the shadows since both Pakistan and Iran are Muslim countries. Many analysts believe that is part of the reason they made the absolutely unprecedented decision to decline a seat on the UN Security Council recently. However, the decision to decline the seat required approval by the Saudi King. He has openly criticized the United Nations for its inability to halt Syrian genocide and Iranian-sponsored terrorism. The Saudi government is so concerned about Iranian and Pakistani nuclear capabilities that, for the first time, they are considering revoking their ratification of the Nuclear Non-Proliferation Treaty in order to develop their own nuclear weapons. They see that as a necessary defensive response to the instability of the radical governments in Pakistan and Iran."

Prime Minister Nayar spent several minutes looking over the folders before responding.

"An Indian attack on Pakistan's military complex at Wah would be seen as an act of war with potentially severe consequences. I'm sure you are both aware that India has the third largest number of Muslim citizens in the world and also the third-largest Shite population. If we attacked our Muslim neighbor, it will almost certainly trigger mass demonstrations in India."

General Melott nodded but then offered a counter argument. "Certainly, Pakistan's allies might characterize an attack as an act of war but most of the rest of the civilized world would be supportive after General Jhalawan's graphic demonstration of what nuclear terrorism really means. After what happened in Paris, many of the developed nations fear that one of their cities might be next."

"Although the UN has passed several resolutions defining and prohibiting acts of war and wars of aggression, the UN has clarified that a nation state's defensive reaction to acts of terrorism is justifiable and is not considered to be an act of war."

"To be clear, General Montreuil and I did not come here today to argue that India should destroy the nuclear weapons stored at Wah. We promised to share our intelligence with you so you would have the facts that might reasonably bear on whatever decision you make. We fully understand that this is a complex question with no easy answers. Thank you again for your support and for taking time to meet with us today."

Prime Minister Nayar and President Gupta both stood to indicate that the meeting was over. "Thank you both for coming," The Prime Minister said as he shook hands with both of the French generals. "You have certainly given us much to consider. Please convey our utmost respect to your President for his courageous leadership of your country in these dark days."

ISI Director General's Office; Islamabad, Pakistan
5:15 PM

Lieutenant-General Zafar Khalil kept Malik waiting on purpose. When his executive assistant finally ushered Malik into his office, he didn't show even the minimal courtesy of standing to shake hands with his most powerful subordinate.

"Sit down."

Malik took the chair in front of the General's desk and waited. After several awkward moments, General Khalil looked up.

"I'm suspending you from your duties. To make sure you don't disappear, you will be taken to the Adiala maximum security prison in Rawalpindi and held in isolation until I decide what to do with you. That may seem harsh but I assure you that the alternative was much worse. When it became clear that you'd been working with the Iranians without my knowledge, the President wanted you executed as a traitor. Only his respect for me has at least temporarily saved you from the hangman's noose. Don't forget that Ali Bhutto, one of Pakistan's most revered Prime Ministers, was hung for what in retrospect seems minor in comparison to what you have apparently done.

"General Khalil, I'm no traitor. I am absolutely loyal to Pakistan and to you personally. My business is intelligence with the necessary ingredient of intrigue. In my eyes at least, much of what I do is so routine that I would not burden you with the details. It is true that I often fabricate elaborate stories to conceal what I am really trying to accomplish. That was the case in my recent trips to Iran and Libya. I was almost certain that an Israeli agent named Nahbi was behind the destruction of the Jamkarān Mosque as part of a plan to destabilize the Iranian government. The story that I told Ali Yazda allowed me to infiltrate some of my best men to set a trap. We were successful and Nahbi is currently under guard in a private villa in Libya. I intended to interrogate him myself. I am convinced that he knows a great deal about the inner workings of the Mossad and its network of agents. My return to Islamabad has at least temporarily delayed that interrogation. What can I do to restore your confidence in my loyalty?"

General Khalil said nothing for several moments but the fierce look on his face made it clear that his anger was barely under control. When he finally spoke, his voice was low and tight.

"Perhaps nothing. On Tuesday, you told me that you didn't have a specific suspect in mind. Now you say you did. Then, a few hours after we spoke on Tuesday, the same phone I called you on was used to place a call to an unlisted number for the Deputy Director of the CIA. I'm not especially surprised that you knew Roger Benson's unlisted number. What is most curious to me is why you would call him shortly after I ordered you to return to Islamabad. I couldn't help but wonder if you were trying to negotiate your defection."

Malik was trapped. He couldn't admit that he'd been trying to negotiate a hostage swap for General Jhalawan. His current boss and the President would both recognize the threat that Jhalawan's release would present to them. Even his supercomputer brain needed several seconds to devise a plausible response.

"General Khalil, just before you were appointed Director General, the President's Chief of Staff asked me to find out where General Jhalawan had been taken. He made it clear that it was in Pakistan's strategic interest that General Jhalawan should meet an untimely end. I learned that one of Benson's agents was a co-commander of the French Special Ops mission that abducted General Jhalawan. The Political attaché at the American Embassy in Malta flew one of the helicopters. The French Constitution was amended several years ago to prohibit capital punishment so General Jhalawan was taken to the American detention center at Guantanamo Bay, Cuba. The purpose of my call was to pretend concern for General Jhalawan's welfare while hoping that Roger Benson might reveal that the General was scheduled to be executed. Unfortunately, I didn't learn much."

General Khalil touched an intercom button to speak to his assistant. "Send them in."

Four heavily-armed soldiers entered the office. One of them was carrying a set of belly chains and leg irons. Another was carrying a set of high-voltage stun cuffs.

General Khalil stood. "These men will transport you to Adiala. It is clear that you aren't yet finished lying to me. Perhaps some time in isolation will change your mind."

Operation White Rabbit Drop Zone; Mediterranean Sea 9:25 PM

The Israeli Air Force C-130J Super Hercules was in level flight heading due west at an altitude of 20,000 feet. The lights of Zuwarah, Libya were plainly visible off the port side of the aircraft thirteen miles to the south.

The Hebrew name for the C-130J was Shimshon which translates as Samson in English. It was the newest version of a legendary aircraft that had been in continuous production for over sixty years. The original design had been so fundamentally sound that decades of improvements had created what was widely acknowledged as the finest tactical cargo plane of its type in the entire world.

The Israeli variant had a range of four thousand kilometers and was equipped with advanced computer systems and avionics including a glass cockpit and dual flight dynamics heads-up displays. Its maximum takeoff load capacity was more than forty percent greater than the older models allowing it to carry the equivalent of twenty-eight paratroopers. That was far more than the current mission required.

Mode S radar transponders are mandatory in most European countries so the Samson's transponder had been intentionally programmed with a false aircraft identity. Some countries require the use of Mode S transponders even in uncontrolled airspace such as over the central Mediterranean. Mode S transponders use a 24-bit binary code that includes the permanent identity of the aircraft. Instead of an Israeli military aircraft, the air traffic controllers in Libya would see the identification code for an Alitalia commercial airliner.

The pilot and copilot of the aircraft were both career IAF officers and had been hand-picked for the mission. When the HUD GPS showed they were nearing the drop zone, the pilot put the aircraft into a steep descent. Leveling off at six hundred meters above the ocean surface, he adjusted the throttles to slow the aircraft to 200 knots and turned on the steady red five minute warning light back in the cargo bay. The copilot released the loading ramp latches and began to lower it to the horizontal position.

In the cargo bay Israel's finest naval commandoes were doing buddy checks and getting ready to go. As the drop ramp opened, the roar of the engines made even shouts over their mission headsets difficult to hear. The aircraft's loadmaster made one last check of the rigging for the RHIB. Moonrise was still almost an hour away but the team would be on the beach before then and would make their attack in the dark. A jump from this low altitude meant they'd be in the water a few minutes after leaving the aircraft.

At 9:28 PM, the pilot banked the plane to port and switched the mission light to a slow pulse to let the raiders know they were two minutes from the drop point. A minute later, he switched the mission light to rapid pulse. Back in the cargo bay, the loadmaster got ready to release the drogue parachute that would yank the Muraena Class rigid hull inflatable boat off the drop ramp. Its three main parachutes would open automatically 1,500 feet above the water.

When the mission light went steady, the loadmaster released the drogue parachute and the RHIB flew off the drop ramp into the night sky. Eleven Shayetet 13 naval commandoes were right behind.

Operation White Rabbit Drop Zone; Mediterranean Sea 9:43 PM

Less than fifteen minutes after their low altitude drop from the Samson, all of the raiders were aboard the RHIB headed toward the Libyan coast at 40 knots. The two turbocharged Caterpillar engines were much quieter than earlier models but running at maximum speed entailed some risk that their approach might be heard at the villa.

Captain Efron looked at his watch. The RHIB would reach their splash point five hundred meters from the beach a minute or two after ten o'clock. All of them were strong swimmers but in full gear it would take ten to fifteen minutes to reach the beach. They needed to be in position outside the villa's walls a few minutes before the moon came up. It was going to be close.

The choppy sea state made for a rough ride so the eight commandos that would go ashore were strapped in. All of them were wearing black wetsuits and Draeger oxygen rebreathers that wouldn't leave a trail of bubbles during their approach. Their masks and swim fins were hanging by straps ready for use.

Four of the Bats were armed with silenced 9mm X95 Tavor submachine gun assault rifles with EOTech laser sights and three thirty-two round magazines. Two of them had Benelli M4 automatic twelve gauge combat shotguns and two had Mini Uzi submachine guns equipped with silencers and Trijicon ACOG combat gunsights. All of them had silenced Glock 19 semi-automatic 9mm pistols with three fifteen round magazines and a Dustar combat knife in thigh holsters. Two of them had pneumatic pistols that could fire tranquilizer darts to quietly immobilize any guard dogs they encountered.

Four of the Bats had conventional thermal night vision goggles hanging from straps. Captain Efron and three of his men were wearing top secret contact lenses that allowed them to see not only visible light but also infrared. Developed by a privately-held Israeli defense company, the revolutionary contact lenses used graphene technologies pioneered in the United States by IBM and the University of Michigan.

All of the Bats were wearing bone conduction amphibious assault headsets that allowed them to communicate with each other without obstructing their ability to hear ambient sounds. Two of them were carrying submersible radios designed for naval commandos.

Of the three commandos that would remain with the RHIB, one was manning the helm and the other two were manning fifty caliber machine guns at the bow and stern in case covering fire was needed.

The helmsman cut the throttles when the RHIB's surface radar and GPS showed that they had reached the drop point. Fifteen minutes later, eight Shayetet 13 naval commandos came out of the water and stashed their diving gear. Faint moonlight was already showing in the eastern sky when Captain Efron gave the signal to advance.

The ornamental wrought iron gate facing the beach was locked but the wall was only two meters high. Two Rottweiler guard dogs were the first casualties of the attack. By the time the moon peeked over the eastern horizon all of the Bats were inside the compound. Adjacent to the garage was a small apartment for a driver or domestic servant but there was no one inside. When they'd completed their search of the compound, Bats Seven and Eight transmitted the all-clear signal of two clicks followed by three clicks.

Giving two of his men time to reach the roof and take up their positions, Captain Efron led three members of his team quietly up the stairs leading up to the deck.

After listening to their conversation and laughter for several minutes, Nathan could make out four distinct men's voices. He understood just enough Urdu to realize they were playing a popular Pakistani card game called Rang. He risked a quick look around the door jamb and saw four men sitting around a table. All of them were in shirtsleeves with shoulder holsters holding semi-automatic pistols. Another man was lying on a couch wearing earphones and watching television. A Czech 9mm K105 pistol was lying on the coffee table beside him. The polymer framed pistol held fifteen 9mm rounds and was capable of both semi-automatic and fully-automatic fire. It was a formidable handgun.

Nathan turned to his men and held up five fingers. When they'd nodded their understanding, he selected an M84 flash bang stun grenade and motioned for the man behind him to do the same. Nodding to coordinate their action Nathan pulled the pin on the grenade and tossed it around the door frame into the living room. The two M84 blasts were almost simultaneous.

As Nathan and his men leaped onto the deck the man on the couch grabbed his pistol and rolled onto the floor. The earphones he'd been wearing had given him some protection from the deafening sound of the grenade's explosion. He got off only one poorly aimed shot before Nathan's mini Uzi sent him to meet his maker. The other four men were on the floor but were quickly recovering. Nathan fired another short burst from his Uzi and shouted at the men to freeze. Three of them obeyed. The one who didn't died instantly as one of Nathan's men squeezed off a full burst from his Tavor.

The three ISI agents who were still alive understood that their lives hung by a thread. They offered no resistance as Nathan's men disarmed them before binding their hands and feet with nylon zip ties.

"Check the other rooms while I cover them," Nathan shouted.

A few moments later, one of his men returned with David Logan.

"Mr. Logan, I'm Captain Nathan Efron in command of this team. As you may have guessed, we are Israeli naval commandos. Are you OK?"

"Yes. I'm fine and I did guess who you were. Where's Nahbi?"

At just that moment, Nathan's two men came back into the room. One of them told Nathan what they'd found.

"The other hostage is in one of the bedrooms but it looks like he's dead. He's just lying on the bed motionless. When we spoke to him, he didn't respond and there's no sign of breathing. There are no visible wounds or blood but he definitely isn't showing any signs of life."

"Let me see him," David said as he pushed past Nathan's men.

As he entered the bedroom at the end of the hallway, David spoke loudly. "Nahbi, it's me, David. Can you hear me?"

When there was no response, David asked one of the commandos to help him roll Nahbi over face down on the bed.

"Please leave me alone and close the door, David said. "There are some security issues here that I don't think you're cleared for."

David raised Nahbi's shirt and opened the almost invisible panel in the small of his back. As he'd suspected, Malik had removed the power supply.

"Captain Efron, can you join me please," David called loudly enough to be heard.

When Nathan entered the room, David asked him to close the door behind him.

"What I'm going to tell you is highly classified by the Israeli government. This hostage is not human. It's an incredibly advanced android robot that has been granted Israeli citizenship by your Prime Minister. Like me, he was abducted by Pakistani S-Wing agents as part of a prisoner exchange strategy designed to free the man responsible for the nuclear terrorist acts against the western democracies. The reason he's not responsive is that the head of S-Wing, a man named Ahmed Malik, has removed his power supply. If we can find it, we can restore Nahbi's capabilities."

“What does this power supply look like?”

“It’s about the size of a brick with a colored, four-wire connector coming out of one end. It looks like it’s coated in white shrink-wrap plastic.”

“OK. I’ll have my men look for it. What if we can’t find it?”

“We’ll have to carry him. He’s far too important to Israel’s security to leave him behind. I hope we can find the power supply. If we can’t, one of the Mossad’s technical geniuses can probably devise a substitute for it.”

Two hours later, it was clear that the power supply was not at the villa.

“I should have guessed that Malik would take the power supply with him as insurance that no one could re-activate Nahbi,” David told Nathan. “What will you do with the men you have tied up?”

“We can’t kill them and we can’t take them with us so I have no choice but to leave them. I’m not especially happy about that because they’ll be able to identify us as Israelis,” Nathan answered.

“Whether you killed them or even if you were able to take them with you, it wouldn’t make any difference. Ahmed Malik is the head of what is arguably the best intelligence agency in the world. Not even the directors of the CIA or the Mossad would dispute that assessment. Malik will know it was the Israelis within minutes of learning that his hostages are gone.”

Nathan nodded and then keyed his command headset to let his men know it was time to go. He signaled the RHIB that they were ready to be picked up and assigned two of his men to carry Nahbi down to the boat.

When everyone was aboard, the RHIB backed away from the beach and then turned to head directly out to sea. The helmsman motioned to Nathan.

“Captain, while you were ashore I got a message from Colonel Mirsky. He asked for a status report. I told him that the mission was successful and that none of our guys got so much as a scratch. He then told me that instead of rendezvousing with the sub, we’re going to be picked up by a Sikorsky CH-53 Super Stallion helicopter from the Theodore Roosevelt. The Roosevelt is about forty kilometers offshore and the helo is already en route. I've been in contact with the pilot and have turned our transponder on so they can find us. The Roosevelt has a Navy SEAL team aboard so they have the lift slings for our RHIB. The plan is to hoist all of us aboard except for two of us in the water to position the slings. Once the slings are in place, we'll be hoisted aboard the helo too. The Roosevelt is moving east and will offload our RHIB to Atlit in a couple of days. The rest of your backup plan is pretty much intact. An Osprey will fly us from the carrier to Sigonella where one of our Samsons will pick us up. We'll be home by late tomorrow afternoon.

Thursday, 25 September

USS Theodore Roosevelt (CVN-71); Central Mediterranean 7:00 AM

David was waiting in the communications center aboard the Roosevelt when Pete Nocona returned his call.

"Good morning. You OK?"

"Yes. I'm fine but I'd like to go to Israel with Nahbi. Malik removed his power supply so right now he's just a pile of heavy hardware that doesn't do anything. Aharon Jacobi can probably create a replacement power supply. I'd like to be there when Nahbi wakes up."

"Sure. Go ahead. Levi called Roger to let him know you were safe but that Nahbi wasn't. Roger predicted you'd want to go to Israel. Stay in touch."

Adiala Maximum Security Prison; Rawalpindi, Pakistan 8:00 AM

Four Special Forces commandos assigned to the Islamabad Capital Territory Police were escorting Ahmed Malik to Pakistan's infamous maximum security prison on the outskirts of Rawalpindi. Malik was securely restrained by leg irons and handcuffs attached to welded steel rings on both sides of the van. Two of the commandos were seated behind him in the third row of the van. Both were holding Baikal MP443 Viking semi-automatic pistols loaded with eighteen rounds of Russian 7N21 9mm armor-piercing ammunition. Both men were fully alert as they approached the security gate.

The commando driving the van slowed as he approached the entrance to the prison and stopped at the guard shack to show their paperwork. The guard directed everyone except the driver to get out of the van to walk single file through a turnstile gate while the driver pulled forward for undercarriage inspection for explosives.

When Malik was out of the van, the commando from the front passenger's seat went through the turnstile first. Malik was next. Just as he stepped up to the turnstile, fifty caliber rounds from multiple sniper rifles struck the two commandos behind him at the base of their skulls, nearly decapitating them. The first commando knelt inside the containment area and ordered Malik to get down on the ground. Before Malik could obey, that commando was also killed.

The guard shack was bullet-proof and the guard sounded the alarm. As sirens began to wail, a white Mercedes cargo van screeched to a halt and six heavily-armed men wearing body armor leaped out. Two of them ran to Malik and helped him to the van as the others opened fire with fully-automatic weapons to suppress any response from the prison guards. In less than thirty seconds, all of them were back in the van as it headed south on Adiala Road.

Two kilometers south of the prison, the van's driver pulled over at the intersection where the Adiala-Rawal road turned south. Three of his rescuers helped Malik out of the van and into a dark blue Toyota Land Cruiser. As soon as they were inside, the Toyota's driver headed due west while one of the men removed Malik's leg irons and handcuffs. Near the center of the small town of Adiala, the Toyota's driver turned north on a secondary road toward the N80 national highway leading west toward Pakistan's border with Afghanistan. Six hours later, they were met by Jalaluddin Haqqani's son at a Haqqani network checkpoint near the small town of Pewar a few kilometers east of the border.

"Āssālam álaykum Ahmed," Siraj Haqqani said with obvious affection for an old friend as the two men hugged and shook hands.

"Wālaikum assalām Siraj," Malik responded. "How is your father?"

"He is well, thank you. We will have dinner with him this evening. He is eager to tell you about our success in bringing Taliban operations under our direct control. Your strategies worked just as you said. He also wants to hear more about your situation and how we can help."

"Siraj, your men were magnificent today. The Capital Territory Special Services commandos are formidable opponents and they've never lost a prisoner before. If I'd been processed into the prison, it would have been all but impossible to escape. I am in your debt."

"No, Ahmed, it is my father and I who are in your debt. Without your protection and financial support, the Americans would almost certainly have captured us by now.

Tari Mangal; Federally Administered Tribal Area, Pakistan 5:30 PM

Jalaluddin Haqqani was seated at a folding table that served as a makeshift desk when his son and Ahmed Malik entered the room. He stood at once and came forward to greet his guest.

""Āssālam álaykum Ahmed," the old warrior chief said as he shook hands with the man who had protected him from the Americans and provided the money to raise a formidable army of fighting men. "Welcome my dear friend."

"Wālaikum assalām," Malik responded. "You look stronger than ever. You are well I trust?"

"Ah, a few aches and pains here and there but they are nothing compared to the joy I feel in seeing you. It has been far too long."

"Yes, too long. Thank you for arranging my escape. Your men made it look almost easy."

"The money you have provided has allowed us to recruit the very best of our young men and to equip them with weapons that are comparable to what the American Special Forces have. I am glad that they were able to show our appreciation in such an important way. Let's have our dinner and then we'll talk more."

An hour later, after a simple dinner of roasted lamb with vegetables, the three men went outside to sit on the porch as twilight shadows lengthened across the valley floor from the mountain peaks just west of the village.

"I was stunned to learn that you had been arrested and sent to Adiala prison. What happened? Jalaluddin asked.

"I was working on some very sensitive intelligence matters and had decided not to brief my new boss. As you might suspect, it is no surprise that he's feeling insecure following in the footsteps of General Jhalawan. Unfortunately, my Iranian counterpart did something very foolish and it was misinterpreted by our President and my new boss as evidence that I might be guilty of treason. The President overreacted and wanted me summarily executed. To his credit, my new boss persuaded the President to send me to the maximum security prison at Adiala until they could decide what to do with me. When the van turned into the prison driveway with still no sign of your men, it seemed my message hadn't reached you in time."

Jalaluddin nodded. "By the time we got your message, your mobile phone messages were being routed to an answering machine. There was no way to tell you our plan. What will you do now?"

"A magic trick," Malik said while smiling mysteriously. I'm going to do something the Americans think impossible. I will need your help. Very early tomorrow morning, I want to slip unnoticed through the Peiwar Pass into Afghanistan. I will need clothing and some temporary identification papers. Once I'm across the border, I need for some of your men to escort me safely into the outskirts of Khandahar."

"Yes. Of course. What else can we do?"

"Do you have an Internet connection?"

"Yes. It's slow but normally reliable," Siraj answered. "You can use my laptop. It has a wireless router connected to the satellite modem and sophisticated encryption software."

Ramat David IAF Airbase; Northern Israel
6:10 PM

Standing on the upper level observation deck of the control tower, Gavriella kept her binoculars trained on the Sampson as it made its approach. A fire truck and ambulance with red lights flashing accelerated to keep pace with the big aircraft as it touched down.

"Levi, I have a car waiting below for you and Miss Adler," Colonel Mirsky said as the C-130 turned off of the runway and began to taxi to its assigned parking space.

"Thank you Colonel. We'll be down in a moment." Putting his hand lightly on Gabe's shoulder, Levi asked, "Are you ready?"

"No, not really. I can hardly bear to think of Nahbi being carried off the airplane lying lifelessly on a stretcher."

"I know but Aharon is confident that he can create a temporary power supply that will work. Are you sure that taking Nahbi to your condo is wise?"

"My condo is the only home Nahbi has known for a very long time. It just seems to me that it's appropriate. I know you had some reservations but I appreciate you agreeing to it." Gabe's smile was a bit forced as she teased her boss. "You're so pragmatic that it probably wouldn't bother you if they laid Nahbi out on a steel table in the hanger while Aharon worked on him with a big wrench. But after living with him, Nahbi's a real person to me."

Levi frowned and pinched his lower lip in thought. "I'm sure you realize he won't have any memories of anything that happened after Malik removed his power supply. I just hope that his non-volatile long-term memory hasn't been compromised."

The strained look on Gavriella's face was unmistakable. Her boss had just said out loud what she'd been trying hard not to think about. "Yes. Of course I know that's a possibility but Nahbi's memories survived a very long time without power before. Surely they'll be OK. What does Aharon think? Has he said anything to suggest otherwise?"

"Not that I know of. I shouldn't have thought out loud. You know I worry about everything. Sorry I said anything. Let's go down. The plane is turning off the runway now."

Captain Efron was the first one off the plane followed immediately by his men.

"Welcome home," Colonel Mirsky said as he shook hands with the team one by one. "The Prime Minister asked me to convey his deepest appreciation for the success of your mission. He's relieved that you got the hostages out without even a peep of indignation from the Libyan or Iranian governments. Well done!"

Two IAF EMTs carrying a collapsible gurney went aboard the aircraft and were out of sight for several minutes. When they reappeared, they were carrying Nahbi strapped to the gurney. David Logan walked down the stairs behind the EMTs to where Gavriella was waiting at the rear doors of the ambulance.

"Hello Gabe," David said as he gave her the customary kiss on her still bandaged cheek. "I'm sorry to bring him home in this condition. As far as I can tell, he's unharmed except for the missing power supply. I called Aharon from the Roosevelt to make sure he knew what voltage Tiera and I used to recharge Nahbi's power supply after we found him. Aharon assured me that he can build a makeshift replacement with limited storage capacity as soon as he can verify the dimensions of the power supply compartment. He said the unusual connectors will be the hardest part but he seemed confident that he could deal with that too."

"It's good to see you David. I wasn't sure we were still friends after my role in abducting Nahbi and sending him to Israel at such a critical time."

David smiled gently. "Gabe, after what we've been through together, we'll always be friends. I understood why your Knesset gave those orders and I certainly never blamed you for carrying them out."

"David, would you mind riding in the ambulance with me? I could use some of your calming influence now. This has been really bad for me."

Gavriella's Condo; Tel Aviv, Israel
6:45 PM

Aharon Jacobi finished putting his tools away and was getting ready to leave. "OK, I've taken measurements and pictures of the power supply compartment and the connector. I think I've got everything I need. I'll be back at nine tomorrow morning to install a temporary power supply.

Gavriella walked the Mossad's resident technical genius to the door. "Goodnight Aharon. Thanks."

As Gabe closed the door, David stood up. "I probably should be going too. I still need to check into my hotel and we could both use a good night's sleep."

"Why don't you stay here? I'm going to make a pallet on the floor in Nahbi's room. I know that probably sounds silly but it's what I want to do. You can sleep in my room. That'll make it a lot easier in the morning."

"Are you sure?"

"Yes. I'm sure. You can make us some sandwiches while I change the sheets and put out some fresh towels."

Friday, 26 September

Gavriella's Condo; Tel Aviv, Israel
9:05 AM

Aharon was punctual and didn't waste time on small talk. Going straight into Gabe's bedroom, he set a small box on the nightstand and asked David to help turn Nahbi over.

When the two men had Nahbi lying face down, David removed the cover to the power supply compartment in the small of Nahbi's back. Aharon inserted the rechargeable power supply he'd created and joined the connecting graphene wiring harness to the makeshift connectors he'd fabricated. David replaced the cover and helped Aharon turn Nahbi face up. Gabe took Nahbi's left hand in hers. All of them saw Nahbi's eyes open and watched him turn his head from side to side.

"Nahbi, I'm here," Gabe said softly. "Are you OK?"

Nahbi just stared at her but said nothing. David took Nahbi's right arm to help him sit up. With that assistance, Nahbi did sit up but still said nothing.

"Gabe, let's get him on his feet," David said as they both pulled on Nahbi's arms to help him stand. Nahbi stood up without hesitation and began to look around the room.

"Nahbi, can you hear me? Gabe asked. "Please say something so we'll know you're OK." Nahbi looked at her but still said nothing.

Gabe didn't see the look that flashed between the two men. Both of them had already realized that Malik had found a way to erase all of Nahbi's memory. Gabe's eyes brimmed with tears as she came to the same conclusion.

"Aharon, it's all gone, isn't it?" Gabe asked softly. "Malik has erased all of his memory hasn't he?"

"It looks that way but I'll need to run some tests. His mobility seems unimpaired probably because it's hard-coded into his graphene circuits. I suspect that his senses and brain are undamaged but that he's like a new-born baby. He's so smart that he'll learn quickly but right now it appears that what we think of as his personality and all of his memories have been erased."

Gabe sat down on the edge of the bed and covered her face with her hands while shaking her head slowly from side to side. After a few awkward moments, she looked up. "I begged him not to take the Iran assignment. None of our agents has returned from Iran in over ten years. I kept telling myself that none of them had his incredible talents and that he'd beat the odds. Look, I know this will seem strange but I wonder if the two of you would mind leaving me alone with him for a few minutes."

David and Aharon nodded and went into the living room. David went to the sliding glass door and stood staring out at the Mediterranean.

After a few moments, Aharan asked, "Did you know that Gabe's first husband was killed by a terrorist bomb?"

"No, she never told any of us on TF3 much about herself."

"Her husband was a handsome young Army Captain and Gabe was madly in love with him. They'd only been married a short time when he was killed by a suicide bomb strapped to a frightened eight year-old Palestinian girl. It detonated while he was trying to disarm it to save her life. Gabe coped with her husband's death by pouring herself into her work. None of us who knew her well thought she'd ever get involved with another man. So far as I know, she's had no social life since. Then she invited Nahbi to stay in her condo until he found a place to live. It wasn't long before some of us realized that Gabe had fallen in love with him. Levi may have suspected their relationship had evolved but he didn't confront them so far as I know. Now Gabe is terrified that Nahbi is also effectively dead to her. Whatever she and Nahbi had together has been erased along with his memory. She was too proud to see a shrink after her husband was killed but I don't think she'll be able to get through this without some professional help."

"Aharon, if that's how it plays out, it'll be hard for me too. Nahbi is the most incredible leader I've ever worked with. In addition to his intellectual brilliance, he has the most appealing personality of anyone I've ever known. We became very close while working on Task Force 3 in London. Is there even a remote possibility that some way can be found to restore his memory?"

Aharon hesitated a moment before answering. "Maybe. Just maybe. It's a long shot and I certainly don't want to get Gabe's hopes up. Just before Nahbi left for Iran, he asked me to figure out how to make what amounted to a backup copy of his entire non-volatile memory. He knew the risks he was facing. But before you start doing high fives in the end zone, you need to know that I'm not even sure we were successful. Let me try to explain the difficulty of what he asked me to do."

"Nearly everyone who uses a modern computer has heard the term terabyte. Some of today's hard drives can store one or more terabytes of information. To put that in perspective, a terabyte is a trillion bytes or roughly the equivalent of fifteen hundred CD-ROM disks. But, Nahbi's memory storage was far greater and that's where I entered uncharted waters."

"A petabyte is defined as a quadrillion bytes of information. That's enough storage capacity to hold twenty million four-drawer filing cabinets filled with text. Yes, you heard me right. *Twenty million file cabinets.* That's obviously a very large amount of storage. Yet, the human brain can store two to three petabytes of information. Think of that for a moment. With ten billion neurons that are each connected to other neurons through approximately ten thousand synapses, *the human brain can store the equivalent of sixty million four-drawer file cabinets of information.*"

"Nahbi's memory uses advanced technology that we can't yet duplicate. We can only dimly understand how his graphene computer brain and its associated memory storage work but one thing is very clear. His memory capacity is many times greater than that of any human. My own estimate is that his memory capacity may exceed one hundred petabytes. And therein was my problem. I couldn't just go on-line to order backup drives that can store one hundred petabytes of information. In fact, Nahbi's memory capacity is so great that in the entire world only the National Security Agency's massive data centers in Utah and Maryland could do a complete backup in any reasonable time."

"Aharon, I followed most of that. But you said at the beginning that you might have a viable backup copy."

"You read too much into what I said. It's more complex than that. There was no way I could validate the copy that was made. Even if it's an accurate and complete copy of Nahbi's memory, it's not under my control and I don't yet have a strategy for transferring the copy back into Nahbi's memory."

"OK. Now you've totally confused me. Why isn't the copy under your control?"

"I had to trick the Americans into making the backup copy for me. Israel is one of the most technologically advanced nations on Earth but we simply don't have the resources that you Americans do. The United States is the only friendly nation with the ability to do it but after we essentially kidnapped Nahbi, I wasn't sure they'd be willing to help."

"The National Security Agency has the largest group of supercomputers in the world and by far the greatest data storage capacity. A recent article in *Forbes* magazine estimated the initial storage capacity of NSA's new data center in Utah at between three and twelve exabytes. The backup copy of Nahbi's memory is probably stored there. I'm sure the NSA operations folks know they have it because it uses a lot of their storage capacity but I'm equally sure they don't quite know what it is or what to do with it."

"The Mossad and the CIA still cooperate to a surprising degree given all of the political obstacles. After 9/11 the CIA and the NSA collaborated on a key word and key phrase dictionary that the NSA's supercomputers could use to scan global communications for terrorist activities. The CIA shared that dictionary with the Mossad. Nahbi memorized it so all of those key words and phrases were in his memory."

"But how did you get the NSA to make a backup copy of his memory?"

"You know that high-pitched scream you hear when you accidentally place a call that connects to a fax machine?" David nodded. "Nahbi's graphene vocal chords are capable of producing extremely high-frequency audio signals that are similar but with far greater data density. Nahbi explained to me how he'd moved his volatile memory into non-volatile storage when his power supply was almost exhausted after his ship sank. Knowing that he could do that gave me an idea. It turned out to be surprisingly simple, at least in theory. Over a period of several days, I had him create a high-frequency digital copy of everything stored in his memory and then archive it in his non-volatile memory so it could be downloaded in segments. Then I figured out a way to feed it all to the NSA knowing they'd record it for subsequent analysis."

"How?" David asked.

"You probably know that the NSA monitors global communications as they sift for clues about potential terrorist threats. It's no big secret that their computers give higher priority to encrypted messages originating in countries known to sponsor terrorism. The Mossad has several embedded agents in Lebanon keeping tabs on Hezbollah. We normally communicate with them using encrypted high-frequency burst messages routed through our Ofeq-10 military intelligence satellite. All I had to do was have Nahbi place calls to our agents that were routed through commercial satellites that we know the NSA monitors. Instead of carrying on a verbal conversation with them, as soon as they answered, he transmitted portions of his digitized memory with the embedded red flag key words and phrases. It took almost a week but he assured me that he'd transmitted everything in his memory up to the moment he terminated the final call. The military satellite phones used by our agents couldn't store what he transmitted but the NSA could and almost certainly did. But, nothing's simple. There was another complication."

"What do you mean?"

"Nahbi left for Iran a few hours after the last data was transmitted to the NSA. Just before he left he pointed out something I'd overlooked. He'd realized that we needed to find some way to archive the additions to his memory that occurred after he left Israel."

"Couldn't he just make subset copies and transmit them to you?"

"Yes, but it created a security issue. We'd gone to considerable effort to devise communication strategies that wouldn't be recognized by Iranian intelligence and thus potentially compromise his mission. Nahbi came up with a brilliant solution. You may know that the human ear can hear frequencies between about twenty and twenty thousand cycles per second. Dogs can hear sound up to about forty-five thousand cycles per second. Musical instruments and singer's voices create an incredibly complex audio signal that is then digitized so it can be burned into commercial compact disks using laser technology. Before transmitting his updates to me via our Ofeq-10 satellite, Nahbi assembled an audio signal that consisted of a popular Iranian song with an embedded signal of much higher frequency that contained his backup data. Anyone picking up the signal would hear the popular song but would not realize that there was something else going on in the background. His incremental backups required relatively little storage space so I was able to store them on our system and then transmit them to our agents in Lebanon. After he deployed to Iran, he sent updates every day until Malik took his satellite phone. I'm confident that the NSA has copies of all of those backups too."

"OK. The Americans have what amounts to a backup copy of everything that was in his brain. So what's the problem? Can't you work with them to restore it?"

Aharon frowned and slowly shook his head. "I can think of several problems. First, the NSA isn't going to be happy that they've been played. Someone will have to do some serious fence mending before they'll even consider helping us. Also, there are quite a few people in senior levels of your government that have never forgiven Israel for abducting Nahbi. I don't think the NSA is going to cooperate. Even if they do, I'm not sure we can get the information back into Nahbi's non-volatile storage. In theory, we ought to be able to reverse the call so to speak. If Nahbi accepted a call from one of the NSA's supercomputers, and they transmitted the binary data back to him, his computer brain should be able to receive and store the data. *In theory*. But Nahbi is the only one who can do it and he's been reduced to the mental state of an infant. I just don't know if it's even possible."

"Aharon, couldn't you teach Nahbi what he'd need to know to receive and store the backup data? You said yourself that he's so smart he'll learn fast. When we were working together on Task Force 3, I saw first-hand his ability to absorb and correlate enormous amounts of information at blinding speed. You said his senses and brain were undamaged. Couldn't you create an electronic tutorial to teach him what he'd need to know to receive the data? "

"Maybe. It would take a little time but, conceptually at least, it's probably possible. What I don't know is whether Nahbi would know what to do with the data once he received it. I don't really understand how he consciously moves data between his volatile and non-volatile memory so I can't teach him how to do it. I don't know if that ability is hard-coded into his graphene circuits or if it was learned behavior from the higher-level programming that's been erased. Remember, he's the one who converted the contents of his memory into ultra-dense, high-frequency data streams. I understood what he was doing but not how."

"Assuming we can get the NSA to play nice, how long would this take?"

"Once the NSA begins transmitting the data, perhaps forty hours. It's a mind-boggling amount of information. The only secure way to do it would be to have the NSA route it through our Ofeq-10 satellite. Our military satellites are arguably some of the best in the world. Israel is one of only ten countries with orbital launch capability so everything associated with the program is national security top secret. We launch our satellites from Palmachim Air Base into retrograde orbits over the Mediterranean. That ensures that a launch failure won't crash in an enemy country where our technology would be compromised. We've never given any other country the access codes to our satellites. The Prime Minister would have to approve it."

David took a few moments to assimilate everything Aharon had told him.

"Aharon, President Marshall is probably the best friend Israel has. He's even got a strong majority in Congress supporting his pro-Israeli agenda. I'm confident that he'll direct the NSA to cooperate if your Prime Minister asks him to."

Millionaire's Mile; Hillsboro Beach, Florida
10:45 AM

Señor Campos, there's a call for you from Chicago on the secure line. It's Mr. Russo himself.

The Chicago Association, headed by Frank Russo, was one of the largest crime syndicates in the United States. Carlos Campos was the head of its Miami operation.

"Hello Frank. What's up?

"CC, I've got a contract for you. I'm going to have Angela transfer five million dollars to your operating account later this afternoon. Our client wants us to purchase an ocean-going yacht licensed in Florida and furnish some muscle for a little cruise. He's already picked out the boat he wants. It's a seventy-five foot Maiora at a yacht brokerage in Key West. He wants four guys that can pass for wealthy businessmen on a Caribbean fishing trip. The client wants the boat to cruise back and forth in the Windward Passage between Haiti and Cuba until it receives a coded message with geographic coordinates for a pickup. After the pickup, the yacht will drop off two men at the small port city of Jérémie on the north coast of Haiti and take the yacht back to a broker in Key West. The contract will be completed at that point."

Saturday, 27 September

Saeed Ghorbani's Shop; Yazd, Iran
9:05 AM

"Why have you come here? This is very unwise," Saeed Ghorbani told the two young men that had just entered his shop.

"There's been no message from Ahura Mazda in over a week. We didn't want to risk having a phone call intercepted by the secret police so we came to see if you knew why we haven't heard from him. If he's been arrested, all of our lives may be in danger. No man can resist torture indefinitely."

"Calm down. Think about it. *If* he'd been arrested, and *if* he'd been tortured, and *if* he'd given our names to the secret police, we'd already be in their custody. Ahura Mazda doesn't even know all of your names. He knows mine but as you can see, I'm still a free man."

"So, we just wait?"

"For now, yes. Just be patient. My sense is that his plan is complex and cannot be executed until all of the critical elements are in place. I feel sure that we'll hear from him soon. Until then, please do not come here again. If you need to communicate with me, send me an encrypted text message using the IAM Smartphones that he gave us."

NSA's National Data Center; Camp Williams, Utah
3:45 PM

"Mr. Jacobi," my team is standing by. We've confirmed that we have four dedicated channels of communication with your Ofeq-10 military intelligence satellite. Is everything ready at your end?"

"Yes. You can start the transfer when you're ready."

With the ability to perform twenty-thousand trillion calculations per second, the NSA's Titan III Supercomputer in Utah had recently reclaimed the title of the fastest computer in the world. It was connected to the world's largest data storage facility that could process and store more than two million gigabytes of data per hour.

The Mossad's supercomputer was appropriately named Tamnun, the Hebrew word for octopus. Developed in Israel specifically for intelligence work, the latest variant of Tamnun used a massive array of EnLight256 electro-optical processors that could each perform eight-thousand billion calculations per second. Each EnLight256 processor was equivalent to a network of previous generation supercomputers. It wasn't as fast as the American Titan but it wasn't slower by much. The slowest link in the connection would be Israel's military satellite.

Aharon was watching a digital counter showing the progress of the data transfer. Nahbi was sitting at a console next to him wearing a special pair of high-frequency stereo headphones while staring at a continuous stream of complex images flashing across a large monitor.

"How long will it take," Gavriella asked.

"I really don't know. It took almost a week to transfer everything from Nahbi's memory to the NSA data center using a much less efficient connection scheme. The NSA can receive and transmit intelligence data at the rate of about two petabytes per hour but they won't dedicate their entire capacity to us. It might take two or three days. I just don't really know. I'm going be here for a while to make sure the transfer is stable. There's really no sense in you and David sticking around. I'll call you if there's a problem."

Monday, 29 September

Gavriella's Condo; Tel Aviv Israel
6:30 PM

By Sunday afternoon, Nahbi's memory had been fully restored. It had gone much faster than Aharon had thought possible. The Israeli Prime Minister's call to President Marshall, and the President's subsequent call to the NSA Director, had leveled several bureaucratic mountains. By the time it was over, the total amount of data transmitted had exceeded a staggering nine petabytes of digital information. Once the Israelis shared the classified details of how the Ofeq-10 processed digital data, the Americans had devised a sophisticated parallel processing algorithm that had dramatically expanded the Israeli satellite's throughput.

Sunday evening, Levi Resnik had allowed Gavriella to take Nahbi home for a couple of days with the understanding that most of Nahbi's time for the remainder of the week would be spent in debriefings and preparing reports.

Gavriella and Nahbi had a late dinner at Baskov's and then returned to her apartment. Hardly daring to believe that he was safe, she nestled against him on the couch lying quietly in his arms until she fell asleep. Sometime after midnight, she barely stirred when Nahbi carried her to the bedroom. For the first time in many weeks, Gabe slept peacefully with Nahbi's strong arms wrapped protectively around her.

Georgetown Pike; Great Falls, Virginia
6:45 PM

Deborah Tillman parked her BMW roadster by the front door instead of pulling into the garage. She and her eight year old daughter Beth would only be home a few minutes before heading for the restaurant to meet her husband for dinner. As they got out of the car, a FedEx delivery truck came down the long curved driveway and stopped behind her BMW. The driver stepped down and said he had a package for her. He walked around to the back of the truck and opened the rear door. In a moment, he returned carrying one of the standard FedEx boxes. Too late, she realized he was wearing a life-like latex mask just as he pulled a semi-automatic pistol from beneath the box.

"Don't move and don't scream or your daughter will die before your eyes."

Two more men wearing latex masks came from behind the truck. One of them was also holding a pistol. The other was holding two black cloth hoods and several nylon zip ties. He pulled the hoods over their heads and bound their wrists behind their backs. In less than two minutes, Deborah and her daughter were lying on their sides in the back of the truck as it headed west on Georgetown Pike. One of the men followed the FedEx truck in Deborah's BMW.

The Capital Grill; Tysons Corner, Virginia
7:20 PM

Richard Tillman looked at his watch again. His wife and daughter were late and Deborah wasn't answering her phone. He was starting to worry that maybe they'd been involved in a traffic accident on the way to the restaurant when his phone vibrated on the table in front of him. Deborah's face filled the screen as he took the call.

"Hello Babe. Where are you?"

"Listen carefully a man's voice said. We have your wife and daughter. They are unharmed but unless you do exactly as you're told, they will die. Do not even think of contacting the police. Leave the restaurant now. Drive west on I-66 to the highway 15 intersection at Haymarket. Go south on highway 15 for a quarter mile then turn into the convenience store and gas station on your left . Park and get out of your car. When we're sure you haven't been followed, a car will pull up to where you are standing. Get in the front passenger's seat and do as you're told. Do you understand?"

"Yes. Let me speak to my wife." There was no response. The caller had hung up.

Gas Station and Convenience Store; Haymarket, Virginia 8:45 PM

Richard Tillman was a wealthy attorney who defended people facing the death penalty for capital crimes. Whether his family had been targeted simply because he was wealthy or because of what he did for a living didn't matter. His wife and daughter were in mortal danger so he did as he'd been told. He'd been waiting in the convenience store parking lot for fifteen minutes when a black Chevrolet Suburban with dark-tinted windows stopped beside where he was standing.

"Mr. Tillman?"

"Yes."

"Get in the front seat and close the door. There's a set of handcuffs on the floor. Fasten one of the cuffs to your left wrist and then pass the other cuff under your right knee before you fasten it onto your right wrist."

When the handcuffs were in place, the driver leaned over to put a pair of wraparound sunglasses on his captive. The inside of the lenses were covered with black electrical tape so he wouldn't be able to see where they were going.

"Where are you taking me?"

"To see your wife and daughter."

Thoroughbred Horse Farm; Middleburg, Virginia 9:15 PM

Thirty minutes after leaving the convenience store, the Suburban pulled into a garage and Tillman heard the garage door closing. The driver got out and a moment later the passenger's side door opened.

"Don't try anything stupid. I will kill you without hesitation if you don't do precisely what I tell you. Be still. I'm going to unlock your right handcuff so you can get out. Don't take the sunglasses off or make any noise."

When Tillman was out of the vehicle, the driver put the right handcuff back on him and then removed the sunglasses. Deborah's BMW was parked beside the Suburban. A hallway into the house was visible through an open door and a second man was standing in the doorway pointing a pistol at him.

"Stop just inside the door. There's a security system screen on the right. You'll see your wife and daughter sitting on a couch on the lower level of the house. In a moment, my associate will take a plate of sandwiches and some soft drinks downstairs. You'll see him enter the room and hear him speak to your wife. This will be your proof that we are not merely playing a recording to trick you into thinking your family is safe. As I've told you, your wife and daughter are unharmed and will remain in good health so long as you do what you are told."

A minute later, Tillman watched as a man put the sandwiches and beverages on a coffee table in front of his wife and daughter. He heard Deborah thank him and watched Beth reach for a sandwich just before the driver switched the security monitor off.

"Go through the kitchen and take a seat in the dining room. My boss will be with you in a moment."

Tillman sat down at an expensive dining table and looked around. He was in a luxury home that was obviously much larger than his own multi-million dollar residence. Original oil paintings of thoroughbred horses hung on the walls and several silver trophies were displayed. He was kept waiting less than a minute.

A slender, dark-headed man in his late thirties or early forties entered the room and took a seat across the table. ""Good evening Mr. Tillman. Let me assure you that you and your family will come to no harm if you do what I ask and make no attempt to contact law enforcement or military authorities."

"Who are you and what do you want from me?" Tillman asked.

"Like you, I am an attorney. I am currently using the name Allan Bradford but of course that is not my real name. What I want from you is a job. Over the next hour, you are going to interview me and then you are going to offer me a job with your law firm, which I am going to accept."

"Let me get this straight," Tillman said bluntly. "You've taken my wife and daughter hostage to persuade me to offer you a job? Surely you don't expect me to believe that?"

"Oh, but I do," Ahmed Malik responded. "Your law firm and you personally are under contract to the Department of Defense under the Military Commissions Act of 2009. You are the lead civilian attorney assigned to defend Lieutenant General Parvez Jhalawan at Guantanamo Bay. You report to the lead military defense lawyer, Army Brigadier General Jason Ross. You and your staff work out of offices in Arlington but you spend a lot of time at Guantanamo in offices that the Department of Defense has set up there for your use. I want you to add me to your staff. I've prepared a formal résumé showing my qualifications. It will probably not surprise you to learn that my résumé is a carefully-crafted work of fiction. However, I assure you that I do possess the knowledge and skills to support my contention that I am an experienced criminal defense attorney. I can also assure you that my assumed identity will pass every security investigation the government conducts."

"That is preposterous! No one could create a false identity as an experienced criminal attorney that would fool some of the best Defense Investigative Service and CIA agents in the country."

Malik smiled. "You are wrong Mr. Tillman. People who work for me are among the most highly-skilled computer hackers in the world. University computers, government systems, civilian law firms, news media archives, and hundreds of Internet web sites have been hacked to create an iron-clad false identity."

"No one could do that. I don't believe you."

"I assumed that would be your reaction. Come with me into the study. I have a computer set up for your use. You can use any search query you can devise to try to find a weakness in my résumé.

After nearly an hour of checking and cross checking the educational credentials, licensing, experience, residences, date and place of birth, news articles in several countries, and every other fact he could think of, Tillman finally admitted that the résumé seemed unassailably verifiable.

"OK. I wouldn't have believed it possible."

Malik smiled again. "Allow me to give you an extemporaneous demonstration of my knowledge and my ability to argue complex legal points. I am fully aware that you have a reputation for an encyclopedic mastery of all aspects of criminal defense, including U.S. and International law. So do I. Ask me any question you like."

Tillman thought for several moments and then posed his question.

"What is the single greatest weakness in the Government's case against General Jhalawan?"

"I'm disappointed that you gave me such an easy challenge. In Boumediene versus Bush, the Supreme Court ruled that detainees at Guantanamo Bay had a right of habeas corpus under the Constitution and that the 2006 Military Commissions Act passed by Congress and signed into law by the President was an unconstitutional suspension of that right. As you know, habeas corpus is a recourse in law that allows any person to report an unlawful detention or imprisonment to a court. If that court finds that the detention was unlawful, then the prisoner must be released. In Hamdan versus Rumsfeld, the Supreme Court held that the U.S. Government did not have the right to establish the military tribunal system trials at Guantanamo because they violated the Uniform Code of Military Justice and Common Article 3 of the Geneva Conventions, to which the United States is a treaty signatory. So, ipso facto, General Jhalawan is being detained illegally because he has not been granted his habeas corpus rights. Since he has been illegally detained, General Jhalawan must be set free."

Tillman nodded. "Yes, that's the crux of the matter. But I'm sure you realize that it's more complicated than that with all of the legal challenges from both sides of the issue. What's the strongest rebuttal to that argument?"

"Again, that's almost too easy. For over one hundred years, ever since the 1907 Hague Convention, international law has consistently held that crimes against humanity are deemed to be part of *jus cogens*, the highest standing in international legal matters. Thus, all nation States can exercise their jurisdiction in prosecuting a perpetrator of crimes against humanity regardless of where the crime was committed. It also means that General Jhalawan cannot claim the defense of '*obedience to superior orders*' and that no statute of limitations applies. In short, all that a competent court need show is that General Jhalawan was responsible for crimes against humanity. If convicted, any State that has him in custody is fully within its international rights to execute him.

This time, Tillman slowly shook his head in awe. "None of my staff could have given such concise and accurate answers to my questions without hours or even days of research. *Who are you?*"

"I am someone who believes that General Jhalawan is being illegally detained and placed in jeopardy of execution for what are essentially political reasons. I believe that joining your staff affords me the best means of contributing to his release."

"So, if I agree to hire you, will you let my family go?"

"When General Jhalawan is free, I will release your family. Until then, your wife and daughter will be held hostage to guarantee that you cooperate fully and do nothing to alert the military or civilian law enforcement agencies. The entire lower floor of this house is a large two-bedroom, two-bath apartment. Your wife and daughter will be made as comfortable as possible."

"That is utterly unreasonable. The trials of the Guantanamo detainees have already been going on for years. With all of the legal challenges from both sides, they will likely continue for several more years. How can my wife and daughter be held for what might easily turn out to be years. Even if General Jhalawan's trial was somehow expedited, it is entirely possible that he will be found guilty and executed. What would become of my family then? If you offer me only a despairing scenario, how can you realistically expect me to cooperate with you?"

"Your points are noted," Malik said calmly. "Your best hope for the early release of your family is to do precisely what I tell you to do. That will begin by you extending a formal job offer to me and by me moving into your house in Great Falls as your guest. Until General Jhalawan is free, I will be your constant companion."

Tuesday, 30 September

Villa Maghur; Principality of Monaco, French Riviera 10:20 AM

It was a beautiful morning. Kalifa Maghur and his wife were enjoying a late breakfast on their deck overlooking Monaco harbor when the phone rang.

"Bonjour. This is Kalifa.

"Kalifa, this is Ahmed Malik. I need another favor."

"Of course my friend. What can I do for you?"

"I want to buy your executive jet. I will pay you twenty percent over its current market value but I want you to leave it in your name for now. I'd also like to hire your pilots for a month. I'll pay their normal salaries and all expenses plus a twenty percent bonus at the end of the job."

"Yes. Of course, Maghur said with obvious enthusiasm. "I've been looking at a newer model for some time so your offer is perfectly timed. When do you need the plane?"

"I'd like for your pilots to fly the plane to Jérémie Airport on the northwest coast of Haiti next Monday. They'll need to make a refueling stop at Cesária Évora Airport on the island of São Vicente in the Cape Verde Islands and another one at Crown Point Airport on Tobago Island. They can rent a nice oceanfront house in Jérémie and put it on my tab. I want them to wait for me and be ready to fly me back to your villa in Libya for a few days while I make some other arrangements. If this is agreeable to you, I will transfer ten million dollars to an account you specify tomorrow morning.

"That's more than fair. Hold for a moment while I get the account numbers for you."

Crowne Plaza Hotel; Tel Aviv Beach, Israel
6:45 PM

David's phone vibrated in his pocket. Phone numbers from CIA offices were blocked as a security measure so he didn't know who the call was from until he heard Pete's voice.

"Hello David. I've just received what amounts to a subpoena requiring you to testify for the prosecution in the Military Commission trial of General Jhalawan at Guantanamo. I had to sign for it since you were still out of the country. Did you know this was coming?"

"No, it's news to me but I'm not surprised. When do they want me there?"

"You're booked on the AMC contract flight out of Joint Base Andrews next Monday morning. It says here that you have a room reserved at the Guantanamo Navy Lodge through Saturday the 11th and that they've got you booked on the return flight that leaves GTMO at fourteen hundred. It's scheduled to land at Andrews around seventeen hundred.

"OK. I'll see what flights are available and try to be back in the office by Thursday or Friday."

Wednesday, 1 October

Mossad Headquarters; Herzliya, Israel
6:05 AM

"Good morning and welcome back. I'm surprised to see you here this early," Levi said as he walked into Nahbi's office and sat down.

"I need to get in touch with some of my contacts in Iran. They haven't heard from me and must be wondering when Iran's secret police will be knocking on their doors. Aharon's programming one of the IAM smartphones for me now."

"Does Gavriella know you'll have to go back?"

"We haven't discussed it but I'm sure she does. It seems cruel after all she's been through."

"What about you? I wouldn't be much of an intelligence chief if I hadn't recognized the signals. It's clear that she loves you."

Nahbi turned away from his computer screen to face his boss. "Does it surprise you that one of your best officers, and possibly your successor, is in love with a machine?"

Levi laughed out loud at the absurdity of the question. "Nahbi, forgive me for pointing out the obvious but you are hardly a mere machine. Quite apart from your superior intelligence and physical abilities, you have quite consciously shaped your own personality to incorporate what most of us think of as the very best of human qualities. So no, I'm not surprised that she's fallen in love with you. What about you? Are you in love with her?"

"Yes. Incredible as that may seem, yes."

Levi nodded thoughtfully before responding. "Where do you see this going?"

"I've got to go back to Iran to finish what we started. It's entirely possible that I won't come back this time. But, if I do, I plan to ask her to marry me. When the PM granted me Israeli citizenship, he made it clear that included the right to get married. Strange and far-fetched as that seemed at the time, it now seems like the right thing to do. Gabe and I have wondered if we could both continue working here if you knew that we were romantically involved."

"Let's cross that bridge when we come to it," Levi said as he stood and turned to leave.

Monday, 6 October

Camp Justice Courtroom; Guantanamo Bay, Cuba
8:30 AM

Two military guards escorted LTG Jhalawan to his seat on the left side of the courtroom beside the defense team's front table. Most of the terrorists being tried at Guantanamo Bay wore handcuffs and leg irons in the courtroom but their leg irons were not typically shackled to rings in the floor unless they became unruly and the Chief Judge ordered the additional physical restraint. In deference to his senior military rank, the Commission Judge granted the defense counsel's request that General Jhalawan not be restrained by handcuffs and leg irons once in the courtroom.

A few minutes after Jhalawan was seated, prosecution and defense attorneys started filtering into the courtroom. Despite the seriousness of their roles, the military and civilian attorneys chatted amicably among themselves. The man who would soon be introduced as Allan Bradford moved easily among the other attorneys smiling and shaking hands here and there. Edging his way over to the front of the prosecution's area on the right side of the room, Malik reached a position where General Jhalawan could see him. Jhalawan caught the movement in his peripheral vision and turned to look. After only a split second of eye contact, the cagey old warrior looked away without revealing any visible sign of recognition.

At precisely 9:00 AM, the Chief Judge of the U.S. Court of Military Commissions, Air Force General Janet Wilson, entered the room and took her seat. She then recognized the chief defense counsel, Army Brigadier General Jason Ross who rose to address the members of the military commission.

"Good morning General Wilson and fellow officers and civilian counsel. I'd like to introduce the newest member of the defense team. Allan Bradford, J.D. will be the chief civilian defense counsel for LTG Parvez Jhalawan. Mr. Bradford's experience fully satisfies the requirement under the Military Commission Act of 2009 that every detainee accused of a capital crime must be defended by a learned counsel with extensive experience in capital crimes. Mr. Bradford's top secret security clearance is active and approved for his current role."

"Welcome to Guantanamo Bay, Mr. Bradford," General Wilson said somewhat stiffly. It was no secret that she had little respect for what she considered overpaid civilian attorneys getting rich defending terrorists. Yet, her reputation for integrity was so great that no one would have thought of accusing her of being unfair in her decisions.

"Admiral Gatlin, does the chief prosecutor have any objections to the addition of Mr. Bradford to the defense team?" General Wilson asked.

"Not at this time Judge."

Leeward Airfield; Naval Station Guantanamo Bay, Cuba 1:37 PM

The AMC flight from Joint Base Andrews in Maryland to Guantanamo Bay was a few minutes late. David grabbed his small bag out of the overhead and followed the other passengers down the stairs. Navy Senior Master Chief Thompson was waiting for him at the baggage claim area. "Welcome to Gitmo sir. How was your flight?"

"A little bumpy on final approach but otherwise fine."

"Is this your first time here?"

"Yes. What do I need to know?"

"Except for the airfield, there's not much here on the leeward side of the bay. The ferry to the windward side leaves in about twenty minutes. I have a government vehicle parked by the dock and I'll take you to the Navy Lodge so you can get checked in. Do you need anything at the Navy Exchange while we're in that area?"

"No, but thanks for asking."

"Most of the Military Commission attorneys are billeted in trailers or hard-backed tents at Camp Justice but they didn't have any available room so I booked you at the Navy Lodge in the support services area. It's actually more comfortable than the trailers and tents anyway. Most of the attorneys hate staying out at Camp Justice."

"After we get you checked in, I've been asked to take you to the Camp to meet the chief prosecutor, Rear Admiral Wallace Gatlin. You'll have an early dinner with him and one or two of the other prosecuting attorneys. They'll brief you on the routine."

"Tomorrow morning, I'll pick you up at zero seven hundred and take you to breakfast. I'll drop you off at Camp Justice at zero eight thirty. One of Admiral Gatlin's staff will meet you at the gate and get you checked in with security. That's apparently all they have you scheduled for tomorrow. I'll wait until you're done and then take you to the Navy Exchange to rent a car. They didn't have any available today but I have one reserved for you the rest of the week. You'll be able to drive yourself to the Camp on the days you're scheduled to testify. The rest of the time, you can enjoy our beautiful tropical paradise. Actually, it's really pretty nice although the beaches are rocky. Are you a scuba diver?"

"Yes. Can I rent gear here?"

"If you have either PADI or NAUI certification you can register with Port Ops and rent gear from them. If you need a diving buddy, I'd be happy to show you our best dive sites. We sometimes see giant Hawksbill and green sea turtles. If we get lucky, we might even spot a leatherback."

"What about renting a boat to go fishing?"

"Sorry. That's probably not an option for you on such a short stay. You have to have a boat captain's license and that takes about a week. Fishing's not that great anyway. With the influx of military personnel for the Joint Task Force, there's been a big impact on the fishing. What the JTF soldiers haven't damaged, the explosion in the Lion fish population has taken care of. Fishing used to be spectacular here but not now."

O'Kelly's Irish Pub; Naval Station Guantanamo Bay, Cuba 5:30 PM

There weren't many restaurant options at Guantanamo Bay. Unlike most American military bases, there was no longer a separate Officer's club. What had once been the Officer's Club dining room had been converted into an All Hands restaurant open to everyone. The All Hands food was marginally decent but the venue wasn't really ideal for the senior officers assigned to the Military Commission. Most of them preferred O'Kelly's Irish Pub and it was common to see both prosecution and defense attorneys there on any given night.

"Mr. Logan, I appreciate you coming," Admiral Gatlin said while they stood waiting for a table. I didn't realize you were in Israel when we served the subpoena. I've since spoken to Roger Benson and have at least some sense of how interesting your life has been this past year. I gather that Roger left out a few details."

David nodded without smiling at the Admiral's understatement. "Yes sir, you could certainly say that. Can you tell me what my role here will be?"

"Your work on Task Force 3 and then the mission to capture General Jhalawan means you are the one person with in-depth knowledge of the General's role in the deaths of so many innocent people. And, before I go any further, please accept my sincere condolences for the murder of your sister by this monster."

"As you probably know, we operate under the Military Commissions Act of 2009. One aspect of that revised set of legally-binding rules is potentially very helpful to the defense attorneys. Under the 2009 Act, information obtained by torture cannot be used against a person accused of terrorism. I understand that you were forced to use physical torture to obtain certain information from General Jhalawan. Is that correct?"

"Yes sir. That is correct but that interrogation was not focused on what the General had already done but, instead, on information that we could use to prevent similar future incidents. I asked the General for information on the location, numbers, and types of Pakistan's nuclear weapons. I asked for information on the location of one of his senior agents who I suspect was the real brains behind the attacks."

"I didn't need further proof of his role in the theft of the nuclear devices and the subsequent detonation of one of them over Paris. If the prosecution's case focuses solely on that sequence of events, you will not be in violation of the restriction on information obtained by torture. I can testify to matters that were available to the United States, the United Kingdom, and Israel prior to the mission to capture Jhalawan. All of that testimony can be verified as being the factual basis of the prosecution's case prior to the joint French and American mission into Pakistan."

"Good. I was hoping that would be your answer. We are still in the prosecution phase of General Jhalawan's trial. We won't need you tomorrow but sometime Thursday or Friday you'll be called. After passing through security, you'll be briefed by our security officer regarding what you can and can't talk about. You'll then be sworn in and then one of my attorneys will ask you questions. Following the prosecution's questions and your responses, you'll be cross-examined by the defense team. Their lead civilian attorney is a man named Allan Bradford. He's already demonstrated his encyclopedic knowledge of the relevant Constitutional and Code of Military Justice rules for capital crimes. He's very aggressive on cross examinations so don't volunteer information he doesn't specifically ask for in his questions."

"What about General Jhalawan? Is he participating in his own defense?"

"Yes, to a considerable degree. As I'm sure you know better than I do, the General is an extremely well-educated man with a superb fluency in English. He frequently suggests lines of inquiry to his lead attorney. Mr. Bradford has filed quite a large number of pleadings to add to the thousands that have already been submitted by the defense attorneys for the 9/11 terrorists."

When their table was ready, they were seated beside some of the defense attorneys so the Admiral kept the dinner conversation light. When they were finished eating, they spent an hour in the bar and then one of the prosecution attorney's dropped David off at the Navy Lodge.

Tuesday, 7 October

Mossad Headquarters; Herzliya, Israel
7:25 AM

"Good morning," Aharon said cheerfully as he walked into Nahbi's office. "I've got one of the IAM Smartphones reprogrammed as you asked. It's ready when you are.

"Thanks, Aharon."

A few minutes later, Nahbi closed his office door so he wouldn't be interrupted and hit the phone's speed dial key for Saeed Ghorbani in Iran.

"Good morning, Saeed. I'm sorry to call you so early in the morning. Can you talk?"

"Yes. I'm alone. Where are you? Why haven't we heard from you? Our associates are getting very nervous and I've had a difficult time keeping them calmed down."

"I had to leave the country to take care of some important business that I could not delegate to anyone else. I expect to return within the next few days but my schedule is necessarily still subject to change. I'll call you when I get back."

"So, what do you want me to tell our associates?"

"Saeed, I'd like for you to do two things. First ask our associates to put up as many of the Ahura Mazda posters as they safely can. I'd like to flood the country with those posters right away to get people talking and maybe generate some TV news commentary."

"OK. I can do that. What else?"

"I also want you to make an appointment to meet with a man named Arteman Hamadani. He's a constitutional scholar who has written a book setting forth a compelling argument that Iran needs an entirely new constitution. He advocates constitutional guarantees for freedom of speech, religious tolerance, and government support for the kind of free market reforms necessary to make Iran more competitive in the global economy."

"I've never heard of him."

"Do some research into his writing so you're prepared. When you contact him, introduce yourself as a small business owner who admires his work and believes Iran desperately needs the kind of reforms he's advocated. Ask if he's ever considered drafting an entirely new constitution for the country. The goal of your meeting is to get him thinking seriously about doing precisely that. He is absolutely the right man to draft a really good constitution for the country."

Saeed chuckled softly. “I should have limited you to one request. What makes you think Hamadani will even see me?”

“He has political aspirations. Tell him that you would like for him to autograph your copy of his latest book. Once you tell him you admire his work, I’m confident that he’ll agree to meet with you.”

Camp Justice Security Checkpoint: Guantanamo Bay, Cuba 8:35 AM

Ahmed Malik and Richard Tillman walked to the building housing the Military Commissions courtroom from the tent that they shared with other members of the defense team. As usual, they had to go through the checkpoint in single file and it always took several minutes to clear security. No weapons, cameras, cell phones, or recording devices of any kind were permitted. Every person that entered the high-security enclosure had to walk through a magnetic scanner and then be sniffed by military working dogs trained to detect explosives. Random pat-downs were also common.

Richard Tillman went through the security checkpoint first. Malik waited for his turn. His entire body consisted of graphene components that would not set off the magnetic scanner and the dogs would not detect the gas stored in a pressurized polyethylene bag he had swallowed before coming to Cuba. The bag’s one liter volume had reduced the capacity of his fermentation chamber so he’d had to increase the frequency of his consumption of whole grain cereal and beans that powered his fuel cell and charged his power supply. He’d invested a lot of money and effort to prepare for this day. The next few hours were critical.

Camp Justice Courtroom; Guantanamo Bay, Cuba 10:45 AM

Judge Wilson and the other Commission members returned to their seats following the morning break.

“General Ross, I’ve decided to grant the defense team’s request that the Commission consider excluding the testimony of the prosecution’s next witness on the grounds that his testimony was obtained using torture. You may proceed.”

General Ross rose to address the Commission judges. "If it pleases the Commission, Mr. Bradford will present the relevant facts that support the defense request that Mr. Logan's testimony be excluded from this trial."

"Mr. Logan, you may proceed," Judge Wilson said with just the faintest hint of resignation in her voice. Every attorney in the room knew what was coming next. The defense teams for the 9/11 defendants had filed literally thousands of pleadings that sought to put the government on trial for using torture. Their obvious strategy was to have their clients found not guilty or at least manage to avoid the death penalty.

Allan Bradford stood and walked to the center of the room to face the Commission members.

"Judge Wilson and honorable Commission members, thank you for this opportunity to present compelling evidence that Mr. Logan's testimony should be inadmissible under the U.S. Military Commissions Act of 2009 because it was obtained using torture. I will be presenting explicit details of the brutal torture methods used by Mr. Logan. In accordance with the Commission's previous decisions in similar matters, it may be appropriate to close the curtains to the observation gallery and turn off the audio feed."

Judge Wilson had anticipated that recommendation and gave the order to isolate the observation gallery. With sober consideration for very genuine national security concerns, and under strong advice from the CIA, State Department, and FBI, most of the explicit details of alleged torture by third party CIA contractors and proxies had been redacted from written records and were not allowed to be disclosed outside of the courtroom and its classified records system.

"You may proceed, Mr. Bradford."

Nodding respectfully to General Wilson, Malik looked around the courtroom as he made his final decisions on how he would deploy his secret weapon. As he turned back toward the bench, he contracted his abdominal muscles to increase the pressure on the polyethylene bag in his digestive chamber. That extra pressure overcame the bag's plastic pressure relief valve and a sophisticated gas began to stream from his mouth. At normal atmospheric pressure, the contents of the bag expanded quickly to fill the room with a top secret gas originally developed in Russia in blatant violation of international treaties prohibiting the development, manufacture, or use of chemical weapons.

In 2002, approximately fifty armed Chechen Islamic separatists seized 850 hostages in the crowded Dubrovka Theater in Moscow. Russian Spetsnaz hostage rescue troops flooded the ventilation system with an incapacitating gas consisting of a dangerous mixture of carfentanil and remifentanil in aerosol form. Carfentanil is a powerful, fast acting tranquilizer used primarily to quickly sedate large animals, such as elephants. Despite its effectiveness, it is not approved for human use because the dosage to incapacitate is only slightly less than the dose that is fatal to most people. Remifentanil is somewhat less potent but longer acting. Both chemicals are derived from a powerful opioid drug called fentanyl but are many times stronger. Combined, they act almost instantly and the incapacitating effect can last for over an hour unless neutralized by a suitable antidote, such as naltrexone.

Within three seconds everyone in the courtroom except Malik was unconscious. Some of them, particularly those who were older, would soon be dead. In Moscow, 129 of the 850 hostages had died along with most of the Chechens Islamists because doctors who rushed to the scene had no idea of the type of gas that had been used or what antidotes might be effective.

Looking quickly around the courtroom to verify that everyone was unconscious, Malik walked over to Richard Tillman and knelt beside him. Pushing Tillman's head back, and pinching his nostrils closed, Malik leaned over the unconscious man and placed his own mouth over Tillman's as though he was going to perform mouth-to-mouth resuscitation. Instead, he exhaled a large volume of the toxic gas directly into Tillman's mouth. He waited a moment to ensure that Tillman was dead and then stood and walked over to General Jhalwan.

Kneeling beside the unconscious general, he removed two white tablets from a small cellophane packet he'd concealed beneath his tongue and moved quickly to General Jhalawan's side. He stuck his left index finger in the general's mouth and placed the tablets under his tongue. Using his right thumb and forefinger, Malik crushed the naltrexone hydrochloride tablets and then pressed the general's tongue down firmly. Within a few seconds, the antidote began to take effect and General Jhalawan opened his eyes.

"Be still general. It will take a minute or two for the effects of the gas to wear off. I must work quickly," Malik told his former boss.

Malik walked around behind the judge's bench and began to undress an Army Colonel whose clothes would fit his own lean body reasonably well. In less than two minutes, Malik had swapped clothes with the Colonel. He then dragged the officer's unconscious body around the bench and over to the end of the defense table. He arranged the Colonel's body in a posture that would suggest he'd collapsed in that spot and twisted his head to one side so his face would not be easily seen. He then undressed Army Brigadier General Ross and walked quickly over to where General Jhalawan had recovered enough that he understood Malik's plan and was already getting undressed. While General Jhalawan put on the uniform removed from General Ross, Malik dressed the limp body of General Ross in the bright orange jumpsuit General Jhalawan had just removed. He then lifted the unconscious officer into the defendant's chair and arranged his body and face so that the first responders would not realize the body in the orange jumpsuit was not the prisoner.

"Are you OK General?" Malik asked.

"Yes. What now?"

"We'll go over by the door to the hallway. There are armed guards posted outside the door. I want you to slump against me and let me seem to be assisting you from the room. After that, just follow my lead."

When they were beside the door, General Jhalawan slumped against Malik as he'd been directed. Malik put his left arm around the General and then pushed the door open and shouted to the guards.

"Help us! We've been gassed! General Ross is still alive. I'm not sure about the others. Call for ambulances and fire trucks. One of you help me get the General to a car. I'll drive him to the Hospital myself. Help the others. Hurry!"

The guard that helped Malik with General Jhalawan was a junior Army Reserve enlisted man assigned to the Joint Task Force. The chaos that was unfolding behind them and the uniforms of a General and Colonel was all he needed to know that he shouldn't ask questions. He helped get General Jhalawan into the passenger's seat of the military police pickup that he and his partner had parked there three hours earlier and handed the keys to the Army Colonel.

"Get back inside soldier. I'll take the General to the hospital. Hurry!" Malik shouted at the scared young soldier.

The guard gave a quick salute before running back to the building. Sirens were already wailing in the distance as ambulances and fire trucks converged on Camp Justice. Malik switched on the MP truck's flashing red lights and floored the gas pedal as he turned onto the road leading to the recreational area at Blind Beach just west of Camp Justice.

Turning left into the rec area, Malik turned off the flashing red lights and drove to the far west end of the beach. He parked the truck on the beach side of one of the picnic shelters where it wouldn't be easily seen from the road.

"Get undressed General. We're going for a swim."

Malik ran over to a grassy area where he'd buried his own custom version of what police and special operations soldiers refer to as a bug-out bag. Inside were a waterproof marine radio, swim fins, snorkel tubes, inflatable life vests, two small waterproof flashlights, and two aerosol cans of shark repellant.

"General, stuff our clothes in this bag and drop it in that trash can over there while I call for our pickup boat."

"Ahmed, I'm not a great swimmer. How far are we going?"

"Several miles sir. The boat can't come too close without drawing unwanted Navy patrol boat attention. But, don't worry, we'll take our time. With the swim fins, snorkel, and the inflatable life vest you'll be fine. If you get tired, we'll inflate our vests and rest until you're ready to go again. There are always a few sharks in the area but we'll each have a can of repellent and I can assure you that any sharks won't stick around once they get the slightest whiff of the chemicals inside these cans. Put on your swim fins while I contact our ride."

When the radio tuned to the guard frequency he'd specified, Malik pressed the transmit button and began to speak in a voice that mimicked the tonalities of typical Haitian speech.

"Monsieur Daudier, this is Henri Jean-Michel. You asked me to let you know if the price of your stock went above your sell limits. Two of your stocks are currently trading slightly above those limits. Haitian Gold Partners is trading at fifty dollars and nineteen cents. Haiti First Bank is trading at seven dollars and seventy-five cents per share. If you want me to sell your stock, please let me know before the market closes at 4:00 PM. Thank you for your business."

General Jhalawan gave Malik a quizzical look. "What was that all about?"

It was a coded message to the men on the boat that's waiting to pick us up. I bought a large yacht in Key West and went through the crime syndicate in Chicago to hire some muscle to support our escape. They're cruising slowly back and forth in the Windward Passage between the east coast of Cuba and the west coast of Haiti while they wait for my signal. They are roughly one hundred miles from here but the yacht is big enough and fast enough to cover that distance in four to five hours. I estimate it will take us that long to reach a point at least three miles off shore. My message was coded to give them the pickup coordinates and the approximate time. The fake price of Haitian Gold Partners provides the latitude of the pickup point. The price of $50.19 per share represents nineteen degrees and fifty minutes north latitude. The price of seven dollars and seventy-five cents represents seventy-five degrees and seven minutes of west longitude. The reference to the market's closing time tells them roughly when we expect to be at the pickup point."

"After the excitement you caused, don't you think the Navy will be boarding every boat that comes anywhere near?"

"Perhaps. But even if the yacht is boarded for inspection it shouldn't arouse suspicion. After all, four men on a multi-million dollar yacht, getting drunk and doing a little fishing, is a pretty common sight in the Caribbean. We'll be in international waters when we reach the pickup point so Navy patrol boats won't have legal authority to board the yacht by then. If one of the Navy P-3 Orion aircraft does an inspection fly-by, you and I won't be on deck where a telephoto lens might help them identify us. I think the Americans have already missed their best chance to catch us. If you're ready General, we need to go."

U.S. Naval Hospital; Guantanamo Bay, Cuba
1:45 PM

General Wilson was wearing a blue hospital gown and sitting by the window in a wheel chair when David entered her room. Two military policmen stood guard outside her hospital room door. After the attack, the base commander was taking no chance that a second attempt might be made on the Commission members.

"Thank you for coming, Mr. Logan. It seems that I need your help."

"General Wilson, I'm glad to see that you're apparently unharmed. What can I do for you?"

"I'm sure you know by now that a man posing as the lead civilian attorney for General Jhalawan released an incapacitating gas in the courtroom this morning and has since escaped with the prisoner."

"Yes Ma'am. I've been briefed."

"We now know that the man posing as a civilian attorney named Allan Bradford was an imposter. But, that's about all we know. Naval Intelligence, the CIA, and the State Department have previously verified that his credentials are valid. There's nothing in his record to suggest that he would have wanted to help General Jhalawan escape. I was hoping you might see something we've missed."

"Actually, I know who the imposter was. When the head of base security briefed me an hour ago, he showed me pictures of all of the attorneys and military officers who were in the courtroom this morning. I instantly recognized the man claiming to be Allan Bradford as Ahmed Malik, the head of the ultra-secret S-Wing of Pakistan's Inter-Services Intelligence Agency. He has been General Jhalawan's strong right arm ever since the General was placed in command of ISI."

"Of course I recognize Malik's name," General Wilson said thoughtfully. "His name appears multiple times in General Jhalawan's case files although to my recollection, there were no photographs of him. I still don't see how he could have managed to fool all of our security vetting."

"General Wilson, such a feat would scarcely tax his ability. Can I ask you a question about a topic that probably exceeds your security clearance?"

"Yes. Of course."

"What do you really know about Ahmed Malik?"

General Wilson gave David a questioning look as she tried to remember everything she'd read in the case files pertaining to Malik.

"I guess I don't really know very much about him. Why do you ask?"

"General Wilson, what I'm about to tell you is classified by the CIA, the Department of Defense, and the State Department as need-to-know restricted. Even though I now work for the CIA, I don't have the authority to read you in on Malik. But given the emergency conditions that exist here today, I'm going to risk my career and possible imprisonment to tell you what I think you need to know. Since I don't have time to ask for permission, I'll just have to ask for forgiveness if I get called on the carpet."

"If it comes to it, I'll put my career on the line right alongside yours. Let's have it."

"Ahmed Malik is not human. He's actually one of two incredibly advanced androids we know about. One like him was thoroughly examined and tested in London recently. That android is so advanced that the best scientists in the United States, the UK, and

Israel all agree that no nation on Earth could have produced him with current technology."

"Are you suggesting that he's an extra-terrestrial alien or some kind of time traveler? That's pretty hard for me to swallow."

"General Wilson, believe me. You're not the only one who thinks that."

"So, where do *you* think he came from?"

"General Wilson, I'm one of only a few people who have at least a rough idea of where the two androids came from. Some pretty smart people would think I'd lost my mind if I told them what I really think about it. I assume you've read my bio and know that I was the leader of a task force formed by Sir Lionel Smythe at MI6 to find some stolen nuclear weapons."

"Yes. Go on."

"Well, I came to be on that task force after my girlfriend and I discovered what we first thought was a dead body partially buried in the sand at Hurd Bank, a few miles east of Valetta, Malta. We soon realized that it wasn't human and we subsequently were able to restore its power. Leaving out a lot of details that aren't germane to the current situation, what we found was an advanced android that is virtually identical, at least physically, to Ahmed Malik."

"The name of the android we found is Nahbi. He claimed to have been in Jerusalem in the year 58 AD and to have been a bond servant to a Jewish religious leader. Later, when Nahbi saw a photograph of Malik while we were trying to locate the stolen nuclear weapons, he recognized him from Jerusalem. That may seem like Hollywood science fiction and I admit that I initially had my doubts too. But, right now, today, I firmly believe that what Nahbi told me is the truth. Never mind that our best scientists can't explain how it could be possible. I believe it is."

"Like Nahbi, Ahmed Malik is an incredibly advanced android robot, virtually indistinguishable from a human. Both androids have been on Earth for at least two millennia."

"I guess I'm missing something," General Wilson said with a puzzled look on her face. "Are these androids from outer space or what?"

"General Wilson, I believe they were built right here on Earth but at some time in the future. I don't know how much you know about Einstein's General Theory of Relativity but one consequence of his equations is that space-time can be distorted to form what scientists characterize as tunnels through time and space. I have gradually come to think that some future scientists figured out how to construct something that allowed them to send things backwards in time. I think Nahbi and Malik were built in the future and then sent back in time. Because they are made of graphene, a carbon allotrope with self-healing properties, I think they never grew old. Unless they are physically destroyed they will last indefinitely."

"You say there are two of these androids?"

"Two we know of. There may be more."

"Mr. Logan, it's clear that Malik doesn't mind killing people to achieve his objectives. Three of the older attorneys in the courtroom this morning, including my friend Admiral Gatlin, died from the gas Malik released. What about the other android? Is it also indifferent to murder?"

"No Ma'am. Nahbi is one of the most decent, kind, and morally upright sentient beings I've ever known. Malik is just the opposite. I believe that the theft of the nuclear weapons and their use to murder millions of innocent French citizens was Malik's idea. I think he manipulated General Jhalawan's enormous ego and lust for power as tools to achieve his own goals."

"Why would Malik go to so much trouble to rescue General Jhalawan?"

"I suspect that Malik intends to help General Jhalawan take over the government of Pakistan. As you probably know, the Pakistani government is pretty shaky. Many observers think the military will soon launch a coup d'état to replace the civilian leadership as they did in 1999. General Jhalawan is arguably the strongest military leader the country has ever produced. If Malik can get him safely back to Pakistan, Jhalawan could be the next leader of the country."

"Do you think Malik and Jhalawan are still somewhere here on base?"

"No Ma'am. I doubt it very much. Malik would have put a masterful escape plan in place. My guess is that they're already off the island and headed back to Pakistan."

"Thank you, Mr. Logan. I need to call my boss so he can brief the Secretary of Defense. I'll need to repeat most of what you've told me. I hope you understand. As I said, I'll stand with you if some politician wants to start a witch hunt but within an hour or so, most of the national defense and national security members of the President's cabinet will have been briefed on what you've told me.

"Yes Ma'am. Understood."

CIA Headquarters; Langley Virginia
2:20 PM

Pete Nocona had just walked back into his office when David's call came in on his secure line.

"Pete? David. Jhalawan's gone."

"What happened?"

"Malik was posing as a civilian defense attorney named Allan Bradford. He used some kind of gas to knock everyone out. He and Jhalawan escaped in the confusion. The base commander acted quickly to seal off all exit portals but Malik probably anticipated that and had a plan to get them off the island quickly before the Navy could react. My guess is that they're long gone. There's something else I need to tell you. General Wilson survived the attack although some of the older officers didn't. She was taken to the Naval Hospital and as soon as she recovered, she asked for my help. I'd just been briefed by the head of base security and had recognized Malik. I know she wasn't cleared for everything we know about Malik but I felt she needed to know so I told her who Malik really is. Technically, that means I divulged classified information I wasn't authorized to reveal. I fully understand the possible consequences of my action and wanted to tell you before you got blind-sided."

"OK. I'll tell Roger but I don't think it's going to be a problem."

"Thanks, Pete. As is so often the case, timing was everything. I was at Camp Justice yesterday morning to register my security clearance and get a briefing on what can and can't be discussed outside the courtroom. Malik was literally just a few yards away from me inside the courtroom making defense pleadings to the Commission. The prosecution team didn't need me yet but if I'd stuck around in the visitor's gallery to observe, I'd have recognized Malik without him seeing me. It was just that close."

"David, don't make travel arrangements until I talk to Roger. I think there's a good chance he's going to want you to go right back to Israel. President Marshall has been working hard to regain Israel's confidence in us as a reliable security partner. He's taken a page out of FDR's play book to provide support to them for a critical mission they're planning. I think Roger and the President would both feel better if you were there to offer any help you can and, of course, to keep us better informed than our secretive Mossad cousins typically do."

International Waters; South of Guantanamo Bay, Cuba 7:45 PM

The sun had set more than a half hour earlier and it was getting dark. For some reason, the boat Malik had arranged hadn't shown up and hadn't answered when he'd tried to contact it using the marine radio. Once they'd reached the pickup coordinates, they'd inflated their life vests to maximum buoyancy so they could rest. A Caribbean reef shark had spooked General Jhalawan when it circled them curiously for a few minutes but they hadn't seen it again after it got a whiff of the shark repellant. Malik had roped them together so the wind and currents wouldn't separate them. After more than eight hours in the water, General Jhalawan was exhausted.

"Ahmed, do you think the boat is still coming?"

"Yes sir. Just hang on. They'll show up soon. My guess is that they were boarded and searched by one of the Coast Guard's

Transportable Port Security Boats. The TPSBs are about ten meters long and can do better than forty-five knots. There are four of them currently deployed here.

Holding the marine radio above his head so the signal wouldn't be attenuated by the choppy waves, Malik tried to contact the yacht again. Pressing the transmit key, he again pretended to be a stock broker trying to contact his client.

"Monsieur Daudier, this is Henri Jean-Michel again. I was able to sell your Haitian Gold Partners stock for fifty dollars and nineteen cents per share. Your shares of stock in Haiti First Bank sold for an average price of seven dollars and seventy-five cents per share. I'm still at the office if you want to call me."

Almost immediately, Malik's message was answered.

"Monsieur Jean-Michel, this is Monsieur Daudier. I was hoping you'd call. I'm pleased with the prices you obtained and I'll stop by your office in about thirty minutes to pay my bill. I know I'm a little overdue but I'll explain when I see you."

Wednesday, 8 October

Public Pier; Jérémie, Haiti
3:50 AM

Two Haitian men were waiting on the pier to help them tie up as the big Maiora yacht approached. A few minutes after the yacht was moored, Malik and General Jhalawan got into the back seat of a Toyota SUV for a short ride to the Jérémie airport five miles to the west. The Embrarer Legacy 450 executive jet Malik bought from Kalifa Maghur was fueled and waiting to take off as soon as they were aboard.

General Jhalawan seemed to have recovered from their marathon swim. He'd showered and slept for a few hours during the high-speed run from where they'd finally been picked up.

Once they were aboard the jet and buckled in, Malik placed a call to the men who were holding Deborah and Beth Tillman hostage near Middleburg, Virginia.'

"Kill the woman and her daughter. Bury their bodies in the pasture on the north side of the barn. Remove all signs that they were ever in the house and then wipe down every possible place where any fingerprints might be taken by a police crime scene team. Drive her car back to her home. Wipe it down and then lock the garage door. Make sure no one sees you."

Navy Lodge; Guantanamo Bay, Cuba
7:30 AM

David had just returned to his room after grabbing a quick fast-food breakfast at the McDonald's a few blocks away when his cell phone range.

"David. Pete. Roger spoke to the Director after your call yesterday and they want you to go back to Israel."

"When do I need to be there? I'd like to take a couple of days of personal leave to see Elena if the timing isn't critical."

"Sorry. They want you there as soon as you can get there. Nahbi's briefing the Israeli PM on Friday morning and Levi has agreed to let you sit in. You can't get there in time if you have to wait for a seat on the AMC flight to Andrews. I've arranged for one of our Gulfstreams to pick you up at the leeward side terminal at noon today. They're already wheels up out of Miami. They'll take you to the executive jet terminal at Dulles International. The pilot estimates they'll land at Dulles around 3:15 PM. I'll pick you up at the terminal and drive you to your apartment to pack some fresh clothes. With a little luck, we'll have time to grab some dinner before your next flight. I've got you booked in business class on the red-eye United-Lufthansa code-share flight to Frankfurt that leaves Dulles at 6:20 PM. It's a non-stop flight that lands in Frankfurt at 7:55 AM tomorrow morning. Your connecting flight will be El Al flight LY358 that leaves Frankfurt at 11:00 AM and gets into Tel Aviv at 4:20 PM. I've got you booked back into the Tel Aviv Beach Crowne Plaza for a week."

"Wouldn't it be a lot easier for the company plane to just fly me to Tel Aviv?"

"Easier? Yes, but the Miami Chief of Station shut me down when I asked. Said he couldn't spare the plane that long."

"OK. Not surprised I guess. Any idea of how long I'll be in Israel this time?"

"Assume a week. We can fine tune the plan once you're settled in. See you at the airport."

Levi Reznik's Office; Mossad HQ; Herzliya, Israel
11:40 AM

"Levi, can you spare a couple of minutes?" Gavriella asked her boss.

"Sure. What's on your mind?"

"I know you really have no choice but to send Nahbi back into Iran. I want to go with him."

"I guess I'm not surprised but don't you think having you with him might just make his job harder? I know you're in love with him and he's admitted that he's in love with you too. But even if I suspended my long-standing rule against inter-office romances, I can't ignore the strategic importance of Nahbi's mission to the security of our country. So, unless you can convince me that letting you go with him would actually strengthen the mission, and he agrees, I'll have to say no."

Gavriella took a deep breath before she began. "I anticipated you'd say something similar so I hope I have my ducks in a row. Here's how I see it."

"Nahbi has already been compromised so going back to Iran is now far more dangerous than his original mission. Malik somehow figured out that he was in Iran and Malik has a close relationship with Ali Yazdi, the head of Iran's Intelligence Agency. So, I think it's all but certain that Yazdi knows that Nahbi is an Israeli intelligence agent and that his men will be watching for him. I doubt that Yazdi knows the details of Malik's abduction of Nahbi and David Logan but we've got to assume that it will be much more dangerous for Nahbi to move around in Iran now."

"As you know, Iraq has used chemical weapons against Iranian soldiers in the past. Today, there are more than seventy thousand disabled Iranian veterans from those poisonous gas attacks. They are given the respectful title *janbez*, which in Persian means those who were willing to lose their lives. Most Iranian civilians don't quite know how to respond to disabled veterans they see on the streets. They usually ignore them if they can."

"I think a very good disguise would be for Nahbi to pose as a disabled veteran. I could pose as his wife to help him in and out of vehicles and push him around in a wheelchair. I'd be wearing a roosari scarf over my head and loose-fitting, full-length dark clothing. I would intentionally dress in the most conservative way possible to signal that we were devout Shiite Muslims. Of course, if we had to fight, I could certainly do my part. As strong and quick as Nahbi is, he's been out-numbered and overcome by force before. Any attackers would be unlikely to suspect that a conservatively-dressed Shiite woman was a Special Forces commando. If it came to a fight, I'd have the advantage of surprise in helping Nahbi."

"There's one last thing. The primary purpose of having Nahbi go back is to mobilize the resistance at just the right time and in just the right way to trigger a popular uprising. Many of the young adults Nahbi has recruited are women. I think I can help motivate and organize them to improve their effectiveness."

"How would you get out?" Levi asked. "Nahbi's talents are such that we always assumed he'd find a way to get out safely. Having you along will make that more difficult and potentially more dangerous for you both."

"I've thought about that too. The Saudi's have been signaling for some time that they would cooperate with Israel to find a way to terminate Iran's nuclear weapons programs that threaten them too. They have a naval support activity on the Persian Gulf at Al Jubail about 150 miles southwest of Iran's Bushehr nuclear reactor plant. They have four Sikorsky MH-60 anti-submarine warfare helicopters based at Al Jubail. The MH-60 has a range of about five hundred miles, which is ample for our needs. I'm sure that when the time comes, Nahbi and I can borrow a boat and get far enough off the Iranian coast for a Saudi helicopter to pick us up."

Levi smiled tolerantly. He'd have been very surprised if Gavriella's rationale hadn't been persuasive.

"Let me think about it. I'll discuss it with Nahbi and let you know."

Baskov's Restaurant; Tel Aviv Israel
8:10 PM

Gavriella and Nahbi had already placed their orders and were sipping some red wine before dinner. After a few awkward minutes of trying unsuccessfully to ignore the gorilla in the room, Nahbi finally broached the subject they needed to discuss.

"Levi is leaving it up to me, Gabe. The PM has authorized you to go with me if and only if Levi and I both think it will improve the mission's success probability."

"Do you?"

"Actually, yes. I think the scenario you proposed makes good sense. It would provide excellent cover in case my identity has been compromised as now seems likely. For Israel's sake, I should probably agree for you to go with me. But, it's not quite that simple, is it?"

"Why? Because one or both of us might be killed?"

Nahbi shook his head and smiled at her obvious disdain for a possibility that would terrify most people.

"No. We both know that's always a risk when soldiers go in harm's way. If we're killed, at least we'll be together and I realize that's a big part of your motive for asking to go with me. I see clearly that you've made your peace with the possibility of your own death and if I'm going to be destroyed you want to be with me. I understand all of that."

"So? What is it?"

"If the mission fails, *and it might*, Israel is going to need every citizen soldier it can muster. Iran will retaliate with everything they have including any nuclear weapons our air force doesn't destroy. Beyond question, you could help me succeed but if I don't, you'll be needed here."

Gavriella's eyes filled with tears and she whispered in a barely audible voice, "If you're killed, I won't be of any use to Levi or anyone else. I've already lost one man I loved. I don't think my sanity can survive if you're killed too."

Nahbi reached across the table and took both of her hands in his. Pausing only a moment, he gave her the answer she wanted to hear. "OK, Gabe. We'll go together."

Gavriella smiled through her tears and quiet sobs of relief as she squeezed Nahbi's hands tightly. "Thank you, my dear."

Nahbi gave her a very tender look and then spoke softly. "Now that's settled, I want to ask you something. I'd like for us to get married before we leave Israel, possibly for the last time."

"Is that a proposal?"

"Yes. Will you marry me?"

"Yes."

"Say it again. I loved hearing you say yes."

"Yes, yes, yes. I'll say it a thousand times more if you need more assurance."

With her happy acceptance, the mood of their dinner changed from somber and tense to joyous.

"Now that you've finally agreed to make an honest man out of me, I have a suggestion. I understand that I may be encroaching on territory that is usually the exclusive domain of the bride."

"In what way?"

"When our Prime Minister granted me citizenship, he said I would have the right to get married. If it's OK with you, I'd like to ask him to perform a simple ceremony with just a few of our friends present. He's already asked Levi and me to brief him on the final mission details before we go. I think we could ask him to do it at the end of the briefing."

"Perfect."

Thursday, 9 October

Crowne Plaza Hotel; Tel Aviv Beach, Israel
7:50 PM

David was exhausted after traveling over fifteen thousand miles in less than a week. After claiming his luggage and clearing customs and security at the airport, he took a taxi to the Crowne Plaza hotel. He'd just stepped out of the shower and was mentally debating whether he had enough energy left to go out for dinner when his room phone rang.

"David, this is Gavriella. Nahbi and I are in the lobby. We've come to take you to dinner. We have something to tell you."

"OK. I just got out of the shower. Give me a few minutes."

La Chaumiere Restaurant; Tel Aviv-Yafo, Israel
8:35 PM

The popular restaurant known as *The Cottage* was crowded and their table wasn't quite ready when they arrived. While they waited, Gavriella was uncharacteristically animated. David suddenly realized that he'd never seen her so obviously happy before. It was a remarkable transformation.

"David, I know you must be tired but thanks for coming. This is my second-best favorite restaurant in Tel Aviv. Nahbi and I had dinner at Baskov's last night so I thought we'd come here for a nice change. The food is excellent. Although it's primarily a French restaurant, the menu is eclectic with Mediterranean and Asian dishes too. I'm sure you'll enjoy it."

The hostess stood quietly nearby until Gabe finished speaking. "Miss Adler, your table is ready. Please follow me."

When they were seated, David's curiosity couldn't wait any longer. "OK. What's the big mystery?"

Before Gabe answered, she took Nahbi's hand and then leaned forward with a radiant smile on her face."

"Nahbi has asked me to marry him."

David was stunned and it showed on his face. "Wow, I didn't expect that. You two are really getting married?"

Gabe laughed. "Yes, David. Really."

"When?"

"Tomorrow morning after the briefing."

Crowne Plaza Hotel; Tel Aviv Beach, Israel
10:55 PM

When David returned to his room, he noticed that the message light on the hotel phone was blinking. He sat on the edge of the bed to listen to the message. It was from Elena, asking him to call her.

"Hello Elena. I got your message. How did you know I was back in Israel?"

"Your personal cell phone kept giving me a message saying your number was not in service. I didn't know what to think of that. I left several messages on the answering machine at your condo asking you to call me back. When I didn't hear from you, I began to worry. I finally called your boss. Pete told me where you were staying. He didn't say much but I sensed that you'd been busy."

"Yes, I have. I'm sorry you couldn't get in touch with me. There wasn't any service for my personal cell phone where I've been. I probably shouldn't give the details over an unsecured phone but I've flown over fifteen thousand miles in the last week. I'd just gotten settled into my hotel room and was planning to give you a call and then catch up on my sleep when I got a call from Gavriella and Nahbi. They wanted me to join them for dinner. That's why I wasn't in the room when you called."

"Is Nahbi back to normal, or at least whatever normal means for him?"

David laughed. "Yes, so far as we can tell, Aharon's backup plan worked perfectly. Now that I think about it, what's normal for Nahbi is about to change dramatically. Nahbi and Gavriella shocked me this evening with the news that they're getting married tomorrow morning. The Israeli Prime Minister will preside over a simple Jewish wedding ceremony at the end of a briefing I'll be attending.

After everything we've been through together, I'm glad I'll get to be there."

"That's pretty incredible. Good for them. Something just occurred to me. What are we going to call them? Somehow Mr. and Mrs. Nahbi doesn't seem quite right."

David laughed. He'd had the same question. "Since Nahbi has no last name, Gavriella asked him how he felt about taking hers. In Israel, almost everyone below retirement age is either on active duty or in the national reserves. The government issues both men and women a rifle and pistol on their eighteenth birthday. That has produced successive generations of fiercely independent, self-reliant women which certainly describes Gabe. It's not uncommon today for Israeli brides to keep their last names although some of them hyphenate their own last names with their husband's. So, to answer your question, the names on their wedding certificate will be Nahbi and Gavriella Adler.

"I like that. Are you going to call Tiera?"

"Yes, I'll call her when we're done. Is she in London?"

"No, she's here for her parent's wedding anniversary. You can reach her on her cell phone. David, I miss you. When can I see you?"

"I'm stuck here in Israel for at least a week. You could come here I suppose but I wouldn't be able to spend much time with you. There's a lot going on and President Marshall wants me involved. I'd hoped to stop in Malta on my way to Israel but the timing didn't work. Maybe I can come to your place before I go home. I'll try. I miss you too."

"It's good hearing your voice at least. Get some sleep. You can dream about me but no X-rated scenes, OK? I love you."

"Love you too. Good night."

After he hung up, David couldn't keep from smiling as he thought about Elena's X-rated dream comment. It was like telling someone not to think of a pink elephant. With her mischievous nature, he wondered if that wasn't precisely why she'd made the comment.

After a few moments, he forced some pleasant mental images from his mind and dialed Tiera's number. She answered his call immediately and seemed genuinely happy for Gavriella and Nahbi. David sensed that she was with someone so he kept the call short.

Even though he'd been careful with what he said to Elena over the unsecured hotel phone line, he'd said too much. His hotel room phone was clean but the electronic bug on Elena's phone that had led to his abduction was still in place. Everything he'd said to her had been recorded.

Friday, 10 October

Mossad HQ Main Conference Room; Herzliya, Israel 9:00 AM

Levi Reznik tapped his knuckles on the table to signal that the briefing was about to begin. "Good morning Mr. Prime Minister. Our briefing this morning will summarize what has already been accomplished in our undercover mission to Iran and what remains to be done. Nahbi, please take over."

"Mr. Prime Minister, We have been successful in our mission objectives. A large number of Iranian citizens now see their own leaders as the primary cause of the economic hardships they have endured under the United Nations sanctions."

"Iran's religious leaders have sown the seeds of their own destruction by constantly evoking the name of Allah to justify their hatred of Israel and the United States. By brainwashing their less sophisticated citizens, they made it easy for me to persuade many of them that the plagues visited on the holy city of Qom were unmistakable evidence of the wrath of Allah against their leaders."

"We never expected the more sophisticated and better educated Iranians to see a supernatural cause, but the message that their leaders had brought hardship on their own people was persuasive even to them. Many of the younger adults already see their elderly, radically theocratic leaders as the cause of most of Iran's problems. I reinforced that view whenever possible.

"By distributing large sums of worthless counterfeit money, we have made Iran's runaway inflation even worse. Many things are now too expensive for Iran's poorer citizens. The economic hardships are fanning the embers of their discontent. It will not take much for those embers to burst into flame."

"An organized network of dissidents is now in place. We can communicate with them via encrypted signals routed through our Ofeq-10 satellite. Many of them are idealistic college students ready to lead public demonstrations."

"We have successfully prepared a sizable minority of Iranian citizens to vent their anger at the medieval mentality that has gripped Iran since the Shah was deposed. Most Iranians will now see the bombing of the nuclear sites as a justifiable international response to their leader's support for terrorism and the threat that Iran will use nuclear weapons. Many, perhaps even most, Iranians will be embarrassed and angry that their country has been attacked but will blame their own leaders. By leveraging that anger, we can trigger widespread riots and protests that will topple the government in much the same way the 1979 revolution brought a sudden end to the Pahlavi dynasty."

The Prime Minister interrupted him at that point. "What remains to be done that cannot be done from here?"

Nahbi had anticipated the question. "All popular uprisings need a leader. Among many of the young adult attendees at Jamkaran Mosque, I am seen as a wise man that has provided answers to many of their questions. Among the growing network of disaffected college students, I'm seen as a sophisticated and educated man who would like to see an end to state-sponsored terrorism and the hate-filled propaganda and obsessive pursuit of nuclear weapons that have made Iran a pariah to the rest of the world. Most of these young college students are technically astute and they yearn for a more open society. I need to return to Iran to provide the necessary leadership to catalyze the widespread protests that will bring down the current regime."

When Nahbi's briefing was finished, he answered a few questions and sat down. The Prime Minister then rose and waited a moment until he had everyone's attention.

"It is my great honor to have been asked to bless the joining of two of Israel's unsung heroes in marriage. Nahbi and Gavriella, please join me."

When Nahbi and Gabe were standing in front of everyone, the Prime Minister spoke again.

"Ours is an ancient land. Many of our customs and traditions are thousands of years old. Marriage was first mentioned in Genesis. Scholars believe that Moses wrote the first book of the Tora and the Christian Bible more than three thousand years ago. By ancient laws and customs, a Jewish marriage has long been an exceedingly simple affair. The bride must accept something of modest value from the groom and the groom must declare his intent to marry the bride and to consecrate their marriage. When these simple acts are witnessed, the couple is married."

"Nahbi, what thing of value do you offer Gavriella?"

"I offer a simple gold ring to symbolize our love for each other. I've had it inscribed with an ancient Hebrew phrase that is especially appropriate for one of Israel's most courageous women. The inscription reads, *Eshet Chayil Ateret Balalah* which translates as *A Woman of Valor is Her Husband's Crown.* Everyone in this room knows that Gavriella is a woman of valor. She will be a crown of honor for our marriage and for me as her husband."

There were murmurs of agreement around the big conference room. Everyone in the room knew of Gavriella's bravery and the horrific wounds she'd suffered in defense of her country. The Prime Minister waited a moment until the room was again quiet.

"Gavriella, do you accept Nahbi's gift and his offer to take you his wife?""

"Yes, with joy overflowing, yes!"

"Nahbi, do you accept Gavriella as your wife from this day forward?"

"Yes, I do so gladly for she is very precious to me."

"And, Nahbi, do you make a vow of consecration of your marriage?"

"Yes. In front of you as the leader of our country, and standing before our friends and protectors of our homeland, I promise to cherish Gavriella and to always be the very best husband for her that I can possibly be."

"Then, it is my great honor and privilege to declare that in accordance with ancient Jewish law and traditions, you are husband and wife in the sight of almighty God. *She'Elohim yevrach otcha* -- may God bless you."

Gardez Airbase and Public Airport; Gardez Afghanistan 4:20 PM

After a grueling flight of over nine thousand miles with five refueling stops, and more than twenty hours in the air, the Embrarer Legacy 450 executive jet touched down two miles north of Gardez, Afghanistan. The small regional airport sixty miles south of Kabul had been used by American military forces as a collection center in the largely unsuccessful program to disarm Afghanistan's local militias. The runway was only one mile long and there were almost no support facilities but it was less than sixty miles to the Haqqani base just over the Pakistani border.

Siraj, Jalaluddin Haqqani's son, was waiting beside a dusty Toyota Land Cruiser when Malik and General Jhalawan emerged from the plane.

"Āssālam ålaykum General Jhalawan," Siraj Haqqani said with evident respect. My father is eager to welcome you to his home.

"Wālaikum assalām Siraj," General Jhalawan responded with the dignity and formality that Siraj would expect from him.

"And Ahmed. Welcome back my friend," Siraj said with a broad smile on his face. "I can hardly believe that you succeeded. I want to hear all of the details. None of us dared imagine that you could penetrate American security and then, against all odds, manage to escape from Guantanamo. But, I am being rude. Please forgive me. I know you must both be very tired after such a long trip. The roads from here to Tari Mangal aren't very good. It will take about three hours, so perhaps we should be going. My father's housekeeper will never forgive me if we're late for the welcoming feast she's prepared."

Tari Mangal; Federally Administered Tribal Area, Pakistan 7:30 PM

Jalaluddin Haqqani was waiting on the porch when they arrived. "Āssālam álaykum General Jhalawan. Welcome back."

Two hours later, after an excellent dinner, Jalaluddin rose from the table and led his guests into the living room of his modest home. His housekeeper brought a pot of the strong black tea known as chai, along with some biscuits and small cakes. There was much to discuss.

General Jhalawan and Malik could not return to Islamabad without risking being locked up in a maximum security prison or put to death. Jalaluddin Haqqani understood that. After his housekeeper left the room, he spoke what was on his mind.

"General Jhalawan, we owe a debt to you that can never be fully repaid. I have taken the liberty of arranging a sanctuary residence for you here in the Tribal Area. I will provide some of my best men to protect you while you are here. How else can I be of service?"

"My old friend, you have done much already. Ahmed has told me how your men rescued him and got him safely out of the country. Neither of us would be here tonight without your help. But, you are right. There is more to be done. It has been clear to me for some time that our President is a weak man who must be replaced. Many senior officers are loyal to me. Ahmed and I will hold quiet discussions with those who can help us replace the current administration. With your help, I intend to take over as the leader of our country. There is one more thing. I want the American David Logan and the Israeli agent Nahbi eliminated. I don't care what it costs. Can you do this for me?"

Jalaluddin glanced at his son before responding. Siraj nodded almost imperceptibly in answer to his father's unspoken question.

"Yes, General. We can do this for you. We will need Ahmed's help with current intelligence and perhaps other key information but yes, Siraj will personally take care of them for you."

Hunting Cabin; Tari Mangal, Pakistan
10:26 PM

Just south of the village, two seasonal streams converged before continuing downhill to the southeast. At the end of the long, hot summer, both rocky streambeds were dry. One of them meandered up the valley west of the village. A mile upstream was a hunting cabin that Jalaluddin Haqqani had built for himself several years earlier. He often let his senior men use the cabin but had kept it vacant for the past few weeks in case Malik's rescue attempt succeeded. Pakistan's Army avoided the tribal area out of a fearful respect for almost ten thousand well-trained and well-equipped Haqqani fighting men. With every house in the village occupied by Jalaluddin's men and their families, the cabin was a perfect safe house for General Jhalawan.

The primitive dirt road to the cabin was very rough. Siraj tried to dodge the potholes but there were so many of them that he couldn't avoid them all. A few hundred meters from the cabin, he stopped the SUV where two of his men were standing guard and got out to speak to them.

"Stay alert. I don't expect trouble but our guest is the most important man in all of Pakistan. Guard his safety with your lives."

At the cabin, two more of Siraj's men were on duty. He issued the same orders to them and then escorted General Jhalawan and Malik into the cabin.

"General Jhalawan, my father and I are embarrassed that we have nothing better to offer you. I hope you will forgive us for such a poor demonstration of our respect. My father told me that you are a passionate hunter. In the gun case are several of his favorite rifles. Please feel free to use them. We sometimes see trophy-size Markhor rams in the area. Perhaps you will find time to do some hunting while you are here. There are several hundred rounds of ammunition in the drawers at the bottom of the case. All of my father's rifles are sighted-in for a zero hold at two hundred yards. Even the American snipers would be impressed with the accuracy of these fine old rifles. I've also placed pistols with extra magazines in the night stands in both bedrooms."

"Siraj, please convey my appreciation to your father for all that he has done. This cabin brings back memories of hunting in the mountains near my home in Janawai when I was a young man. Ahmed and I will be very comfortable here while we make our plans."

Malik nodded in agreement. "Yes. I also thank you and your father for all you've done for us. Let me walk with you to your vehicle."

As they walked toward the SUV, Siraj had a question. "Ahmed, do you know where Logan and Nahbi are?"

"No. Not yet. I have some calls to make. I'll know in the morning. Shab bākhair, Good night my friend.

Saturday, 11 October

Hunting Cabin; Village of Tari Mangal, Pakistan 7:20 AM

Ahmed Malik and General Jhalawan were sitting on the porch talking when Siraj parked and walked up to the cabin.

"Subha bākhair. Good morning. My father's housekeeper sent you some breakfast."

Inside the cabin, Siraj unloaded a basket onto the small table. It was a typical Pakistani breakfast of boiled eggs, flatbread, minced meat, yoghurt and strong black tea.

While they ate, Malik shared the results of several calls he'd made.

"Logan and Nahbi are both in Tel Aviv. Logan is staying at the Crowne Plaza Hotel on Tel Aviv beach. Nahbi lives with a Mossad agent named Gavriella Adler in her condo at 36 Retsif Herbert Samuel Street in Tel Aviv. All of them spend most of their waking hours at the Mossad headquarters but a take-down there is out of the question. The security is considered impregnable by anything less than an armored assault supported by helicopter gunships."

"Security at the hotel is better than what is typical in most western cities but you would have little difficulty in penetrating it if you chose to make the hit on Logan while he's in his room. Security at Adler's condo is similar and can be easily breached."

"Adler is a senior Israeli Special Forces officer in addition to being a senior Mossad agent. If she is present when you make the hit, do not underestimate her. My advice is to put her out of action immediately. However, Nahbi is even more formidable. I don't think Adler and Nahbi can be taken out by one man. I think it will take a team of two or more specialists to put them down."

"Logan is somewhat less formidable than Nahbi or Adler but he too has Special Forces training and served with the American Army in Afghanistan. The French Special Forces considered him fully qualified to join their mission. You should not underestimate him either."

"Adler, Nahbi, and Logan sometimes have dinner together at local restaurants. Two of their favorites are Baskov's and La Chaumière in central Tel Aviv. Both restaurants have security cameras and alarm systems for night protection but nothing that would make a hit at either location especially difficult."

"Logan has a girlfriend in Malta. I have an active bug on the phone in her apartment in Valletta. Thursday night, a call from Logan to Elena Scarlatti was recorded. Logan told his girlfriend that he'd try to come to Malta when his assignment in Israel was completed. We don't know when that might be but if you cannot take him down while he's in Tel Aviv, Malta would offer a better opportunity than waiting until he's back in the United States. I'll provide updates to you when anything changes. I'd like for you to leave misleading forensic evidence to suggest the hits were carried out by Iranian agents."

General Jhalawan waited until Malik was finished before speaking. "Siraj, I have arranged for the equivalent of five hundred thousand U.S. dollars to be transferred into the account I set up to support you in the past. Let me know if that's not enough."

"General Jhalawan, I am astonished at your generosity. That is a far greater sum than I imagined."

"If you can do what I've asked, it will be one of my best investments. Both of these men have been a great deal of trouble to me."

Mossad Headquarters; Tel Aviv, Israel
10:30 AM

Levi's assistant called Nahbi, Gabe, Aharon, David, and Avi Ben-Shimon to ask them to join him in the conference room for a short meeting. When they were seated, Levi asked Avi to close the door.

"I've just gotten off the phone with the Prime Minister. He approved the revised mission schedule so Nahbi and Gabe's deployment to Iran is scheduled for next Wednesday night. It'll be a HALO drop from one of our Bardelas aircraft. We've all got a lot to do to help them get ready so I'll keep this short. I'm going to host a belated wedding dinner next Tuesday night at Baskov's in honor of our newly-weds. I've made reservations for the six of us at 8:00 PM."

Crowne Plaze Hotel; Tel Aviv Beach, Israel
10:05 PM

When he got back to his room after another long day, the first thing David did was call Elena. She answered on the second ring.

"Hello, Elena. Sorry I couldn't call sooner."

"It's OK. I've been expecting your call. I know you can't tell me what you're working on but do you know if you'll be able to come to Malta anytime soon?"

"As things stand right now, it looks like I'll leave here next Thursday morning. That's still somewhat tentative, but I think I'll arrive in Malta sometime Thursday afternoon. I haven't checked the flight schedules yet but once my departure date is confirmed, I'll make reservations and call you."

"That's wonderful. Do you know how long you'll be able to stay?"

"I've asked for three days of personal leave. Pete gave me a hard time for being gone so much. He finally said OK on the condition that I have to be back in my office no later than the following Tuesday. That means I'd have to leave Malta Monday morning."

"I'm disappointed that it can't be longer but I understand. You must be exhausted from all the travel and long hours. I won't make big plans while you're here so you can take it easy for a change."

"Thanks. That sounds great. Oh, before I forget, I might not be able to call you next Tuesday evening. Levi has invited a few of us to have dinner with Nahbi and Gabe at Baskov's as a sort of belated wedding celebration. Fortunately, he's made the reservations for 8:00 PM which is fairly early by Israeli standards. I'm sure there'll be a little food and a lot of wine. I probably won't get back to the hotel before midnight. I don't think Levi is much of a party animal but he might surprise me."

Hunting Cabin; Village of Tari Mangal, Pakistan
1:20 AM

Malik didn't need to sleep but he maintained the illusion that he did by feigning fatigue and retiring to his room. Saying good night, he closed the door and lay down on the bed. He was pulling a blanket over himself as part of the charade when his satellite phone vibrated.

"What do you have?" He listened for a few moments and then hung up. He immediately placed another call.

Tuesday, 14 October

Baskov's Restaurant; Tel Aviv, Israel
8:00 PM

Levi Reznik was not a sentimental man. In his profession, being sentimental could get people killed. Yet, he knew that there was a high probability that Nahbi and Gavriella would not return from their mission so he wanted the evening to go well.

The food and service were superb and wine flowed freely. Levi offered the first of many toasts to the new couple. Sometime around nine-thirty, the restaurant's owner came to their table carrying two bottles of expensive red wine that he placed in front of Nahbi and Gavriella.

"Miss Adler, these are a wedding present from a gentleman who was listening to your toasts. After he paid his bill, he went to his car and brought these back. He asked me to give them to you after he left. He told me that he's the director of marketing for the Golan Heights Winery in Galilee and that this Cabernet Sauvignon is arguably the finest red wine ever produced in Israel. I can tell you that his claim was no exaggeration. This particular vintage has won more international awards than any other Israeli-produced wine. It has become quite rare as wealthy people bought up most of the vintage for their private cellars. Would you like for me to have the sommelier open a bottle to let it breathe?"

"Yes, please, but only one tonight. Would you set the other bottle aside in your wine cellar for us to have with dinner on our first anniversary?"

"Yes. Of course. The sommelier will bring it back to your table when he's satisfied that it's ready to enjoy."

Three minutes later, a massive explosion killed everyone in the kitchen. If the bottle had been opened at tableside, as is customary in most restaurants, General Jhalawan's contract would have been fulfilled. Sitting in a rented car across the street, Siraj Haqqani at first thought his job was complete when he heard the deafening sound of the explosion. Then he saw his targets emerge from the restaurant covered in dust but apparently not seriously injured.

The two wine bottles were a type of improvised explosive device similar in concept to the improved Claymore mines developed at the Army's Picatinny Arsenal. Each bottle contained two pounds of C-4 plastic explosive and three hundred steel pellets. The C-4 explosion exploded when the cork was extracted, triggering the embedded detonator. The initial velocity of the steel pellets was more than four thousand feet per second but the masonry wall between the kitchen and dining area absorbed most of the blast.

Beirut Rafic Hariri International Airport; Beirut, Lebanon 2:20 AM

It had taken Siraj Haqqani two hours to drive from Tel Aviv to a petrol station near the Bar'am kibbutz located three hundred meters south of the Lebanese border. With a suitable financial inducement, Hezbollah had created the explosive wine bottles for him. The Shiite Muslim terrorist organization's massive network of tunnels under the Israeli border had allowed him to escape. A taxi had taken him from the small Lebanese village of Muhaybib, near the tunnel's exit, to the Beirut airport where Malik's executive jet had been waiting. As the sleek business jet accelerated for takeoff, he called Malik.

"Hello Siraj. I've been waiting for your call. Has your visit been productive?"

"No, Ahmed. I was watching the restaurant when the bomb went off. I was about to head north to my exit point when I saw the targets come out of the building covered in dust but apparently unharmed. The blast was very powerful so I cannot explain how they survived."

"That's very unfortunate. Where are you now?"

"We just took off from Beirut. I've instructed your pilot to return to Gardez."

"No. Tell him to take you to Zuwara, Libya instead. David Logan will stop over at his girlfriend's place in Malta on his way back to Washington. Kalifa Maghur's Pershing 88 yacht is still moored at an oil terminal just east of Zuwara. It's registered in Italy and you can use it to get you into Malta. I'm sure the Greek passport you used to enter Israel has been compromised and won't be any good within a few hours. Do you have another EU passport with you?"

"Yes. I have several, including one from Spain. My father always insisted that I get a college education so I have a degree in mechanical engineering from the Polytechnic University of Catalonia. I'm fluent in Spanish so the passport should work just fine."

"Good. When you get to Malta, go immediately to the reception dock in Msida marina. With an EU passport and EU registration on the yacht you should have no difficulty in getting an entry permit from the Immigration Police and Customs office at the marina. The marina is less than a mile from the University of Malta campus. Get a hotel room in the area and let me know where you're staying. I'll call you when I know Logan's schedule. A silenced pistol would be ideal if you can find one but I'll leave the details to you."

Wednesday, 15 October

Main Conference Room; Mossad HQ; Herzliyya, Israel
8:00 AM

For those in the conference room who had been in the restaurant, it had been a long night. After the explosion, ambulances had taken them to the Sourasky Medical Center. When the doctors finished examining them and cleared them to be released, the national police had questioned them for over two hours. It was three o'clock in the morning when they were finally allowed to go home. By then, the Prime Minister had arranged for Shayetet 13 Naval Commandos to provide protection and asked Levi to schedule a meeting to discuss the attack and its possible ramifications for planned missions.

The large conference room was packed with Mossad agents, Israeli Defense Force officers, and senior civilian government agency officials. David was the lone CIA representative. The Israeli Prime Minister and several of his cabinet members entered the room right on schedule. When they were seated, Levi called the meeting to order.

"Good morning. Thank you all for coming on such short notice. I realize that many of you in the room have had very little sleep in the aftermath of the attack last night. Our objective this morning is to summarize what is known about the attack and to evaluate what it may mean for several planned intelligence and military missions."

The head of the national police bomb squad gave the first part of the briefing.

"The bomb that detonated was one of two improvised explosive devices commonly referred to as IEDs. They were disguised as one-liter bottles of wine. The explosive material was C-4 triggered by a standard military detonator connected to a battery and a switch activated by removing the cork. Although innovative in several ways, the devices borrowed heavily from R&D work done on Claymore mines at the U.S. Army's Picatinny Arsenal, particularly with respect to the size and type of steel pellets used. Forensic testing is still going on so I can't say for certain where the IEDs were made but our early assessment is that they were probably constructed in southern Lebanon by Hezbollah bomb technicians using materials supplied by Iran. A plastered rock wall between the kitchen and the dining area contained most of the blast. If that wall had been of typical interior construction, most of the people in the dining room would also have been killed or seriously wounded."

The next person to brief was the director of the national police force known as the Mišteret Yisra'el.

"The bombs were delivered by a man claiming to be the Marketing Director for the Golan Heights Winery. We've confirmed that he was an imposter. Security cameras at Ben Gurion airport recorded a man fitting his description entering Israel yesterday on an El Al flight from Athens. We've shown his EU passport photo to the restaurant owner and he's confirmed it's the same man who gave him the wine bottles. His forged passport gave his name as Demetrius Christos. He rented a car that was found abandoned early this morning. Initially, we didn't get a hit from our facial identification search. We sent a copy to the American Embassy. They tapped into the Army's database of known bad characters from Iraq and Afghanistan. That search determined that the man who delivered the bombs is Siraj Haqqani. He and his father have been supported and protected by LTG Parvez Jhalawan, formerly the head of Pakistan's intelligence agency and more recently escaped from the American detention center at Guantanamo Bay, Cuba. We've issued a national search warrant but don't expect to catch him."

The Prime Minister stood to speak.

"Does this attack suggest that the security of our mission plans has been compromised? I'd like to hear your views, Levi."

"That is certainly possible Mr. Prime Minister but Mr. Logan suggests otherwise. David, would you explain your analysis for us?"

"Mr. Prime Minister, I'm all but certain that Parvez Jhalawan and Ahmed Malik are currently being protected by Jalaluddin Haqqani in the Federally Administered Tribal Area of western Pakistan. I think the attack last night was a retaliatory contract hit ordered and financed by Jhalawan. We've interfered with his plans and humiliated him by abducting him from his home and sending him to a bare concrete cell in Cuba. Although the Iranians would be more than happy to kill Mossad and CIA agents, I don't think they ordered this hit. My assessment is that they are only indirectly involved through technical and material support of Hezbollah. Siraj Haqqani almost certainly ordered the fake wine bottles from Hezbollah and had them brought into Israel through one of the tunnels that hasn't yet been discovered. I see no reason to believe that the security of the upcoming missions has been compromised."

In the heated discussions that followed, it was clear that opinion about a possible mission security compromise was sharply divided. Finally, the Prime Minister cut through the fog of disagreement.

"We do not have the luxury of time. I've listed carefully to the opinions and rationales for both sides of the question and it does not appear possible to know for sure if our mission security has been compromised. What we do know, without ambiguity, is that Iran has nuclear weapons and that the Supreme Leader's threat to destroy Israel must be taken at face value. With those thoughts in mind, my decision is that we proceed as planned. I think we have no choice."

Thursday, 16 October

Drop Zone 3; Kavir Desert, Iran
1:57 AM

The skies over the new drop zone Nahbi had specified were clear. The waning moon had risen two hours earlier and was in its third quarter.

The Israeli Bardelas was streaking across Iran's night sky at Mach 3. With its stealthy design, high altitude, and unmatched speed, the Iranian air defense radars could not detect it. Piloting the incredible aircraft, Eli Wolff scanned his instruments and then keyed his microphone.

"Two minutes to target. Please acknowledge that you are ready for release."

In HALO pod one, Nahbi keyed his throat mic. "Pod One's board is green. Ready for release."

"Roger, Pod One," Eli confirmed.

Gabe then keyed her mic, "Pod Two's board is green. Ready for release."

"Roger, Pod Two."

As soon as she heard Eli's confirmation, Gabe keyed her mic again.

"Well, Mr. Adler, this wasn't exactly what I had in mind for our honeymoon. You might want to find a new travel agent."

Dowlat Abad Garden; Yazd, Iran
Noon

Saeed Ghorbani saw them sitting in the shade and walked toward them. When he was still a few steps away, Nahbi greeted him.

"Salaam, Saeed. Thank you for coming. This is my wife, Delbar."

"Salaam. I am honored to meet you Delbar. Forgive me for asking a rude question, Ahura Mazda, but are you well? I did not expect to see you in a wheel chair."

Nahbi assured him that he was fine and that the wheel chair was merely a useful disguise.

Saeed nodded his understanding. "You obviously found the van. I must apologize that I could not find the model you asked for on such short notice. I rented a Mercedes Sprinter camper van in my name. It has 4-wheel drive and GPS navigation. The water and propane tanks are full. It is practical for short-term camping as you specified.

"You did well, Saeed. We are in your debt," Nahbi assured him. "Were you able to arrange for us to meet with Professor Hamadani?"

"Yes. He will meet with you at his home in Nowshahr tomorrow morning at 9:00 AM. It's a small oceanfront villa just east of Chalus. I've written his address on this card and programmed it into the GPS."

"Good. Did you give Professor Hamadani my name?"

"Yes. He recognized it as the same name that's on all of the posters we've put up and asked if you were the same Ahura Mazda. I told him that I'd wondered about that too but didn't really know. He then asked me to tell him a little about you. His facial expression suggested that he was already putting the puzzle together in his mind but he asked me no further questions."

"Can we rely on him to keep our discussions confidential?"

"Yes. Several of my close associates know him well and consider him absolutely trustworthy," Saeed assured him. "Look, I don't want to rush you but you still have a long drive ahead of you. It's over six hundred kilometers from here to the campground I reserved for you in the Mazandaran Forest. It will be dark by the time you get there. I've already put the coordinates into the van's GPS. Tomorrow morning, you'll need to leave the campground no later than 6:30 AM. It's a two-hour drive from the campground to Professor Hamadani's villa but there are often construction delays along the coastal highway."

"Saeed, there's one last thing. I'd like for you to schedule an encrypted broadcast to our network for tomorrow evening at 10:00 PM."

Malta International Airport; Valletta, Malta
2:10 PM

The Alitalia flight was over a half-hour late taking off from Ben Gurion International due to an air traffic hold. David called Elena from the first class cabin of the Airbus A320 when it finally pushed back from the gate to let her know his estimated arrival time.

Elena was waiting for him just outside the baggage claim area. He set his bags down just in time for her to wrap her arms around his neck and give him a passionate kiss that caught him off guard. It was the first time she'd ever French-kissed him. Taking his cue from her and responding in kind, the thought occurred to him that maybe she needed rating controls on her dreams too. Laughing, as the kiss finally ended, he teased her.

"Wow! Who are you? What have you done with my girlfriend?"

"Laughing, she teased him right back. "It's all hormones honey buns. Some days are good. Some days aren't. That's why wedding vows say for better or worse. You men aren't the only ones who have libidos, you know."

"OK. You win. I can't top that."

A few minutes later, Elena turned left out of the airport onto Triq San Tumas and headed north toward her apartment.

"I put a casserole in the oven before I came to the airport. You can shower and we'll have an early dinner. I'm sure you're exhausted and could use some sleep."

"That does sound good but I need for you to stop by the American Embassy first. There's something I need from Emily."

American Embassy; Ta' Qali, Malta
2:40 PM

David had called ahead so Emily Harrison was waiting for him at the Embassy's security checkpoint.

"Welcome back. Last I heard you were in Israel. What brings you to Malta?"

"I'm heading back to DC but I wanted to see Elena. There's a lot going on. Can we go to your office? Some of what I need to tell you is classified."

When they got to her office, Emily closed the door and sat down on the sofa to hear what David had to say.

"After Malik rescued Jhalawan from Guantanamo, Pete and Roger sent me back to Israel. You probably already knew that."

"Yes, go on."

"Did you know that Nahbi and Gavriella were married last Friday morning?"

"No. How interesting. Was it romance or trade craft?"

"Romance. Henceforth, you can refer to them as Mr. and Mrs. Adler. Nahbi has taken Gabe's last name. The trade craft part of their relationship officially began sometime after midnight this morning when they did a HALO jump into Iran to finish Nahbi's mission. Most of that is classified top secret by the Mossad so you didn't hear what I just said."

"Understood."

"Tuesday evening, Levi Reznik hosted a small dinner at Gavriella's favorite restaurant. Officially, it was a belated wedding dinner but the six of us who were there all knew there was a dark undercurrent. Nahbi is the only Mossad agent to return from Iran in over ten years and he almost didn't either. This time, Levi decided to send Gabe with him. We all knew they might not come back."

"After we finished eating, we were trying to keep things light-hearted with some risqué toasts to the new couple when the restaurant owner brought two bottles of wine to our table. He said they were a gift from the Marketing Director of an Israeli winery. Actually, they were IEDs packed with two pounds of C-4 and several hundred steel pellets. One of them detonated in the kitchen when the wine sommelier tried to uncork it. Everyone in the kitchen was killed. We were spared by a plastered rock wall that separated the kitchen from the dining area. I'm convinced that the attack was ordered by Jhalawan and Malik. Since it failed, I'm sure they'll try again. That's what brings me to your office. I need a pistol and a combat shotgun. I can't afford to risk having Elena killed or abducted because my presence puts her in danger. Malik may not know I'm in Malta yet but even if he does, they haven't had much time to adjust their strategies or get assets in place for another attempt. It's possible I'm being overly cautious but I've learned not to underestimate Malik. I don't want to take a chance. Can you help me?"

"Yes. Of course." Emily walked over to her desk and hit a key on her communications console to connect her to the Embassy's military security officer. "Lieutenant, I need a few things from the armory. One of my CIA colleagues is here in my office and he needs some weapons while he's in Malta. I'd like for you to take care of this personally. You can list me as the ordering officer. Pick out one of our Sig Sauer MK25 9mm pistols and mount a combination tactical and laser light on it. Put it in one of the Embassy's lockable diplomatic brief cases along with a silencer and two spare magazines. Load all three magazines with alternating bonded hollow points and the Norwegian M39B armor-piercing rounds. Mr. Logan also needs one of our black Remington twelve gage pump shotguns with a tactical light and a side saddle for extra shells. Put a box of double-aught shells in the briefcase. When you've got it all ready, give me a buzz and we'll meet you at the front entrance."

Elena's Apartment; Valletta, Malta
3:25 PM

"Can you tell me what's in the long box and the briefcase we picked up at the Embassy?" Elena asked.

"Yes but there's some things I need to tell you first. Let's go in the living room."

When they were seated on the sofa, David reached over to take Elena's hand. "You shouldn't repeat what I'm about to tell you. Some of it is classified but you have a right to know. The reason you couldn't reach me on my personal cell phone is that I was in Cuba. I was subpoenaed by the prosecution in the Military Commissions trial of General Jhalawan at Guantanamo Bay. The day before I was scheduled to testify, Ahmed Malik helped Jhalawan escape. Jhalawan and Malik have a history of having people killed who interfere with their agenda so I may now be a target. That means that anyone I'm with could be in danger too. That's why I stopped at the Embassy."

"The long box has a shotgun in it. A pistol and ammunition are in the briefcase. Something bad happened Tuesday night while I was in Israel. I didn't want to tell you over the phone because I knew it would scare you. Nahbi, Gabe, and I were having dinner with three Israeli intelligence officers when someone tried to kill us with a bomb. All of the people in the restaurant's kitchen were killed. We were spared by a masonry wall that absorbed most of the blast."

"I'm sure General Jhalawan and Ahmed Malik ordered the attack to destroy Nahbi and kill me. By now, they know the attack failed. I don't think they know where I am but I can't be absolutely sure. Malik is one of the world's most talented spies. He can find out just about anything he wants to know within a day or two. It's been two days since the attack so I have to assume that Malik may know I'm here. The guns are an insurance policy. I'm probably being overly cautious but I would never forgive myself if something happened to you because of what I do for a living."

Elena was terrified by what David had just told her. With tears brimming in her eyes, she asked, "Can't the CIA protect us? What about the Malta police?"

"Elena, I can't go to the Malta police without compromising the fact that CIA agents work undercover inside the Embassy with innocuous titles like *Political Officer* or *Cultural Officer*. That diplomatic cover is absolutely essential to them being able to do their jobs. The State Department and CIA both go to great lengths to protect that cover."

"I work for the CIA and I'm a former Special Forces soldier. My training was very similar to what the Navy SEALs get. With the weapons and my experience and training, I'm quite confident that I can protect us from any attack. I know this is scary stuff but try not to worry. Trust me to do what I've been trained to do. We'll be fine."

Elena's wasn't so sure. As the daughter of a quietly conservative university professor, and a life-long resident of one of the safest countries in the world, it was almost impossible for her to assimilate the brutal realities of international terrorism. "David, I do trust you but you're just one man. What if they send several people?"

"Elena, you know I grew up in Texas. There's a true story about what the right man can do even when outnumbered. A little over a hundred years ago, a Texas Ranger was sent to quell a riot in Dallas.

When he stepped off the train by himself, the mayor asked where the rest of the Rangers were. Ranger Captain McDonald's answer has become legendary. He said, '*Hell, ain't I enough? There's only one riot isn't there*?' At most, Malik would only send two men. We're prepared and ready for them if they come. With my training and the weapons from the Embassy, I can handle two men."

5:35 PM

Elena was holding dinner until 6:00 PM to give David time to take a shower. He was getting dressed when his CIA phone buzzed.

"David, Emily. I've authorized some overtime work for our roving security contractor. They drive dark blue, unmarked compact sedans. A security patrolman will drive through Elena's neighborhood every hour from 6:00 PM until 6:00 AM. They'll call you on your CIA phone if they see anything suspicious. You didn't say how long you'd be in Malta but my discretionary budget can support the extra patrols for a few days."

"Thanks. Elena is scared stiff. This will help."

"There's something else. I'm sure you realize how messy it could be from a diplomatic perspective if you get into a gun fight. That's why I included a silencer for the pistol. I'm hoping that if you have to use it, the neighbors either won't hear it or won't recognize what it is. Any hit man Malik sends will have entered the country illegally and will be working under deep cover. If I have to send a cleanup crew, we might be able to quietly get rid of a body that no one will miss anyway. Do what you have to do but I'm hoping you don't have to use the shotgun. It would make so much noise that the police would be on the scene in minutes. You and I would have a lot of really awkward questions to answer, both here in Malta and back in DC."

"Understood. Thanks."

9:45 PM

Elena was watching David with a puzzled look on her face. "Can I ask what you're doing?"

"Don't you read spy novels?" I'm installing some early warning devices."

"Pardon me, but that looks exactly like a table knife."

David smiled at her dry sarcasm. "Darn. I thought I could fool you. But, seriously, since you don't have a security alarm system and your front door doesn't have a dead bolt, I'm improvising. The knives work as simple alarms. I'm hanging them by a thread from the top of the window trim to dangle at the side of the window just barely touching the glass. Because they are long and slender, they won't be visible in the dark to anyone trying to come through a window. If they are jostled, they'll make a faint metallic chiming sound that will wake me instantly. Sensitivity to slight noises, even when asleep, is a survival skill I acquired in Afghanistan. It's sort of like the proverbial *sleeping with one eye open.*"

"What about the doors?"

"They're easy too. A table knife dangling from a thread tied to the inside door knobs will alert us if the door is opened.

"But last time, the man who abducted you just knocked on the door and waited for you to open it."

"That's true but last time I wasn't expecting an attack. Now we're ready and anyone Malik might send will know that. Have you ever handled a shotgun before?"

"Yes, but it's been several years. My father inherited a classic Luigi Franchi double-barreled twelve gauge from my grandfather. I used to go to the skeet range with him when I was a teenager. I even won a couple of club tournaments. I can handle a shotgun if I need to."

"Good. I don't think you'll have to but a shotgun is absolutely the best home defense weapon there is. In close quarters, just pointing it in the general direction of the bad guy will get the job done. Also, shotgun pellets won't penetrate walls the way a pistol or rifle bullet can. That means a shotgun is less likely to hurt someone you didn't intend to hurt. The pump shotgun I borrowed from the Embassy is very easy to use. I'll show you what you need to know. If it's OK with you, I think we should both sleep in your bedroom while I'm here."

Friday, 17 October

Arteman Hamadani Villa; Nowshahr, Iran
9:00 AM

Professor Hamadani's waterfront villa overlooked the Caspian Sea. It was a two hour drive north of the University of Tehran main campus via the Chalus River valley road. Professor Hamadani was watering his rose garden when Gavriella turned into his driveway and parked. He turned off the water and walked over to meet his guests. Gavriella removed a folding wheelchair from the back seat and helped Nahbi get into it.

"Welcome to my home. I trust you had no difficulty in following my directions."

"Good morning sir," Nahbi replied. "Your directions were excellent. Allow me to introduce myself. I am Ahura Mazda Amirmoez. This is my wife, Delbar."

Professor Hamadani looked at Gavriella as though he'd seen a ghost. "Please forgive me if I reacted strangely to your introduction. My wife's name was also Delbar. She was killed in a car accident last year. We were married for almost forty years and I was blessed beyond measure for her love and devotion. I'm sure you know that the name Delbar means a woman who is charming, a sweetheart, and a heart-ravisher. My wife was all of those and so much more. But, enough of an old man's reminiscing. Please come into the house."

After a few minutes of polite small talk, Nahbi steered the conversation to the purpose of their meeting.

"Professor Hamadani, you are a respected constitutional scholar. Your views closely parallel my own. If I may speak freely, I believe that Iran's current constitution is an impediment to all manner of progress including human rights and economic prosperity for our people. Iran's constitution has been characterized as a hybrid blend of theocratic and democratic elements but that is misleading. It is not really a hybrid because all rights are subordinate to the Guardian Council and the Supreme Leader. Iran's government is a democracy in name only. All real power is in the hands of the

Supreme Leader. I do not have to enumerate the weaknesses of our current constitution. You know them far better than I do. However, I wanted to ask you if you believe gradually amending the constitution is a viable pathway to something better."

Professor Hamadani said nothing for a few moments while he considered how he would respond.

"Sadly, no. I do not think gradual change will work. The theocratic influence is simply too entrenched. The Supreme Leader and his close confidants will never willingly relinquish or even diminish their near absolute power. I have reluctantly come to the conclusion that it will take a popular uprising, not unlike the one that toppled the Shah in 1979, to bring about the significant changes that are desperately needed. My greatest concern is that our leaders' obsession with the development of nuclear weapons will force the western powers to destroy much of our country to prevent those nuclear weapons from being used by radical Islamic theocrats."

"I agree with your bleak but realistic assessment," Nahbi told him. "Your writing suggests to me that you have a clear vision of what a new constitution would include. Have you actually drafted such a document?"

"Yes. Would you like to see it?"

While Professor Hamadani chatted amicably with Gavriella, Nahbi pretended to spend an hour studying the draft constitution. It actually took him only a few minutes to read and memorize the entire document.

"Professor Hamadani, this is an impressive piece of work. Are you the sole author?"

"No. Although I wrote most of it, I asked a few close associates to read it. They provided some excellent suggestions, many of which I incorporated. If it were presented to a legislative body, I have no doubt that it would be subjected to some passionate debate and likely revisions. After all, that's the messy nature of democratic government. Yet, I think the draft is substantially sound."

"Do you envision that you might take an active role in the formation of a new government and the adoption of a new constitution?"

"What do you mean by an active role?"

"Would you be willing to serve in a political capacity, perhaps as a democratically-elected President to guide the nation through its transition to an entirely new form of government?"

Arteman Hamadani bowed his head for several moments before he looked up and gave his answer.

"Yes. I would serve in a high-level political role but only if I was democratically elected. Yet, that seems highly unlikely. I am not well known outside of academic circles and I do not have a network of wealthy and influential people to advance and finance my candidacy."

"As an alternative, would you be willing to serve as an appointee of a democratically-elected President? For example, what if former President Mohammed Khatami were elected? He remains very popular and seems to share many of your views. What if he were to nominate you to serve on a new Supreme Court organized and empowered as you describe in your constitutional draft?"

"That would be the crowning achievement of my career. Yes, I would gladly serve on the Supreme Court where my advocacy of constitutional law would have an impact each time I voted."

A few minutes later, Professor Hamadani walked with them to their car and waited while Gavriella helped Nahbi into the passenger seat. He shook hands with Nahbi through the open window and expressed an obviously sincere invitation for them to visit again.

When they were heading west on the coast highway, Gavriella asked Nahbi for his assessment of their meeting.

"It went extremely well. Professor Hamadani's draft constitution is just what the country needs. Let's see what we can do to help make it happen."

Public Campground; Mazandaran Province, Iran
10:00 PM

More than four thousand of the special IAM Smart Phones had already been distributed to college students and other young adults in the major cities of Iran. Encrypted and linked to the Israeli Ofeq-10 satellite, they provided a secure way for Nahbi to communicate with the steadily increasing numbers of recruits.

"Good evening. This is Ahura Mazda. Thank you all for joining me. Earlier today, I reviewed a draft of a new constitution. It was prepared by several constitutional law scholars and it is superb. Tonight I will explain how we are going to create an entirely new government within a matter of weeks."

"Most of you were not yet born when the 1979 revolution took place. Civil resistance, labor strikes, and public demonstrations preceded the actual collapse of the Pahlavi dynasty. At that time, unlike today, our country was enjoying relative prosperity. Then, as now, the country's leadership had become increasingly unpopular with our citizens. Yet, at the time of the revolution, Iran was a respected member of the civilized world instead of the pariah that it is today."

"Since the 1979 revolution, our leaders have obsessively pursued the development of nuclear weapons in violation of the nuclear non-proliferation treaty that Iran, and nearly every other nation on Earth, is a signatory to. In an effort to spread their medieval vision of an Islamic caliphate, our leaders have sponsored unspeakable acts of terrorism that have included mass murders and using innocent children as suicide bombers. Quite apart from the economic hardship that their evil policies have created, they have made the rest of the world hate us."

"With a new constitution, and new leadership, we can again become a respected member of the world community. We can create the kind of economic conditions that will bring an end to the privation our people have suffered for decades."

"Some of you may wonder if it is possible to replace Iran's current government without major bloodshed. The answer is yes. Several surveys conducted over the past decade have shown that the vast majority of Iran's citizens do not support the totalitarian government that exists today. One survey put that number at ninety-two percent of our citizens."

"Another survey found that more than two-thirds of Iran's adults would support what has been called a '*Velvet Revolution*' to remove the government's leaders. The phrase velvet revolution refers to what happened in Czechoslovakia in 1989. It was coined by Rita Klímová who later became the ambassador to the United States. In a matter of days, street protests in Prague grew to more than a half-million people. In November, a general strike involving all citizens was held. On November 24th, the entire leadership of the Communist Party resigned. On November 28th, the Communist Party announced that it would relinquish power and dismantle the one-party state."

"Yet another survey here in Iran found that fifty-eight percent of Iranian citizens were so dissatisfied with the country's leaders that they would even support foreign military action to bring about regime change. You can be absolutely confident that most or our citizens will be sympathetic to your protests. Your courage will instill courage in them and the movement will expand rapidly, just as it did in Czechoslovakia."

"We will begin by paralyzing Tehran's major railroad stations next Monday. At 6:30 AM, just before the morning rush hour begins, hundreds of you will suddenly appear in the train stations wearing bright red tee shirts with the words *Nuclear Free Iran* printed in bold letters on both the front and back. Your cell leaders will have the tee shirts ready for you to pick up Sunday afternoon. The national news media will be notified in advance so you'll get plenty of attention. Your cell leaders will provide you with packets of printed handouts explaining the linkage between the economic hardships our country is suffering and our national pursuit of nuclear weapons."

"Some of you have been selected to act as spokespersons to be interviewed by TV and radio correspondents. Others have volunteered to lie down across the tracks to prevent the trains from leaving the stations. When thousands of people can't get to work, your voice will definitely be heard. We want this to be a non-violent protest, the opening salvo in our own velvet revolution. Our national leaders will be furious of course and they will order the police to break up the demonstrations. In doing so, they will play right into our hands. When the news programs show pictures of young, idealistic college students being carted off to jail for having the courage to oppose our country's pursuit of nuclear weapons, many of our citizens will be proud of you. This is how the revolution in 1979 began. Your courage will spread. Soon literally thousands of your fellow countrymen and women will join in public demonstrations of their opposition to our current government."

Nahbi's broadcast continued for almost another hour as he outlined the protests and demonstrations that would occur over the following two weeks.

Saturday, 18 October

Elena's Apartment; Valletta, Malta
1:30 AM

The previous night had passed quietly. For Elena's sake, David had downplayed the risk of an attack but after what had happened in Tel Aviv, his instincts said otherwise. The 9mm pistol was under his pillow and the shotgun was leaning against the night stand on Elena's side of the bed. Both weapons were loaded, cocked, and ready to fire.

Always a light sleeper, David came instantly awake at the faint sound of one of the table knives bumping softly against the front door. Elena heard it too. She slipped quietly out of bed and grabbed the shotgun. She hid in her clothes closet and pulled the door almost shut as they'd planned. David crouched down beside Elena's dresser where he'd be out of sight and have some protection. He didn't have long to wait.

A man dressed in black stepped silently into the bedroom's open doorway. He was wearing night vision goggles and holding a silenced pistol. When he realized there was no one in the bed, he looked quickly around the room. His night vision goggles limited his peripheral vision so he didn't see David at first. When he did, he and David fired almost simultaneously.

David's aim was good but the first round of his combat double-tap was a bonded hollow point that didn't penetrate the assassin's body armor. His second shot punched through the Kevlar vest and drilled a hole through the assassin's heart. The man fell to the floor and lay still after a few spasms. The assassin's only shot glanced off David's left clavicle bone and nicked his carotid artery. Blood gushed from his wound but his combat training saved his life. Keeping his gun aimed toward the assassin, he made a fist with his left hand and pressed it against the wound. He knew he was bleeding badly and needed medical help soon or he would die.

"Elena! It's over. Turn on the lights and get me a towel. Hurry!"

When Elena knelt beside him, David used the towel to make a temporary compression bandage while telling her what to do.

"My CIA cell phone is on the night stand on my side of the bed. Call Emily. Just hold the five key down until the speed dial places the call. When she answers, hit the speaker button and hold the phone where I can talk to her."

While Elena was placing the call, David removed the man's night vision goggles and balaclava. He instantly recognized Siraj Haqqani.

Elena held the phone where David could hear it when Emily answered. "David, you wouldn't be calling me at this hour unless something happened. Have you been attacked?"

"Yes, by Siraj Haqqani. I put two rounds into his chest. He's dead. I took a round in my left carotid artery. I've managed to slow the bleeding but I've already lost a lot of blood so I could lose consciousness at any time. Can you send medical help and maybe a couple of military personnel?"

"Yes. It'll take them fifteen or twenty minutes. Is the situation under control?"

"Yes. Can you let Pete know?"

"Will do. I'll be there in ten minutes. I need to ask if there was enough noise to wake the neighbors."

"I don't think so. I fired two silenced rounds and Siraj fired one. I doubt if the shots were heard outside Elena's apartment."

2:20 AM

Even with David's prompt efforts to stop his own bleeding, he'd lost a lot of blood by the time he was brought to the embassy's infirmary. The embassy's medical staff consisted of one nurse practitioner assisted when necessary by one of the security guards. The guard's sole medical claim to fame was that he'd once been an ambulance driver.

The NP had the guards put David on the examination table. Finding almost no pulse, she told her ad hoc assistant to get two bottles of freeze-dried blood plasma from the refrigerator and talked him through the process of adding sterile water and shaking the bottles to reconstitute the plasma. When it was ready, she hung one of the bottles from a hook over the examination table and began an intravenous drip as a temporary replacement for some of the blood David had lost. His blood pressure was dangerously low.

The bullet had fractured David's left clavicle bone before plowing a deep furrow in his neck and shredding a section of his carotid artery. The broken bone wasn't life threatening but the neck wound was so she focused on it first. By then David was unconscious so no time was lost waiting for a general anesthetic to take effect.

David's wound was well beyond the scope of the NP's training but he would be soon be dead if she couldn't stop his bleeding without cutting off the blood supply to his brain. Even if she was successful, the risk of brain damage was very real. After making a three-inch long incision, she clamped the artery in two places just long enough to insert a makeshift nylon catheter that would serve as a temporary shunt around the damaged section. When she released the clamps, she watched carefully to verify that the loops of surgical thread she'd used to secure the catheter weren't leaking. The temporary repairs seemed to be holding. Then David's heart stopped.

It took two jolts from the embassy's portable defibrillator to restart his heart. A few moments after his heart started beating again, David stopped breathing. Keeping the defibrillator paddles in her hands in case they were needed again, the NP coached the guard on how to use the bag valve mask to manually resuscitate him. By a little after 3:00 AM, they'd done all they could do. David was still unconscious and his blood pressure was critically low. He was alive only because the embassy's nurse practitioner had simply refused to let him die.

"Is he going to make it?" Emily asked from the doorway.

"Too soon to know for sure," the NP said with evident and understandable fatigue in her voice. "He's still unconscious so there's no way to know if the interruption of blood flow to his brain has caused irreversible damage or not. We've done all we can until you can arrange to have him airlifted to Landsthul."

"Ambassador Stevens made arrangements for him to be transported to the Naval Hospital at NAS Sigonella instead. It's much closer than Landsthul. A Navy MEDEVAC Huron piloted by two CIA agents from the American Consulate in Palermo will land at Malta International in the next half hour. Can Mr. Logan be transported that soon?"

"Yes. I'll put together an emergency travel kit and go with him. We almost lost him twice. If you'll get me a good phone number, I'll brief the doctor while we're airborne so they can be ready.

4:15 AM

The Navy's designation for the Beechcraft King Air 200 turboprop is the UC-12M Huron. It is arguably the finest aircraft in its class with more than three thousand built over a continuous production run spanning forty years.

The Huron was configured as a MEDEVAC aircraft with two medical stretcher stations. The port stretcher station was located just forward of the aft main cabin door. The starboard station was offset to the aircraft's rear about two feet to make room for a medical equipment rack. An extremely narrow aisle ran between the stretcher stations. Forward of the stretcher stations and immediately aft of the cockpit were five passenger seats.

Cruising at three hundred miles per hour, the Huron would land at NAS Sigonella just before 5:00 AM. An ambulance was already waiting to take David to the naval hospital five miles north of the airfield.

Twenty miles due north of Malta, the pilot slowed the aircraft to just above its 75 knot stall speed and descended to an altitude of five hundred feet. He then depressurized the cabin so the copilot could open the main cabin door located aft on the port side of the aircraft. Emily Harrison helped the copilot drag the weighted black body bag from the port stretcher station to the open doorway. A few minutes after being shoved from the Huron, the body of Siraj Haqqani came to rest in an unmarked grave eight hundred feet below the surface of the Mediterranean.

6:05 AM

Four minutes after David was transferred to the operating table in the basement of the Naval hospital, he went into cardiac arrest again because of the massive amount of blood he'd lost. It took two high voltage jolts from a defibrillator to restart his heart.

Navy Captain Marilyn Walker was the senior Medical Service Corps surgeon. She was a fifth generation descendent of Alvah H. Walker, brother of Dr. Mary Edwards Walker, the only woman to have ever won the Congressional Medal of Honor. Marilyn Walker was a brilliant surgeon. She was also one of the world's foremost experts on the use of super cooling to minimize the risk of irreversible brain damage due to serious traumas, such as gunshot wounds.

Already prepped for surgery when the assisting doctor used the defibrillator, she realized the patient wasn't going to survive without extraordinary measures. Turning to the senior OR nurse, she issued terse instructions.

"Prepare the patient for a cold saline flush. We've got to lower his body temperature to buy some time or he's not going to make it."

Over the next half-hour, chilled saline was pumped through David's veins until his body temperature was cooled to ten degrees Centigrade. At that low temperature, almost all metabolic activity ceased except for a process called *anaerobic glycolysis* that would produce energy when the body's oxygen supply was severely limited due to blood loss. Anaerobic glycolysis would keep David's body cells alive for only a couple of minutes at normal body temperature but at the super cooled temperature the cells could live for several hours. Dr. Walker hoped that would give her enough time to repair the artery. With her surgical skills, and a lot of luck, the patient had a fighting chance.

Ninety minutes later, Dr. Walker was finished repairing David's carotid artery. She'd used a high magnification headlamp to perform the delicate surgery that involved removing the damaged section of the artery and then using micro sutures to reconnect the ends. She'd placed a nylon mesh sleeve over the damaged artery to provide mechanical support to allow the damaged tissue to heal.

"OK. Let's start replacing the saline solution with whole blood. Keep a close watch on his core temperature. We don't want to warm him up too fast."

9:55 AM

David was still unconscious when he was moved from the operating room to the hospital's emergency care unit. His vital signs were within acceptable limits but the next few hours would be critical. Emily Harrison and the embassy's Nurse Practitioner were in the nearby waiting room when Dr. Walker came to give them an update on David's condition.

"Mr. Logan is alive. The surgery to repair his carotid artery was successful but we won't know until he regains consciousness if there's been brain damage. He lost an awful lot of blood. I've decided to keep him sedated for several more hours as a precaution. My admin assistant has reserved rooms for you both at the Navy Lodge. The hospital's duty driver is waiting for you at the main entrance. I'll have the ECU nurse call you if Mr. Logan regains consciousness."

10:05 AM

Emily called Ambassador Stevens on the way to the Navy Lodge. She waited until she got to her room to call Elena.

“Elena, this is Emily Harrison. I wanted to let you know that David is out of surgery and has been transferred to the emergency care unit. The chief surgeon is keeping him sedated for the next few hours. I’ve just checked into the Navy Lodge here at NAS Sigonella. I’m going to grab a little sleep and then call David’s boss. There’s a six hour time difference to Washington so it’s still too early to call him. You can reach me on this number.”

“Thank you for calling. I’m at my parents. We’ve been holding a family prayer vigil and hoping for a miracle. With so much blood all over my apartment, and so much of it being David’s, I’ve been terrified that I’d never see him again. What does the doctor think? Is he going to be OK?”

“Elena, Dr. Walker is one of our best military surgeons and a world-renowned expert on the use of super cooling to reduce the risk of brain damage. David was lucky to be brought to Sigonella but we just won’t know until he wakes up and his doctors can evaluate his condition. Right now, the best thing you can do for David is to keep praying.”

Sunday, 19 October

44000 Constitution Avenue; Islamabad, Pakistan
1:00 PM

Since the creation of Pakistan in 1947, there had been three successful military coups in 1958, 1977, and 1999. Civilian government had last been restored in 2008.

Just after midnight on the 19th of October the fourth successful military coup d’etat, led by Lieutenant General Parvez Jhalawan, again removed the country’s civilian government from office. Mistrust of the civilian government was so pervasive among the military officer corps, and respect for General Jhalawan was so widespread, that it had taken him just two weeks to organize and execute the coup after his return.

Seated behind the Prime Minister's desk, General Jhalawan was about to address the nation. The room was brightly lit for the television cameras of the government-owned Pakistan Television Corporation. At precisely one o'clock, the PTC director looked up from his watch and pointed to General Jhalawan signaling that the broadcast was live. Impressive in his full dress uniform, General Jhalawan looked directly into the center camera and began to speak.

"Good afternoon. For those who do not recognize me, I am Lieutenant General Parvez Jhalawan. I am the senior military officer of Pakistan's armed forces. At midnight, the civilian government of Pakistan was abolished by our nation's armed forces as a direct result of the civilian government's failure to protect the national security interests of our homeland. The Prime Minister and President, and all of their senior cabinet ministers, have been placed under house arrest. None of them have been harmed. A military tribunal will be established in the next few days to decide if these civilian leaders should be tried for willful neglect or misconduct."

"As of Midnight, martial law has again been established to protect our great nation. The National Assembly of Pakistan and all provincial assemblies are hereby dissolved. Their functional responsibilities will be carried out by the military for the foreseeable future. I have also temporarily suspended the Constitution to enable your military leaders to act promptly to secure our borders and protect our civilian population. I will announce the interim military provincial governors at six o'clock this evening."

"In closing, I want to remind everyone of the words of Pakistan's founding father, Muhammad Ali Jinnah. His words are immortalized on the State emblem of Pakistan. In our national language, these words are *Īmān, Ittiḥād, Naẓm.* In English, these well-known Urdu words translate as Unity, Faith, and Discipline. We must be united in our love of our country. We must have faith in each other and we must have the discipline to do what is best for our people." General Jhalawan looked steadily into the center camera until the red light beside the lens blinked off signaling that the broadcast was over.

General Jhalawan's assurance to the nation that none of the deposed *civilian* leaders had been harmed was disingenuous although technically accurate. At four minutes after midnight, Lieutenant-General Zafar Khalil, had been taken from his home and summarily executed by soldiers assigned to Pakistan's Special Services Group based at Tarbela Cantonment. General Jhalawan appointed Ahmed Malik to take his place as the Director General of the Inter-Services Intelligence Agency.

Levi Reznik's Office; Mossad HQ; Herzliya, Israel
4:25 PM

With the three hour time difference between Tel Aviv and Islamabad, Levi had just turned the television in his office off at the end of General Jhalawan's broadcast when the blinking red light for his direct line to the Prime Minister lit up on his desk. He didn't even have time to say hello when he picked up the handset.

"Levi, what did you make of General Jhalawan's address?"

"Mr. Prime Minister, this is very bad. Now we have two nuclear-armed enemies who want to destroy us. Between them, Iran and Pakistan have more than thirty times our population."

"Our best estimate is that Pakistan may have as many as three hundred nuclear weapons, which would put them third behind only the United States and Russia in total devices. Pakistan also has four plutonium production reactors giving it the ability to produce at least twenty new warheads each year."

"The latest version of Pakistan's Shaheen-III strategic missile has sufficient range to hit any target in Israel. Making matters worse, the CIA believes Pakistan has already retro-fitted some of them with multiple independently targetable re-entry vehicle warheads. Missiles armed with up to five MIRVs each would overwhelm our Iron Dome missile defense systems. In a massive attack, some of their warheads would reach their targets. Half of Israel's population could be killed outright with many more dying later from burns and radiation overdoses."

After a moment's reflection, the Prime Minister said, "Yes. I agree. We didn't need this new threat just as we're preparing to deal with Iran. Until today, I was convinced that Pakistan's focus was on gaining nuclear superiority over India as a counter balance to India's much larger military forces. Now, I'm not so sure. Jhalawan has already made it clear that he's willing to use nuclear weapons. Even the United States has found it all but impossible to execute their national defense strategy of being able to fight two major conflicts, anywhere in the world, at the same time. Israel simply can't do that. Just dealing with Iran will stretch our resources to the absolute limits."

"Mr. Prime Minister, what do you want the Mossad to focus on?"

"Before I answer that question, let me pose another. If we carry out our attack on Iran's nuclear facilities, do you think Jhalawan would use that as a pretext for launching an attack against us?"

"I've been wondering that myself," Levi answered. "Jhalawan has an enormous ego. We've been convinced for some time that he sees himself as the eventual top dog in the Muslim world. If we are successful in defanging the Iranian cobra, we might conceivably be doing Jhalawan a favor by weakening his strongest competitor for Islamic leadership. Then, by attacking Israel in retaliation, Jhalawan could position himself as the defender of Islam against the hated Zionists and Americans."

"Levi, as always, your analysis is cogent and succinct. You may very well be right that Jhalawan will at least be tempted to use our attacks on Iran's facilities as a pretext for attacking us. But, I think we'll just have to take that chance. We have to deal with Iran now and hope that we'll have time to deal with Pakistan later if necessary. Fortunately, the new American President is more willing to work with us than his predecessor was. I would also think that India's Prime Minister cannot be happy to see Jhalawan take over Pakistan's government. Levi, do you think the Indians would risk a pre-emptive attack against Pakistan before Jhalawan can attack them?"

"Logically, you'd think so but I have my doubts. Even with the bitter dispute over Kashmir, my sense has always been that India's nuclear capability was primarily a sort of mutually assured destruction doctrine similar to what existed between the United States and the Soviet Union during the cold war. I think India's leaders want Pakistan to understand that they would bring about their own destruction if they ever used nuclear weapons against India."

"OK. I'll answer your question now. We are a small country. Iran and Pakistan are enormous by comparison. We simply don't have the manpower or military resources to engage them in a conventional way. So, I want the Mossad to develop contingency plans to eliminate Jhalawan, Malik, and several other senior Pakistani military officers in one bold stroke. I want to cut off all of the Hydra's heads at once. What I have in mind is something similar to the Mossad's 'Operation Bayonet' that tracked down and killed the Palestinian terrorists that murdered eleven of our athletes in the 1972 Munich massacre. Let's refer to these contingency plans as *Operation Stiletto.*"

Monday, 20 October

Tehran Railway Station; Rahahan Square, Iran
7:05 AM

"What do you and your fellow students hope to achieve by interfering with rush hour rail service?" The reporter asked a young college student wearing a bright red tee shirt. The TV cameraman zoomed in on the student's face and then panned to the *Nuclear Free Iran* slogan printed on the shirt.

"Just what our tee shirts say. We believe that our nation's pursuit of nuclear weapons is the primary cause of the UN sanctions that have crippled our economy and made civilized people everywhere hate us. We love our country and want it to regain the international respect that it previously had. We strongly believe that if our leaders persist in their obsessive pursuit of nuclear weapons, the western powers will eventually be forced to take military action against us. If that happens, many innocent people will die needlessly. It is time to stop this madness. Iran has proven oil reserves of more than one hundred and fifty billion barrels. With so much oil, we do not need nuclear power production. We certainly don't need crippling economic sanctions resulting from our pursuit of nuclear weapons. Again, it is time to stop this madness. "

The Islamic Republic of Iran Broadcasting cameraman zoomed in on the TV reporter's face as the newsman began his summary statement.

"That was one of the leaders of the student protest here at the central Tehran Railway Station. This protest is one of many similar large-scale demonstrations that have occurred in all of Iran's major cities this morning. Railroad workers all over the country have refused to risk injury to the protestors lying on the tracks. The morning commuter trains are filled with passengers trying to get to work but the trains are stuck at the loading platforms. Police and military personnel are beginning to make arrests so the trains may be moving again soon. Tune in tonight when IRIB's six o'clock news will cover the day's protests in depth." As the news reporter finished his summation, the final TV image was of riot police in black SWAT gear clubbing and dragging student protestors away.

44000 Constitution Avenue; Islamabad, Pakistan
8:00 AM

General Jhalawan had taken over the Prime Minister's residence and office. Since he'd announced the military coup on national television the previous day, a steady stream of senior officers had kept him busy giving orders and dealing with myriad details he hadn't foreseen. One thing needed his attention immediately so he'd asked Malik to come to his office.

"Ahmed, we need to ensure that our nuclear weapons are secure. With our takeover of Pakistan's government, I feel sure that India, Israel, and the United States are very nervous and might decide to strike. So long as our nuclear capability is intact, they won't dare attack us for fear of retaliation. I'm concerned that a massive strike on Wah could wipe out all of our nuclear weapons in a single blow."

"I want you to assemble some special trains like the Americans use to move their nuclear weapons from where they are produced to their ready stockpile locations at various military installations. Pakistan Railways has more than seven thousand miles of track. Roughly half of that total is on main lines rated for up to one hundred kilometers per hour for freight trains. As an interim measure, I want you to identify a sufficient number of refrigerated freight cars that can be used to keep our nuclear arsenal in constant motion. The weapons do not need especially cold temperatures but they do need the dehumidification that the refrigerated cars can provide."

"Spread the weapons among several trains and coordinate the train's movements so they are only rarely in close proximity to each other. Provide personnel cars for Special Forces soldiers to guard the trains. Provide me with a detailed plan and schedule as soon as possible."

U.S. Naval Hospital; Sigonella, Sicily, Italy
9:10 AM

Emily was checking her messages in the waiting room when a nurse came to tell her they'd moved David to a private room on the second floor and that she could see him for five minutes. He was awake and watching television coverage of the military coup in Pakistan when she came in. The left side of his neck was bandaged and his left arm was in a sling to immobilize his collarbone until it healed.

"How are you feeling?"

"Lucky."

Emily nodded in agreement. "Yes. The embassy's nurse practitioner unquestionably saved your life and bought you enough time for the trip to Sigonella but you can thank Dr. Walker that you didn't suffer permanent brain damage. I'd never even heard of super cooling a critically-injured person. Turns out she's one of only a handful of doctors in the entire world who have successfully used the technique. The lead OR nurse told me that without it, you'd have suffered massive brain damage or died."

"The nurse also told me I had to be out of here in five minutes so I need to update you on a few things. I sent a cleanup crew to Elena's apartment and had a couple of my men search and sanitize Siraj's hotel room. They found a well-worn leather-bound copy of the Quran in his suitcase. It's printed in Urdu. There are two hand-written entries on the title page indicating that it was given to Jalaluddin Haqqani by his father. Jalaluddin then wrote a personal note to Siraj. It's apparently an old family treasure that has been passed from father to son. I know you are fluent in Urdu so I thought you might like to have it."

David opened the old book and read the handwriting on the title page. The first inscription, written by Siraj's grandfather in a scrawling style, was *'lā ʾilāha ʾillā-llāh.'* David knew that was a common Islamic creed known as the first shahada that translated as '*there is no god but God.*' The second, written by Jalaluddin Haqqani translated as '*fear Allah and be his obedient servant.*' David recognized the remarkable similarity to what King Solomon had written in Ecclesiastes 12:13 fifteen hundred years before the birth of the Islamic Prophet, Muhammad. '*Let us hear the conclusion of the whole matter: Fear God and keep his commandments, for this is the whole duty of man.*'

"Yes. I would like to keep it. Thanks."

"The cleanup crew also recovered Siraj's cell phone," Emily continued." We've made copies of all of the phone numbers and text messages. Siraj had Ahmed Malik in his speed dial directory. We have no further use for it so you can have it too if you like."

David reacted visibly to what Emily had just said and took the phone from her hand.

"No one tried to place any calls from Siraj's cell phone did they?"

"No, not so far as I know. Why"

"I'd bet my next paycheck that Malik has Siraj's phone tapped. It's the kind of thing he'd do. Not only would he want to know everything that Siraj said over his phone, he'd use the phone's GPS feature to keep track of his location. I'm glad your men retrieved it but it's potentially very dangerous to whoever has it in their possession."

"I hadn't thought of that but you're probably right. Listen, I need to go. My time's up. Is there anything you want me to do for you?"

"Yes. The Quran and the phone have given me an idea. Set up a conference call with Roger and Pete for later today. I'd like for you to sit in on the call."

"OK. Anything else?"

"Yes. Persuade Dr. Walker to release me."

"I'll try."

3:00 PM

Emily's secure CIA phone rang right on schedule. She answered the incoming call and put it on the speaker.

"Hello. This is Emily Harrison and David Logan. I have you on the speaker."

"It's just Roger and me at this end," Pete told them.

David didn't waste any time. "I'll be as brief as possible. I think I may know how we can eliminate Malik and Jhalawan."

"That certainly gets our attention," Roger replied. "What do you have in mind?"

"There's still a lot of detail to be worked out but when Dr. Walker releases me, I need to go back to Israel. I want Aharon Jacobi to build something for me. When he confirms that it's done, I'll use Siraj's cell phone to call Jalaluddin Haqqani to offer my sincere condolences for the death of his son. I'll tell him that I have his family's leather-bound Quran and that as a matter of honor I would like to return it to him if he will guarantee my safety."

Pete didn't think much of that idea. "David, aren't you forgetting something? Jalaluddin's son just tried to kill you. Not once, but twice. It was almost certainly a contract hit paid for by Jhalawan. Jalaluddin Haqqani owes a lot to Jhalawan. What makes you think he won't just rat you out once you're on his turf?"

"I don't think so Pete. Although Jalaluddin Haqqani's code of honor is very different from ours, it's at the very core of everything he's fought for. There's no question that Jalaluddin is intensely loyal to General Jhalawan but his loyalty to Allah is absolute. Haqqani knows the Quran inside out. In Islam, the hospitality relationship is triangular. It consists of host, guest, and God. In Islam, hospitality is a God-ordained right, not merely a gift bestowed by the host. Islamic doctrine makes it clear that the host has a duty to treat his guests well. If Jalaluddin Haqqani guarantees my safety, he will honor that commitment with his own life if necessary. My offer to return his family Quran will touch him deeply. It's not Haqqani I need to worry about."

"In order to pull this off, I'm counting on Malik monitoring all calls to or from Siraj's cell phone. He didn't leave capturing Nahbi to his subordinates. After two failed attempts, I don't think he'll leave capturing and killing me to his subordinates either. I fully expect that Malik will come to Jalaluddin's home to deal with me personally. In fact, I think General Jhalawan will insist on it."

Pete was exasperated with what he clearly viewed as a really dumb idea. "David, you've lost your mind. If I understand what you're proposing, you'll be alone in a mountain cabin half-way around the world with no weapons and no backup plan. With one arm in a sling, you'll face an android with twice your strength and who knows how much more brain power. What makes you think you can take him down by yourself?"

"Actually, I will have precisely the right weapon. Nahbi told me that Malik had a high-voltage stun gun in Iran in case his men's guns weren't enough. We know that Malik's entire body is composed of graphene. Like graphite, one of graphene's outer shell electrons is not involved in crystal bonding and is free to move between atoms, which allows it to conduct electricity. Nahbi and Malik can be totally immobilized by a high-voltage stun gun. A company in Israel makes a twelve million volt stun gun that looks exactly like a cell phone. It fires two thin-wire electrodes up to eight meters. I'll also have a pistol with me as a backup."

"Malik won't try to kill me in Jalaluddin's home. I'm counting on getting a chance to immobilize him with the stun gun. While he's down, I'll open his power supply compartment and remove the rechargeable battery pack. I'll replace it with the device Aharon will make for me."

"What makes you think Jalaluddin will let you do all of this?" Emily asked.

"I'm counting on him being embarrassed that Malik has violated his guarantee of my safety and the Quran's admonition to treat guests well. I'll tell him that I suspected that Malik might show up so I came prepared. I'll explain that to respect his home, I did not intend to harm Malik, only to temporarily disable him to give myself time to get away. That will satisfy Jalaluddin's sense of honor."

Roger asked the obvious question. "What's this mysterious thing you're going to have Aharon build for you?"

"A very specialized bomb."

Tuesday, 21 October

Port of Shahid Rajaee; Bandar Abbas, Iran
7:00 AM

Ninety percent of Iran's container ship traffic goes through the port of Shahid Rajaee. Located strategically at the mouth of the Strait of Hormuz, the port is a hundred miles northeast of the United Arab Emirates. The contrast between the vibrant, wealthy city of Dubai and the bleak economic conditions in Iran could not have been sharper. Employment at the port was only a fraction of what it had been before sanctions imposed by the United Nations decimated international trade with Iran. Many of the port's stevedores had been out of work for several years and their families were suffering. The day after the railroad protests, the stevedores staged their own *'Nuclear Free Iran'* protest. None of them showed up for work. As stevedores at the other Iranian container ship ports heard about the strike at Shahid Rajaee, they too walked off their jobs. By noon, virtually all of Iran's port facilities were shut down.

With no containers being off-loaded, the container truck drivers were idled too. The flow of food and other goods simply stopped. The strikes were the lead story on the six o'clock news.

Navy Lodge; Sigonella Naval Air Station, Sicily, Italy
3:30 PM

David hit the speed dial for Jalaluddin Haqqani on Siraj's cell phone. After a few moments, the call was answered by the old man.

"Siraj, I've been wondering when you'd call. Where are you?"

"Mr. Haqqani, this is David Logan. I'm sorry to be the one to tell you that your son is dead."

After a long moment of stunned silence, Jalaluddin asked, "Mr. Logan, why would you call me to tell me this?"

"Mr. Haqqani, after the bombing in Tel Aviv, I was afraid there would be a second attempt on my life so I was prepared. Siraj tried to kill me while I was sleeping. I heard him enter the apartment and was waiting when he came into the bedroom. He and I fired simultaneously. He died instantly and I almost did too. I lost a lot of blood and required emergency surgery to save my life. When my associates searched his hotel room, they found his cell phone and what appears to be an old family copy of the Quran. Although our religious beliefs are quite different, I respect your commitment to your faith. As a sign of one soldier's respect for another, I will return your family Quran to you if you will agree to guarantee my safety. Think of it as a battlefield truce."

Jalaluddin Haqqani was silent for several moments before he responded.

"You are saying that you would come all the way to Pakistan simply to return our family's Quran? I find that hard to believe."

"Mr. Haqqani, I am quite serious and sincere. If you will guarantee my safety, I will do as I've offered. I would fly into Kabul Friday and then drive to your village Saturday, arriving around noon."

Again, Jalaluddin took several moments to consider what David had said.

"Why would you trust my promise to guarantee your safety? Surely you must at least suspect that I am the one who accepted General Jhalawan's contract on your life."

"Yes sir, I did assume that it was General Jhalawan who ordered me killed. Perhaps you know that I was a military intelligence officer in Afghanistan until I was wounded. I know a great deal about you, sir. I know that your men revere you and that you are widely seen as a man of honor. This matter of your family's Quran is about honor. It is not about hatred or retribution. My family Bible is precious to me as I believe your Quran is to you. I speak and write Urdu fluently and I was touched by the handwritten notes from your father and then from you to your son. These are not trivial things in my eyes. So, I tell you again that I am sincere. I will bring the Quran back to you as a man of honor who bears no personal malice toward you or even toward your son."

In a whisper, Jalaluddin agreed. "Mr. Logan, you have my blood promise that I will guarantee your safety and protect your life with my own if need be."

4:10 PM

David called the flight operations section at the airfield to confirm the arrangements he'd made earlier in the day and then called his boss.

"Pete, Jalaluddin Haqqani agreed to my conditions so I'll head for Pakistan Thursday morning. I spent most of this morning trying to come up with a plan to get the package Aharon made for me into Afghanistan without having to pass through civilian airport security anywhere."

"I went over to the airfield and talked to the Chief Master Sergeant in flight operations. He said NAS Sigonella is a refueling stop for Air Mobility Command shipments to Afghanistan and that a C-17 Globemaster out of McGuire Air Force Base will land here Thursday morning en route to Kabul. It's carrying several air-tight shipping containers of spare engines for three Israeli F-16 squadrons based at Ramon AFB. It will land there to off-load them and refuel. The crew will spend Thursday night at the American Camp at the north end of Ramon's main runway. Aharon will drive down from Tel Aviv sometime Thursday afternoon and spend the night at the Bachelor Officer Quarters. He'll meet me in the air ops terminal Friday morning to give me our surprise package. After a short hop to Ali Al Salem in Kuwait to refuel and off-load some more cargo, the C-17 will land in Kabul late Friday night. I've already made hotel and rental car reservations. I'll drive to Tari Mangal Saturday morning and meet with Haqqani around noon."

"David, you are certifiably crazy. I just hope it works."

LTG Jhalawan's Residence; Islamabad, Pakistan
10:10 PM

"Why are you calling me this late, Ahmed?"

"Siraj is dead. Logan killed him. I just intercepted a call Logan made to Jalaluddin using Siraj's cell phone. He offered to return the Haqqani family's old leather-bound Quran if Jalaluddin would guarantee his safety. Jalaluddin agreed. Logan will be in Tari Mangal Saturday morning."

"It must be a trick. Why would Logan take such a chance just to return an old family Quran?"

"Actually, I think Logan is sincere. He's a devout Christian. He made a point of telling Jalaluddin that he'd read the handwritten notes that Siraj's grandfather and Jalaluddin had written on the title page when they passed the holy book down to their sons. He told Jalaluddin that even though their faith was quite different, and they could never agree on many points, he could still respect him for his life-long commitment to his religious faith. Bringing the family Quran back to an old man whose only son he's just killed is fully consistent with Logan's character profile. My guess is that Logan has CIA business in Kabul so it's probably not a special trip just to return an old Quran."

"I still have my doubts but it doesn't really matter. I want you to be in Tari Mangal when Logan gets there. Make sure this time, Ahmed."

Iranian Intelligence Ministry; Tehran, Iran
10:25 PM

The Deputy Director of the Iranian Intelligence Ministry was in his living room watching the late news when his cell phone rang.

"Ali, this is Ahmed. I have some information that I think you'll find interesting. The man responsible for the destruction of the Jamkaran Mosque is an Israeli agent using the name Ahura Mazda. I have reason to believe that he is back in Iran or soon will be. I'm sending you a recent picture of him. I feel sure that your Supreme Leader would find some way to reward you if you can capture or kill this man.

Wednesday, 22 October

Iranian Parliament Building; Baharestan Square, Tehran
10:00 AM

At the specified time for the demonstration to begin, more than four thousand college students and other young adults had taken their positions around the Iranian Parliament Building. Many of them were wearing bright red '*Nuclear Free Iran*' tee shirts while others had cloth banners and posters. A blast from an aerosol marine horn was their signal to begin marching around Baharestan Square.

Holding hands and walking three abreast, the demonstrators sang the popular Iranian protest song '*Be of a Strong Heart*.' Major Iranian television network vans from IRIB, IRNA, IRINN, PNN, and several local TV stations were parked around the square and the demonstrations were being carried live on most of Iran's networks. The demonstrations were peaceful until police in SWAT gear tried to break up the demonstration. Several of the demonstrators resisted. Within minutes, dozens of demonstrators had been brutally beaten and several were bleeding profusely. A few were lying unconscious on the sidewalks. TV cameramen recorded all of it.

The White House; Washington DC
2:00 PM

"Good afternoon Mr. President. I think I can guess the purpose of your call," the Prime Minister of the United Kingdom began.

"Mr. Prime Minister, I have no doubt that you can. We are both aware that the Israelis intend to bomb all of the nuclear facilities in

Iran. The date has apparently not yet been established but I'm told that it will be soon. I'd like to propose that the U.S. and the U.K. each support that mission with five of our B-2 bombers. With the ten that the Israelis have, that would double the strike force and dramatically improve the probability of success. The Israelis have not asked for help even though I'm told that their total strike capability is at best only marginally adequate to get the job done."

"The Israelis will be doing the rest of the world an enormous service if they are successful. I think the combined support of the U.S. and the U.K. would virtually guarantee that all of Iran's nuclear production and weapons facilities can be destroyed with minimal collateral damage and none whatsoever to civilian population centers."

"You may know that I authorized twelve of our latest F-35 stealth fighters to be flown to Israel for what was billed as a joint training mission. The Israeli's need to neutralize several surface-to-air missile defense sites to protect the bombers on their final target runs. Their F-16 Sufas aren't stealthy so the Iranian missile defenses would take a heavy toll on them."

"The Saudi King has authorized Israel aircraft to fly over Saudi territory. Certainly he knows the risk that a nuclear Iran poses so it may be a classic case of '*the enemy of my enemy is my friend.*' The Saudis have three operational Airbus A330 aerial refueling tankers. I believe they'd be willing to top off the F-35 escorts just before they reach the Persian Gulf. The F-35s have a combat radius of about six hundred nautical miles so they'll need to be topped off again on their way home. We have refueling capability in Sicily and at Diego Garcia but my sense is that the Saudis might actually want us to include them in the mission."

"Mr. President, I need to confer with my cabinet and senior military officers but I think their approval can be safely assumed."

"I thought that would be your answer but I admit that I am once again humbled and deeply grateful for your leadership and our personal friendship. I look forward to your call."

Thursday, 23 October

Highway Rest Area and Restaurant; Qom-Tehran Freeway
7:45 AM

Several hundred students wearing the red 'Nuclear Free Iran' Tee Shirts were in the parking lot of a highway rest stop and restaurant twenty miles north of the Mofid University campus in Qom. From the rest area, it was eight miles to the main gate of the massive underground uranium enrichment facility at Fordow. Walking peacefully along the access road, it would take the students three hours to reach the entrance to Iran's worst kept secret. Television stations had been notified in advance of the protest march and TV cameras were already recording the event. Several hundred soldiers were also strung out along the road but so far were making no effort to stop the protest.

The popular song '*Be of a Strong Heart*' had quickly become the theme song for the anti-nuclear protests and the students were singing it as they marched along the road.

American Camp; Ramon AFB; Negev Desert Israel
10:20 PM

Aharon Jacobi showed his Mossad identification to the security guard at the entrance to the American Camp and asked to see the duty officer. The guard told him to wait outside the gate. A few minutes later, An Air Force Captain wearing the Air Force's BDU version of camouflaged clothing came to the gate and asked him the nature of his business.

"I'd like to see David Logan on official U.S. and Israeli national security business. The matter is classified. If you'll call his room, he will verify my identity and sponsor me into the compound."

Three minutes later, David came to the gate to sign for his guest and walked with Aharon back to his room on the officer's side of the transient quarters.

"I didn't expect to see you until tomorrow morning. Is there a problem?" David asked when they were inside his room.

"No. I wasn't sure how much time we might have in the morning and there are a couple of things I wanted to go over with you. It won't take long. Let me start with what you euphemistically referred to as the surprise package."

"I didn't bring it with me in case the American Camp security included explosive-sniffing dogs. Does the aircrew of the C-17 know you'll be carrying explosives?"

"Yes. The pilot knows I work for the CIA. I told him the target was the mastermind of the nuclear attack on Paris. That was good enough for him. He's expecting you to bring a package to me just before we take off."

"Good. As you specified, I've created a substitute for Malik's power supply that reduces the volume of the battery by half to make room for a high explosive and a radio-controlled detonator. When we talked, you suggested I use C-4 or Semtex. I considered using HMX instead. It's even more powerful than C-4 and is generally regarded as the state-of-the-art in military explosives. HMX is one of the most powerful chemical explosives available. But, there's an even more powerful alternative called Octanitrocubane. At the atomic level, it consists of a cube of eight carbon atoms with nitro groups consisting of nitrogen and oxygen attached to each carbon atom. It's insensitive to shock and at least twenty-five percent more powerful than HMX."

"The modified power supply contains two pounds of Octanitrocubane. If Jhalawan is in the same room with Malik when the bomb is detonated, they will both be instantly destroyed."

"What will trigger detonation?" David asked.

"I've built some artificial intelligence into the detonation device. Part of the design is a digital listening bug that will transmit any conversation it picks up to the Mossad's Tamnun supercomputer where it will be recorded for subsequent analysis. The device also does a voice-print comparison of any conversation it picks up to verify that Malik and Jhalawan are in close proximity to each other. When that proximity condition is verified, a signal will be transmitted to Israel's Ofeq-10 military satellite that will cause a red light to glow steadily on the Mossad duty officer's panel and sound an audible alarm. The voice recognition routine will run three times in quick succession. If all three tests confirm that both Malik and Jhalawan are present, the bomb will detonate automatically. When the signal to the satellite ceases, your CIA phone will automatically receive a text message with just the words 'ding dong' displayed."

David smiled at Aharon's humor. "As in ding dong the wicked witch is dead?"

"Yes. I knew you'd catch it. There's something else I need to go over with you. I modified the fake cell phone. It still fires two darts trailing micro-thin wires from a twelve million volt stun gun but it also does something else you didn't ask for. It occurred to me that Jalaluddin doesn't know Malik is an android. When you press the trigger the first time, it will immobilize Malik. But, my modification provides a second shot. In essence, I created a double-barreled version. If Malik doesn't go down, shoot him again. But, if he does go down with the first shot, I think you should hit Jalaluddin with the second shot. That way, he won't observe you opening the power supply compartment. By the time he and Malik recover, you'll be gone."

The White House; Washington, DC
1:00 PM

The secure video conference began on time. In Washington, the President, Vice President, Secretary of State, Secretary of Defense, Director of the Central Intelligence Agency, and the Chairman of the Joint Chiefs of Staff were in the cabinet room in the West Wing of the White House.

In London, the Prime Minister, Deputy Prime Minister, Security Cabinet members, and Air Chief Marshall were in the cabinet room on the second floor of number 10 Downing Street.

In Jerusalem, the Prime Minister and all members of the Security Cabinet were in the in the cabinet room of Israel's government complex in the Givat Ram district of central Jerusalem. The Chief of the General Staff of the Israeli Defense Force and the Director of the Mossad were also present.

As the acknowledged leader of the free world, President Marshall hosted the conference call.

"Thank you all for participating in this historic alliance. To paraphrase something Golda Meir once said, '*We will only have peace when radical Islamic terrorists love their children more than they hate us.*' Not so very long ago, Iran was a modern, prosperous member of the world community. Today, Iran is a poisonous viper that must be dealt with. There is no discernible difference between Adolf Hitler's determination to wipe out the Jewish people and Iran's Supreme Leader's vow to do precisely the same thing. Diplomacy failed spectacularly in the case of Nazi Germany. It has similarly failed to contain the evil forces emanating from Iran. So, reluctantly, we will do what must be done. We can take solace in knowing that we will strike only military targets, using only conventional explosives. But, our strike will be a massive blow. When our pilots return home, Iran will no longer be a nuclear power."

Israel's Prime Minister responded to what President Marshall had said.

"On behalf of the Israeli people, allow me to thank the United States and the United Kingdom for their friendship. More than two hundred years ago, the American Declaration of Independence set forth certain self-evident truths. One of those truths was that the right to life is an unalienable right given to us by the Creator. Thomas Jefferson understood that the right to life could not come from any form of government because if granted by a government, it could be denied just as easily. Our fight with radical Islam is not about where certain houses are built, or even where certain boundary lines are drawn. No! This fight is about every Israeli citizen's right to live. If those who follow the Islamic religion would stop killing our children, we could have peace. But, until radical Islam lays down its swords, we dare not lay ours down.

The Prime Minister of the United Kingdom then spoke.

"Mr. President and Mr. Prime Minister, as we have previously agreed, the United Kingdom will provide five of our B-2 bombers to support the Israeli attack on Iran's nuclear facilities. After considerable discussion, it has been decided that our aircraft will not deploy to Ramon Air Force Base in Israel prior to the mission. Instead our bombers will take off from RAF Fairford in Gloucestershire and return there after one aerial refueling over Saudi airspace."

President Marshall then explained how the United States would support the mission.

"Thank you, Mr. Prime Minister. The American aircraft will be pre-positioned at Diego Garcia and then return to Whiteman AFB in Missouri after the mission. Although all targets are within the range of the B-2 bombers when staged from Diego Garcia, we have asked the Saudis to top them off before they head back to their home base. Each country will maintain control of its own aircraft. By mutual agreement, the overall mission has been divided into specific target objectives that each country will be responsible for. It is my understanding that five of the Israeli bombers will be held in reserve in case some of the targets are not destroyed by the first wave. All aircraft and air crews will be on alert as of midnight tonight UTC. The decision as to mission timing rests with the commanding general of the Israeli Air Force subject only to final approval by Israel's Prime Minister."

The video conference wrapped up a few minutes later. For the first time in over sixty years, Israel would not have to fight alone.

Friday, 24 October

Mehrabad International Airport; Tehran, Iran
7:00 AM

"Good morning, this is Hamid Askari reporting live from Mehrabad International Airport where yet another anti-nuclear protest is underway. About six o'clock this morning some two thousand demonstrators sat down on the sidewalks in front of every entrance to the terminal buildings. They are sitting so close to each other that no one can get past them into the terminal building. Their demonstration has already caused several flights to be cancelled and airport authorities tell us that the demonstrations are now starting to affect flight schedules in other countries as connecting flights are being impacted. Sergeant Soleymaini of airport security is with me now. Sergeant, can you tell our viewers why the protest hasn't yet been disbanded?"

"Yes. I'd be glad to, Hamid. The demonstrators are peaceful. Airport legal counsel has told us that the Iranian Constitution allows peaceful demonstrations that are consistent with and supportive of the Quran and the country's leadership. As you know, Supreme Leader Ali Khamenei issued a fatwa several years ago that stated that the acquisition, development, or use of nuclear weapons was prohibited by the Quran. It is the opinion of the airport authority lawyers that we cannot legally disband this protest. Simply stated, the protestors are supporting a fatwa issued by the Supreme leader so it is hard to imagine how we can legally or ethically disband them."

Turning to face the camera, the newsman told his audience that a similar protest was currently underway at Imam Khomeini International Airport twenty miles south of downtown Tehran.

By nine AM, news coverage of the protests had gone globally viral. News commentators in the western democracies were already making comparisons to the 1979 Iranian revolution and the 1989 Velvet Revolution in Czechoslovakia.

Imam Khomeini International Airport; Tehran, Iran
9:10 AM

Nahbi was sitting in his wheel chair with a blanket over his legs watching the protest from the far side of the terminal access road. Gabe was standing a few feet away. Airport security guards were monitoring the protest but doing nothing to break it up. News reporters and cameramen were interviewing one of the protest leaders when a black government SUV pulled up to the curb and stopped directly in front of Nahbi. Four men in full SWAT gear got out and walked toward him while covering him with MP5 submachine guns.

"Ahura Mazda, you are under arrest! Put your hands in the air and don't make any sudden moves."

Nahbi and Gabe had always known that something like this might happen so they were prepared. Israel's Prime Minister had emphasized that allowing the Iranians to capture two Mossad agents was not an acceptable option. The leader of the SWAT team was the first to fall as Nahbi shot him between the eyes with the 10mm pistol concealed under the blanket. In quick succession, Nahbi and Gabe killed the other three men who were caught totally by surprise. They had not expected Ahura Mazda to be armed or to have an armed partner. It was all over in a matter of seconds.

The SUV's engine was still running. Nahbi picked up two of the submachine guns and shouted to Gabe to get the others as he ran around to the driver's side and got behind the wheel. Gabe climbed into the passenger side just as Nahbi floored the accelerator and headed for the airport exit. Two miles to the east, he turned south on the Persian Gulf Freeway before he said anything.

"It's almost five hundred miles to Bushehr. We need to find another vehicle quickly. Within a few minutes, police helicopters are going to spot us."

Gabe was studying the touch-screen GPS on the dash. "Take the next exit. It's about two miles. Head east toward the Shamasabad Industrial Park. It looks like it's a huge complex of warehouses and parking lots. We ought to be able to hide the SUV beside one of the warehouses so it won't be easily seen from the air. I'm guessing that we can find another ride there too."

Four minutes later, Nahbi turned right onto Bustan Boulevard, heading south into the industrial park.

"There's our answer," Gabe said as she pointed to an Islamic Republic Trucking Corporation's long-haul tractor-trailer rig with a sleeping compartment behind the cab. The young driver was walking around the rig checking his tires when Nahbi pulled up beside him. Gabe got out and walked over to say hello to him.

"Salâm. I'd like to hire you and your truck for a trip to the coast. I will pay you five hundred thousand Iranian rials to drive me and my partner to Bandar Bushehr."

"I'm sorry. We're not allowed to carry passengers."

"I understand. So, let me try again." Gabe was still wearing a black chador so the young man was surprised to see a 9mm pistol suddenly appear in her hand. "I'm sorry but you do not really have a choice. If you do as I say, you will be able to tell your employer that you were hijacked. You don't have to tell him that the hijackers gave you the equivalent of fifteen thousand U.S. dollars for your trouble. Otherwise, I'll simply gag you and tie you up and then take your truck anyway. The choice is yours."

To his credit, the young man wasn't easily intimidated.

"Show me the money."

Whiteman Air Force Base; Missouri, United States
6:00 AM (1200 UTC)

"AV-1, you are cleared for takeoff."

With those words from the control tower, the pilot of the Spirit of America released the brakes and advanced the throttles to full power for the four GE turbofan engines. With its fuel tanks full and two of the 30,000 pound massive ordnance penetrator bombs in its belly, the two billion dollar stealth bomber was near its maximum design loading of 336,500 pounds. Whiteman's main runway was over two miles long but the bat-wing aircraft was airborne forty seconds later with plenty of runway to spare.

Three minutes later, AV-3 was cleared for takeoff. The Spirit of New York had been included in the mission out of respect for the passionate support of Israel by the state's large number of Jewish-American citizens. AV-7, the Spirit of Texas followed three minutes later. AV-17, the Spirit of Florida was next and AV-18, the Spirit of Oklahoma was last off the runway.

With two refueling rendezvous, the eleven thousand mile flight to Diego Garcia would take twenty-two hours. Once the bombers reached their cruising altitude of 50,000 feet, only one of the pilots had to be in the cockpit at a time while the other took a nap on an inexpensive folding chair. On some of the early missions to Afghanistan, pilots had tried to sleep on the deck behind the cockpit but the vibration was so significant that sleep had been all but impossible. A folding chair from a discount store had solved that problem and had quickly been adopted as standard equipment by the seventy pilots who flew the giant stealth bombers. The autopilot was so sophisticated that there was little for the pilots to do but monitor the instruments once the aircraft was in level flight at its cruising speed of 560 MPH. Touchdown on Diego Garcia's twelve-thousand foot runway would be about 4:00 PM local time on Saturday.

Hotel Inter-Continental; Kabul, Afghanistan
11:40 PM

David was exhausted and his neck and collarbone were bothering him as he rode the elevator to the sixth floor in Kabul's only five-star hotel. Although the hotel's security was very good, it hadn't prevented Taliban gunmen from killing seven guests a few years earlier. As a precaution, he'd parked his rental near the front door under a parking lot light in full view of the curbside baggage valet station. A Sig Sauer P-239 compact 9mm pistol was tucked into his waistband under his shirt.

As soon as he set his roll-aboard bag down on the bed, he called Pete.

"Hello David. I've been expecting your call. I'm guessing you're in the hotel by now."

"Yes. Just got in. Any intel I need to know about before my meeting with Haqqani?"

"Not really. Get some sleep and call me tomorrow when you head back to Kabul. I still can't believe you volunteered for this assignment. The Comanche Code Talkers in World War II would have called you *posah-tai-vo*. It means crazy white man. Good luck."

Saturday, 25 October

Jalali Marina; Near Bushehr, Iran
1:20 AM

The trip from Tehran to the coast had taken thirteen hours but had been mercifully uneventful. Nahbi told the driver to park in a gravel parking lot near the public marina in the small town of Jalali. Unwilling to risk having the driver sound an alarm, Nahbi told him to walk ahead of them as they looked for a boat they could use to escape . Near the seaward side of the northern jetty, they saw a man standing nude on the stern diving platform of a large yacht with twin three-hundred horsepower Yamaha outboard engines. The man was taking a shower and was surprised to see anyone on the pier at that hour. He put the shower head down and grabbed a towel to cover himself as they walked up.

"Is this your boat?" Nahbi asked.

"No. I work for the owner. He runs a charter fishing business and I handle the boat for his clients. I sleep on the boat to protect his investment. There's no shower on board so I wait until everyone is asleep or gone home before bathing. I'm sorry if I offended the lady."

"We need to hire you for a short trip. You'll be back before daylight and we'll pay you one hundred thousand Iranian rials for your time," Nahbi told him.

"I'd have to clear it with the owner first. Can it wait until morning?"

"I'm afraid not," Nahbi replied as he drew his pistol. "You see, we're in a bit of a hurry. Get dressed. We need to leave immediately."

Five minutes later, the twin Yamaha engines were purring like two big kittens as the boat idled its way out of the marina past the breakwater.

"Where to, mister?"

"Head southwest on a heading of 225 degrees. As soon as you're clear of the jetty, bring the boat up to its top speed."

Gabe called their mission contact at the Saudi port of Al Jubail to request a helicopter pickup at 28 degrees 30 minutes North Latitude and 50 degrees 26 minutes of East Longitude in one hour.

Twenty Miles South of Jalali Marina; Persian Gulf
2:05 AM

The SWAT team's stolen SUV hadn't been discovered until late in the evening. The connection to the missing tractor-trailer rig took another hour. By midnight, Ali Yazdi knew that Ahura Mazda had hijacked the truck. Like most commercial trucks, it had a GPS signal device that allowed it to be located in case of theft or accident. When the truck's location was determined, it was readily apparent to Yazdi that Ahura Mazda and his partner intended to escape by boat. Thirty minutes after midnight, two twin-rotor Boeing CH-47 Chinook helicopters took off from Khatami Air Base near Isfahan. It was two hundred and eighty miles from Isfahan to Jalali. Despite their age, the Chinooks could cruise at 180 MPH with a maximum combat radius of two hundred miles. Two pintle-mounted thirty caliber machine guns were installed inside both shoulder windows and another was mounted at the rear of the aircraft on the loading ramp. Ali Yazdi's call to the Khatami base commander had made it clear that the Chinook's mission was to capture or kill Ahura Mazda and his partner.

The Chinooks reached Jalali just before two AM and were about to land in the parking lot beside the truck when the lead copilot noticed the bioluminescent wake from the yacht that had left the marina just minutes earlier. A quick check of the radar screen showed a fast-moving contact fifteen miles off the coast. Both Chinooks gave chase.

Eight minutes later, the Chinooks caught up with the yacht. The lead helicopter fired a warning burst from the port shoulder gun but the yacht didn't slow down. Dropping down to only fifty feet above the surface, the copilot used the aircraft's forward-looking infrared camera system to take a closer look at the people on board the yacht. Just as he confirmed that he'd identified Ahura Mazda, the helicopter came under automatic weapons fire from two submachine guns wielded by Mazda and his partner from the stern of the boat. The helicopters responded with fully-automatic fire from their thirty caliber machine guns. A few seconds later, the firing from the yacht stopped.

As the lead Chinook approached the yacht for a closer look, two Saudi Sikorsky MH-60 helicopters suddenly appeared. Firing short warning bursts at eleven hundred rounds per minute from their fifty caliber machine guns, the Saudi aircraft made it clear that the Iranian helicopters should break off their attack. Badly out-gunned and over international waters, the Iranian pilots got the message and headed for home.

By then, the yacht had drifted to a stop. As the Saudi helicopters hovered overhead waiting to take their passengers aboard, Nahbi knelt beside Gavriella. She was mortally-wounded but still conscious. She'd taken multiple thirty-caliber machine gun rounds in her torso and was bleeding badly. Gabe knew she had only moments to live but she was trying to say something. Nahbi had to lean close to hear her last words.

"I love you so very much. Kiss me, my darling."

Nahbi's eyes could not produce tears and he did not have a human heart. Yet, the hopeless sense of utter loss he experienced as he kissed his wife for the last time would never go away. Unlike humans, where the passage of time would eventually ease the pain, the intense memory of the incredibly sad moment would never fade for him. "I will always love you," he whispered.

And then she was gone.

Border Crossing; One Mile West of Tari Mangal, Pakistan 11:40 AM

Seventy miles east of Tari Mangal, the main highway border crossing at Torkham followed the legendary Khyber Pass route that Marco Polo had used seven hundred years earlier. Officially, the secondary border crossing a mile west of Tari Mangal was closed but the Pakistani Army was not welcome on Jalaluddin Haqqani's turf so various kinds of illicit traffic flowed freely across the pass in both directions.

Near the abandoned border crossing station, several of Haqqani's men stopped David's SUV and ordered him out of the vehicle. They were Pashtuns of the Mangal Tribe that had lived in the area for four hundred years. Two of the men kept him covered with AK-74 rifles while the third searched him for weapons. When he found the pistol in David's waistband, he pulled it out and waved it in front of David's face as if to say, 'what's this?' David just shrugged and smiled. The man searched David's soft-sided briefcase but found only a laptop computer and what appeared to be a reserve computer battery. The men were surprised when the American spoke to them in Urdu.

"I am a guest of Jalaluddin Haqqani. He has given his personal guarantee that I would not be harmed. I have a meeting with him at his home in the village at noon. Only a fool would wander around in these hills unarmed. I am no fool so please return my pistol. Just as Jalaluddin has guaranteed my safety, I am honor-bound not to harm my host while I am under his roof."

With David's brief speech, he established himself as someone important. The man who had searched him gave his pistol back and told him that he would guide him to Jalaluddin's home. Ten minutes later, David parked and got out. Jalaluddin was waiting for him on the porch.

"Mr. Haqqani, I am David Logan."

"Mr. Logan, please come into the house. We will have some food and then we will talk.'"

The meal was awkward for both men. David praised the cooking and Jalaluddin gave credit to his housekeeper. When they had finished eating, Jalaluddin invited David to sit on the front porch and stood aside to let David chose where to sit. David expected Malik to appear at any moment so he picked the chair near the west end of the porch where he'd have a clear view of the front door. As he sat down, he reached inside his briefcase for the old Quran and handed it to his host. He unclipped the fake cell phone from his belt and held it as if expecting a call. Jalaluddin laid the Quran on his lap and opened it to the title page. His face was stoic but David could tell that he was struggling to hide his emotions when he looked up and spoke.

"Thank you Mr. Logan. You are clearly a man of honor. All of us die eventually. My son was a soldier and the direct descendent of

Pashtun warriors. I think he would be satisfied that his earthly life was ended by a worthy adversary who fears God."

At just that moment, Ahmed Malik stepped out of the house onto the porch. A 9mm Russian Makarov pistol was visible in his right hand alongside his leg. In one smooth motion, David aimed the fake cell phone and pressed the trigger button before Malik had time to shoot. Malik fell convulsing to the porch as twelve million volts surged through his body. Jalaluddin stood in shock and stared at David.

"Mr. Haqqani, Ahmed Malik is my mortal enemy. He has tried multiple times to kill me and he murdered my only sister. Unlike you, he is not a man of honor. To respect you and your home, I will not kill him but I must search him to see if he is carrying something important. Please go inside. I do not want you to see what I do. I give you my word that I will not harm Ahmed while he is your guest but I must leave quickly to keep him from killing me."

Jalaluddin did as David asked and closed the front door behind him. David rolled Malik face down and lifted his shirt to expose the power supply compartment in the small of his back. Working quickly, he used double nylon ties to bind Malik's hands and feet before replacing the power supply with the one Aharon had constructed. Spasms were still wracking Malik's body but they were already fading. David needed to leave. Standing to go, he knocked on the front door.

"Mr. Haqqani, I'm leaving now. Ahmed Malik is unharmed. He will recover from the shock in a few minutes. I have bound him to ensure that he couldn't harm me if he recovered more quickly than I expected."

Marketplace Roundabout; Gardez, Afghanistan
4:20 PM

Twenty miles west of Tari Mangal, David turned off the Chamkani-Gardez highway to get gas near the village of Mirazi Kalay to see if he could spot anyone following him. He couldn't so if someone was following him, they were no amateur. Even so, he was nervous. He fully expected Malik to send someone after him so he took many detours and doubled back several times. Just north of the roundabout near the center of Gardez, he stopped at a small neighborhood grocery store to buy a bottle of water and some raisins. Before he got out of the car, he checked his cell phone again and saw that he finally had decent signal strength. The connection was slow and he was about to hang up and try again when Pete answered.

"Pete. David. It worked. Malik is now a fully-equipped, mobile IED. I'm in Gardez on my way back to Kabul. I didn't check out of my room at the hotel this morning thinking I'd use it again tonight. But, after thinking about Malik's intelligence network and his operational reach in this part of the world, I called and made reservations using one of my alternate identities. I'm going to return the rental car when I get back to Kabul and take a taxi to the hotel in case Malik knows the license number and type of vehicle I'm in. I don't think I was followed but I'm still in a dangerous neighborhood where Malik has lots of resources. I've already snagged a space-available seat on the Air Force C-17 return flight to Sigonella Monday morning. I hate to ask, but could I persuade you to let one of your poor wounded employees take a few days of personal leave? I really owe it to Elena to see her before I come back to the States. I wouldn't blame her for telling me to get lost after what I just put her through but she was a real trooper. Not many young women would have grabbed a shotgun and been ready to back me up."

Eleven thousand miles to the west, Pete was smiling to himself and shaking his head in amusement.

"You know, David, when Roger offered you a job, we sort of thought it was a full-time position but I've hardly seen you since. Just when do you think this poor wounded employee of mine might deign to grace us with his presence again?"

"What would you say to five days of personal leave and me being back in the office on Monday, November 3rd?"

"How far are you in the hole on your leave?"

"I hope you're kidding because I don't have a clue about where my leave balance stands. I don't remember annual leave even being mentioned when you and Roger offered me the job. How about it? Can you manage without me for a few more days?"

"And what, may I ask, would be different about your continued absence? Still, under the circumstances, I guess this poor, wounded employee you're referring to has earned a little down time. We'll treat next week as medical leave. Say hi to Elena and Emily for me. I know it'll be hard, but you might try to set a new personal record by staying out of trouble for a whole week. By the way, your new company nickname is *posah-tai-vo*.

4:40 PM

David's next call went immediately to voice mail so he kept his message short. "Aharon, this is David. I thought you'd like to know that our package has been delivered. I've got a seat on an Air Force C-17 out of Kabul Monday morning. It stops at NAS Sigonella on its way back to McGuire AFB in New Jersey. I'm taking a week's leave in Malta but I'll be watching my text messages. Thanks for all your help."

David was about to get out of the car when the passenger side door opened suddenly and Ahmed Malik slid into the seat and pressed a silenced 9mm pistol against his ribs.

"Hand me your gun and cell phone then start the car and head back toward Tari Mangal. Keep both hands on the steering wheel where I can see them."

IRGC Garrison Headquarters; Tehran, Iran
6:00 PM

The Islamic Revolutionary Guard Corps, or IRGC as it is more commonly known, is considered a terrorist organization by most of the western democracies. Like the legendary Roman Praetorian Guard, one of the IRGC's primary duties is to protect Iran's Supreme Leader. The IRGC is also Iran's primary interface with its terrorist proxies in Iraq, Lebanon, Syria, and the Gaza Strip. IRGC soldiers are the elite of Iran's military forces and they receive superior weapons and training. Although the IRGC is feared by both its military counterparts and the civilian population, the elite unit is also bitterly hated for its eagerness to crush any alleged opposition to the Supreme Leader, no matter how innocent or minor and often without evidence of subversive intent. Operating much like the infamous Gestapo of Nazi Germany, they are a brutal military force answerable only to Iran's Supreme Leader.

Just before four o'clock, the commanding officer of the IRGC, Major General Qasem Hosseini, and several of his senior officers were just concluding a briefing about the protests when ten heavily-armed soldiers wearing full battle gear burst into their conference room.

"General Hosseini, you and your officers are under arrest. Put your hands on the table in front of you and don't make any sudden moves."

Two of the soldiers went around the room disarming the officers before handcuffing them with their hands behind their backs.

"Who are you?" General Hosseini demanded.

Smiling grimly, the leader of the rebel soldiers answered. "We are your worst nightmare. Over the past two years, your command has been infiltrated by a few genuine patriots who passionately love what Iran once was and bitterly despise what it has become. We have bided our time. The outbreak of student protests and demonstrations this week was the catalyst we've been waiting for. About an hour ago, some of my men arrested the Supreme Leader and the President. They are under house arrest and safe for now but we will kill them if we are attacked. Your fate will be the same if you make any attempt to have your loyal troops overpower us. At six o'clock, all of you will board a chartered civilian aircraft that will take you into permanent exile in Zimbabwe. Your families will follow in a few days. If you ever set foot on Iranian soil again, you will be executed by a firing squad without trial.

Jalaluddin Haqqani's Home; Tari Mangal, Pakistan
6:45 PM

Jalaluddin's men had called ahead to let him know that Malik was coming. The old man was waiting on the porch when the rental SUV turned into his front yard.

Keeping the pistol pointed at David's side, Malik removed the ignition keys and told David to get out of the vehicle and walk slowly toward the front porch. At the foot of the steps, Malik told David to stop.

"My old friend, I must impose on your hospitality again by asking you to let us spend the night here."

"Yes, of course, Ahmed. What will you do with Logan? You know that I guaranteed his safety."

"Do not concern yourself. I will not harm him while he's at your home. But, I need your help. General Jhalawan will be saddened by the death of your son but he will not be pleased to learn that Logan is still alive. I've decided to take Logan to Islamabad so General Jhalawan can have the satisfaction of killing the man who disfigured him and sent him to the American prison in Cuba. I want you to go with me to Islamabad to help guard Logan. We'll leave early in the morning."

Sunday, 26 October

44000 Constitution Avenue; Islamabad, Pakistan 8:00 AM

General Jhalawan was in a meeting when Malik asked to see him. Unlike a civilian secretary, the general's enlisted aide was armed with a 9mm Beretta 92F pistol. If he was surprised by the fact that one of the men with Malik was handcuffed and wearing leg shackles, he didn't show it.

"The General's meeting will be over in a few minutes. Please take a seat and I'll call you as soon as the General is free. Can I tell him what this is about?"

"Just tell him that Ahmed Malik is waiting to see him."

A few minutes later, several senior military officers walked out of the office and the General's aide went in. In less than a minute, he came out and told Malik that the General would see him.

"I told General Jhalawan that you had two other men with you but he said he wanted to speak to you alone first."

"Wait here for me," Malik told Jalaluddin Haqqani. I'll tell General Jhalawn why we're here. I'm sure he'll want to see you and Logan immediately."

General Jhalawan's aide closed the heavy oak door behind Malik and returned to his desk. Twenty seconds later, a massive explosion shook the entire building as Aharon's voice recognition feature confirmed that Jhalawan and Malik were in the same room.

It was only by the grace of God that David wasn't in the room when the bomb he'd installed in Malik's power supply compartment detonated. Alarms were going off all over the building as General Jhalawan's aide forced the door open and saw that there was very little left of the two men that had been in the room. The small remaining fragments of Ahmed Malik's body and clothing were actually on fire. The general's body was horribly mutilated.

Jalaluddin Haqqani's swift reaction belied his age. He'd seen the damage that IEDs caused and instantly realized the bomb that had just been detonated was far more powerful. He knew without further evidence that Jhalawan and Malik were both dead

"Hurry, follow me!" he shouted at David. The scene outside the building was utter chaos. Ninety seconds after the bomb detonated, Haqqani unlocked the door to the rental SUV and helped David into the passenger seat. Less than a minute later, they were on their way out of Islamabad heading west toward Tari Mangal. Haqqani handed David the keys to his handcuffs and leg shackles.

"Why are you helping me escape?" David asked.

"Mr. Logan, I guaranteed your safety. It was very awkward when Ahmed Malik showed up at my home with the obvious intention of having you killed. He assured me that my promise to you would not be violated but I felt that his actions said otherwise. His death and that of General Jhalawan released me from my obligations to them. Now, I can fulfill my promise to guarantee your safety by helping you escape from Pakistan."

Hotel Inter-Continental; Kabul, Afghanistan
6:30 PM

After leaving Jalaluddin Haqqani at his home in Tari Mangal, David drove back to Kabul. He returned the rented SUV and took a taxi to the hotel. As soon as he was in his room, he called Pete's cell phone.

"Well, if it isn't my wandering, poor wounded employee. Feeling guilty?" Pete razzed.

"No. Not especially. I need to confess that I wasn't able to stay out of trouble for a whole week as you suggested. I do think you're going to need a new nickname for me though. Somehow, '*crazy white man*' doesn't seem quite right after what's happened. Maybe '*lucky white man*' would be more appropriate."

"OK. You've got my attention. Let's have it."

David told him everything that had happened since they'd spoken on the phone the day before. He heard Pete's soft whistle when he heard about his narrow escape.

"I see why you said 'lucky white man' should be your new nickname. If you'd gone into Jhalawan's office with Malik, you'd have been killed too. Are you OK?"

David couldn't resist a small barb. "Well, I guess I'm OK for a poor, wounded CIA employee who was hired for what was supposed to be a desk job."

Monday, 27 October

Israeli Defense Force Headquarters; Camp Rabin, Israel
10:00 AM

As soon as his morning security briefing was finished, the Prime Minister called the commanding general of the Israeli Air Force.

"Good morning general. I assume you've seen this morning's satellite photos showing what appears to be pre-launch activity at the Badr and Imam Ali missile sites. Levi thinks the Iranians are preparing to launch an attack on Israel sometime within the next forty-eight hours. His analysts believe they have at least ten active

Fajr-3 ballistic missiles at both sites and that all of them are configured with multiple independently targeted reentry vehicles. Given Iran's frequent threats to destroy Israel, it's entirely possible that some of the missiles are armed with nuclear warheads. We know that some of the Shahab-3 missiles at Badr are armed with biological weapons including anthrax. The satellite photos show condensation plumes over the silos that are characteristic of pre-launch transfer of liquid propellants. It looks like we're out of time. Is the strike force ready?"

"Yes sir. With your approval, I'm ready to notify the Americans, Brits, and Saudis that Operation Thor will be executed tonight."

"It's a six hour flight to the targets from Diego Garcia and just about the same from RAF Fairford. Iran's time is three and a half hours later than UTC and two and a half hours earlier than Diego Garcia's time. In order to be over the targets by 0200 local Iranian time, the Americans will take off from Diego Garcia at 2230 local time. The Brits will take off at 2230 UTC tonight and be over their assigned targets at about the same time.

The F-35 fighters will take off from Ramon AFB at 2230 local time. They'll top off their fuel tanks ninety minutes later from Saudi A330 tankers two hundred miles west of Al Jubail before crossing the Persian Gulf. As soon as the fighters are airborne, five of our B-2 bombers will take off. Their role in the mission is to destroy all of Iran's strategic missile sites. They won't need to be refueled until they're on their way home.

"Eight of the F-35 fighters will knock out the surface-to-air missile batteries at Fordow, Isfahan, Natanz, and Parchin just before the bombers make their runs. The American's new target recognition system is so advanced that the Iranian SAM batteries probably won't have time to launch a single missile before they are destroyed. Each of the fighters will be carrying a total of eight GBU-53 precision-guided small diameter bombs. The other four F-35s will stay close to the bombers just in case any Iranian F-14 Tomcats or Mig-29 Fulcrums show up. The mission commander will be Lieutenant Colonel Ezra Feldstein piloting Gibor Ben-Gurion. With your permission, I'll release the mission operational order as soon as I hang up."

"Yes general. You are authorized to execute Operation Thor." He concluded the call with the first line of a traditional Jewish prayer, "*May we see the day when war and bloodshed cease.*"

Diego Garcia; British Indian Ocean Territories
2230 Local Time/1630 UTC

All five of the American B-2 bombers were very near maximum takeoff weight with full fuel tanks and two of the 30,000 pound MOP bunker buster bombs. The Spirit of America was first off the runway followed at three minute intervals by the Spirit of New York, the Spirit of Texas, the Spirit of Florida, and the Spirit of Oklahoma. Their mission call signs were Hammer Forty-One through Hammer Forty-Five. All Operation Thor call signs that ended in a one signified a squadron commander's aircraft.

Royal Air Force Base Fairford; Gloucestershire, England
2230 Local Time/2230 UTC

The B-2 bombers from the United Kingdom were named after British Royal Air Force aviators that had been awarded the Victoria Cross for gallantry in the face of the enemy during World War II. Like their American counterparts, each of the bombers was very near their maximum takeoff weight. The mission call signs for the bombers from the United Kingdom were Hammer Fifty-One through Hammer Fifty-Five.

First off the runway was VC Eric Nicolson. It was a six hour flight to target but neither pilot was in the mood for a nap. Their mission was something both of them had long hoped they'd one day get a chance to do. They were closely followed off the runway by VC Guy Gibson, VC Cyril Joe Barton, VC John Cruickshank, and VC Leonard Cheshire.

The first wave of U.K. B-2 bombers were responsible for destroying the runways at the airfields and airports where Iran's fighter aircraft were based. Each of them was carrying eighty MK-82/GBU-30 JDAM bombs weighing five-hundred pounds each. The

JDAM guidance kit converted so-called dumb bombs into all-weather smart bombs. Guided by an integrated inertial guidance system coupled to a GPS receiver, the JDAM bombs could be dropped up to fifteen nautical miles from their target and reliably land inside a circle twenty feet in diameter. Some of the airfields had more than one runway but eighty five-hundred pound bombs, individually targeted would still do so much damage that aircraft launching would be impossible after the attack.

Ramon Air Force Base; Negev Desert, Israel
2230 Local Time/2030 UTC

The twelve F-35A Lightning II stealth fighters on loan from the United States would use the Israeli call signs of Hammer Twenty-One through Hammer Thirty-Two. Each of the fighters was carrying a maximum weapons load that included AGM-158 air-to-surface missiles, GBU-39 bombs, and a 25mm GAU-12 Equalizer cannon with 182 rounds of ammunition. The four-barrel cannons were mounted internally as part of the stealth technology. Like the bombers, they were all very near their maximum takeoff weight of 70,000 pounds. With their maximum range of twelve-hundred nautical miles, they would reach the refueling rendezvous in Saudi Arabia's airspace with roughly one-third of their fuel remaining. After being topped off, they would proceed to their targets and then return to the refueling point to be topped off again before returning to Ramon AFB. The actual top speed of the Lightning II aircraft was classified but most military aviation experts estimated the top speed of the supersonic fifth-generation fighters at over twelve hundred miles per hour. With their takeoffs spaced only two minutes apart, and using both of Ramon's main runways, all twelve fighters were airborne in less than fifteen minutes.

At 2245, Israel's B-2 bombers Gibor Ben-Gurion and Giborah Meir were cleared for takeoff on parallel runways. Four minutes later, Gibor Rabin and Gibor Dayan released their brakes and began their takeoff runs. Four minutes and fifty seconds later, Gibor Sharon was the last of the Israeli B-2 bombers to become airborne. Their call signs were Hammer One through Hammer Five. Five of Israel's B-2 bombers would be held in reserve in case some of the targets weren't destroyed in the first wave. Their call signs were Hammer Sixty-One through Hammer Sixty-Five.

Bandar Ganaveh; Iranian Coastline on the Persian Gulf
1:15 AM Local Time

Eight of the F-35 Lightnings crossed the Iranian coastline fifty miles north of Iran's nuclear power plant at Bandar Bushehr en route to their respective targets. All of them were equipped with the top secret AN/AAQ-37 Distributed Aperture System that gave them 360 degree, spherical situational awareness. Using six electro-optical sensors, the systems provided missile detection and tracking, launch point detection, and situational awareness of the entire battlespace by linking the sensors of all of the aircraft together. The systems also provided unprecedented day and night precision navigation capabilities.

Shortly after crossing the coast the fighters separated into four two-aircraft attack groups but kept their speed subsonic to avoid the tell-tale sonic booms. The waxing moon had set five hours earlier. The skies were dark but clear. The F-35's very low observable stealth allowed them to safely enter the defense airspace around Iran's nuclear facilities without being detected by Iranian radar. Iranian air search radar might get a few random pings off their stealth coatings but it wouldn't be enough to recognize the threats that were approaching their nuclear facilities. For all practical purposes, the Lightnings were invisible.

Tapeh Air Base; Tehran, Iran
1:45 AM Local Time

VC Eric Nicolson's target was the Tapeh Air Base near Tehran. Almost six hours after leaving RAF Fairford, the B-2 was still ten miles from its target when its F-35 escort detected an Iranian MiG-29 Fulcrum taking off on what was probably a routine patrol flight. Long before the pilot of the Fulcrum knew the F-35 was in his neighborhood, the F-35 pilot opened his weapons bay doors and launched an AIM-120 AMRAAM 'fire and forget' air-to-air missile. Keying his secure tactical radio, he transmitted 'Fox Three' to indicate that he'd just launched an active, radar-guided missile.

Using its own on-board radar for terminal guidance, the missile killed the Fulcrum while the F-35 was still well out of visual range. A moment later, the F-35 pilot's situational awareness system detected a second MiG-29 taking off from Tapeh Air Base so he sent a surprise package to that unsuspecting enemy pilot as well. "Fox Three," he again announced. Less than two minutes later, VC Eric Nicolson dropped forty-thousand pounds of MK-82 bombs on the airfield from an altitude of fifty-thousand feet. It would be weeks before any other aircraft were able to take off.

At almost the same time, the military runways at Tabriz International Airport, Bandar-e Abbas, Shahroki Air Base near Hamadan, and Vahdati Air Base near Dezful suffered similar fates.

Natanz Centrifuge Facility; 25 Miles SE of Kashan, Iran
1:55 AM Local Time

The pilot of Hammer India Twenty-One began his run on the uranium enrichment plant with the weapons bay doors closed to maintain a clean, stealthy configuration. Hammer India Twenty-Two was two miles back. Both F-35 aircraft locked onto the surface-to-air missile batteries and dropped their loads of eight GBU-53 precision-guided small diameter bombs before the Iranians even knew they were under attack. The missile defense for the Natanz Centrifuge Facility was utterly destroyed by over three thousand pounds of precision-guided high-explosive shaped-charge bombs. What General John Gordon had said about the top secret target recognition systems was true. It was the best surface-to-air missile killing system ever used in combat.

By 2:00 AM, the surface-to-air missile defense systems at Fordow, Isfahan, and Parchin had also been destroyed to prepare the way for the bombers.

Badr Missile Base; Semnan Province, Iran
2:00 AM Local Time

The Israeli B-2 named Gibor Ben-Gurion arrived over the secret Badr missile base twenty-five miles southeast of Semnan at precisely the planned time-over-target. The bomber was carrying eight MK-

84/GBU-30 JDAM bombs in each of its two bomb bays. To maintain the aircraft's clean configuration, the pilot waited until the last few seconds before opening the bomb bay doors. It took the precision-guided, two-thousand pound bombs less than four minutes to reach the ground. The B-2 bomber named Giborah Meir was over the target five minutes later and dropped an identical bomb load. After the attack, a few of the missile silos were damaged but not completely destroyed. Even so, the destruction of its power and control systems effectively ended the missile complex's ability to launch strategic missiles against Israel.

At almost the same time, the Imam Ali strategic missile launch complex was destroyed by the other three Israeli B-2 bombers. Without political change in Iran, the attacks could not indefinitely guarantee that Israel was safe. But, for now, the threat of Iranian missiles was over.

Fordow Centrifuge Facility; 20 Miles NE of Qom, Iran
2:10 AM Local Time

The huge, heavily reinforced underground nuclear enrichment site at Fordow was the reason the United States had developed the thirty-thousand pound Massive Ordnance Penetrator bomb. Without it, only a nuclear weapon could have destroyed the Fordow site. Each B-2 could carry two of the enormous bunker-buster MOP bombs. The Fordow site was so important, both from a technical perspective and for political reasons, that the mission's operational plan had committed two of the American B-2 bombers to its destruction.

Inside the mountain were over three thousand cascaded IR-1 centrifuges in two underground halls. Access to the halls was via four tunnels that opened to the northwest. The Iranians had told the International Atomic Energy Agency that the facility would only be used to produce medical isotopes. That was a transparent lie. The real purpose of the facility was to produce enriched uranium for nuclear weapons. As a state sponsor of terrorism, and an implacable enemy of the United States and Israel, Iran's government could not be trusted with nuclear weapons. The Fordow facility had to be destroyed.

The Spirit of America's pilot released two of the precision-guided MOPs right on schedule. Five minutes later, the Spirit of New York's pilot dropped two more. The sophisticated detonation-delay mechanisms allowed the giant bombs to penetrate two-hundred feet into the hardened underground facility before exploding. Two of the MOPs penetrated the northern centrifuge hall before exploding and destroying the fifteen-hundred centrifuges installed there. A few minutes later, two more penetrated the southern hall and destroyed all of the centrifuges there. The combined explosive force was so great that the extensive system of tunnels began to collapse, burying what was left of the centrifuge complex under hundreds of tons of rock and dirt.

By 2:30 AM local time, The Spirit of Texas had destroyed Iran's nuclear facility at Natanz. Spirit of Florida took care of Isfahan and Spirit of Oklahoma destroyed the facilities at the Parchin military complex outside Tehran.

An hour and a half before sunrise, Israel's five reserve B-2 bombers put an exclamation point on the attack by dropping another eighty MK 84/GBU-32 JDAM precision-guided bombs on secondary targets. Collectively, those eighty bombs represented one-hundred-sixty thousand pounds of high explosives. The Arak heavy water production facility was totally destroyed. Sixteen of the JDAM bombs destroyed the yellow cake uranium production mill at the Ardakan site. The Lavizon-3 uranium enrichment facility outside Tehran was destroyed as well.

After refueling over Saudi Airspace, the F-35 fighters and Israel's ten B-2 bombers returned to Ramon Air Force Base. The F-35 squadron would be given post-mission maintenance attention and then be flown back to the United States.

The American B-2 bombers were refueled by the Saudi A330 tankers and then headed back to their home base in Missouri. Israel's five reserve bombers were the last to refuel from the Saudi tankers before returning to Ramon AFB.

Despite the maniacal boasts of Iran's leaders, the combined airpower of the United States, United Kingdom, and Israel destroyed or severely damaged all of Iran's major nuclear facilities in less than three hours of high-tech aerial warfare with no civilian casualties and without the loss of a single allied aircraft. With that bold stroke, the Iranian nuclear threat was ended.

After thirty-five years of theocratic repression and economic hardship, the people of Iran would soon have a new constitution and a new democratic government.

Epilogue

The national cemetery of Israel is located two miles west of the heart of Jerusalem on top of Mount Herzl.

Israel's highest military honor is the Medal of Valor. It is awarded for '*Performing a supreme act of valor while facing the enemy and risking one's life.*' Gavriella Adler's first husband was awarded the Medal of Valor posthumously for his heroism in trying to save the life of an innocent Palestinian girl. He was buried on Mount Herzl in the National Military Cemetery.

Gavriella was also posthumously awarded the Medal of Valor for leading the Kidon team that killed Abdul Muqtadir, the Hamas commander in Gaza. She received a second award for the mission to Iran that resulted in her death.

On a bright, sunny day in mid-November, at Nahbi's request, Gavriella was laid to rest beside her first husband.

Proof

66686135R00273

Made in the USA
Charleston, SC
25 January 2017